BIOCHEMICAL SOCIETY SYMPOSIA

No. 60

MOLECULAR BOTANY:
SIGNALS AND THE ENVIRONMENT

BIOCHEMICAL SOCIETY SYMPOSIUM No. 60
held at The University of Leeds, Spring 1993

Molecular Botany:
Signals and the Environment

ORGANIZED AND EDITED BY

D. J. BOWLES, P. M. GILMARTIN, J. P. KNOX
AND G. G. LUNT

PORTLAND PRESS
London and
Chapel Hill

Published by Portland Press, 59 Portland Place, London W1N 3AJ, U.K.
on behalf of the Biochemical Society
In North America orders should be sent to Portland Press Inc.,
P.O. Box 2191, Chapel Hill, NC 27515-2191, U.S.A.

ISBN 1 85578 050 X ISSN 0067-8694

British Library Cataloguing in Publication Data
A catalogue record for this book is available from the British Library

Typeset by Unicus Graphics Ltd, Horsham, Sussex
and Printed in Great Britain by Whitstable Litho Printers Ltd

Contents

Preface

This book summarizes the lectures given during the Annual Symposium of the Biochemical Society, held at the University of Leeds in the Spring of 1993.

We chose the term 'Molecular Botany' as our title for the Symposium to reflect our enthusiasm and conviction for the subject in its own right, and to fly the flag for those responses which only plants exhibit to environmental change.

All organisms respond to their environment, but, significantly, plants routinely use environmental cues as developmental signals. Plants have evolved a sedentary lifestyle, and have maximized their surface areas above and below ground, to gain maximum opportunity for absorption of sunlight, water and nutrients. The down-side of this evolutionary strategy is that they are vulnerable to whatever the environment offers: whether climate changes or attacks by pests and pathogens.

Signalling pathways lead from environmental stimuli to end-effects within the plant. The end-effects range from exquisitely sensitive and rapidly reversible fluctuations in ion movements — such that, for example, stomata open or shut — through to changes in gene activation and triggering of co-ordinate transcriptional events that ultimately lead to new developmental programmes or multi-faceted defence responses.

We wanted the Symposium to reflect the widespread interest of the research community in signalling events and the pathways that lead from molecular recognition of an environmental stimulus to a defined change within the plant. We also wanted to highlight the importance of the plant cell surface in these recognition and signalling events. The cell surface comprises the plasma membrane and the cell wall. Increasingly, the apoplastic space within plants, with an external boundary to the outside environment and an internal boundary at the surface of the cell, is recognized to play a central role in signalling. To understand fully the functioning of the apoplast, it will be essential to understand its three-dimensional dynamic organization in terms of the structure of its component polysaccharides, glycoconjugates and proteins, and in terms of its role as a reservoir of regulatory molecules, ions and defence-related products.

We anticipated that the Symposium would interest a broad spectrum of researchers and this was also reflected in financial support from two diverse groups within the Biochemical Society: the Carbohydrate Group and the Molecular Biology Group. We would also like to thank the companies who sponsored guest lectures, including Advanced Technologies, Courage, Gatsby Charitable Foundation, Monsanto, Schering, Stratagene and Unilever. The financial support and goodwill of the Society, its research groups and the industrial section provided the financial basis to invite a superb group of specialist speakers for the Symposium from laboratories throughout the United States, Europe and Australia. The lectures were well-received by a large audience and discussion on all the topics was wide-ranging and highly useful. All in all, we believe that the Symposium was a great success and certainly enjoyed by all the participants. We hope readers of this text will gain a flavour of the excitement and enthusiasm of the science presented and join with us in supporting the future success of Molecular Botany.

DIANNA BOWLES

University of York
June 1994

Abbreviations

ABA	abscisic acid
ACC	1-aminocyclopropane-1-carboxylate
AGP	arabinogalactan protein
AGT	appressorial germ tube
APS	adenosine-5′-phosphosulphate
B	blue
BSM	basic staining material
$[Ca^{2+}]_c$	cytosolic free calcium concentration
Cab	chlorophyll *a*/*b*-binding protein
CAT	chloramphenicol acetyltransferase
Cdi	cathepsin D inhibitor
CHS	chalcone synthase
CMV	cytomegalovirus
ConA	concanavalin A
COX	cyclo-oxygenase
cyt. b_6f	cytochrome b_6f
2,4-D	2,4-dichlorophenoxyacetic acid
DP	degree of polymerization
EFE	ethylene-forming enzyme
EPS	extracellular polysaccharide
ER	endoplasmic reticulum
Fab	fragment antigen binding
FBP	fucose-binding protein
FITC	fluorescein isothiocyanate
FR	far-red
GA	gibberellic acid
GRP	glycine-rich protein
GUS	β-glucuronidase
HIR	high irradiance response
HMG I-Y	high mobility group I-Y
13-HPLA	13-hydroperoxylinolenic acid
HR	hypersensitive response
HRGP	hydroxyproline-rich glycoprotein
IgM	immunoglobulin M
Ins(1,4,5)P_3	inositol 1,4,5-trisphosphate
IPTG	isopropyl β-D-thiogalactoside
JA	jasmonic acid
KLH	keyhole limpet haemocyanin
LA	α-linolenic acid
LAC	light-absorbing compound

LFR	low fluence response
LPS	lipopolysaccharide
mAB	monoclonal antibody
MeJA	methyljasmonate
MGBG	methylglyoxal bis(guanylhydrazone)
NSAID	non-steroidal anti-inflammatory drug
OG	oligogalacturonide
OMT	*O*-diphenol methyltransferase
ORF	open reading frame
12-oxo-PDA	12-oxo-phytodienoic acid
PA	propionic acid
PAL	phenylalanine-ammonia lyase
PAPS	3′-phosphoadenosine-5′-phosphosulphate
PCR	polymerase chain reaction
PE	promoter element
PEG	polyethylene glycol
Pfr	far-red-absorbing phytochrome
PG	polygalacturonase
PGIP	polygalacturonase-inhibiting protein
PGT	primary germ tube
pHc	cytosolic pH
PHYA	phytochrome A
PHYB	phytochrome B
pin	proteinase inhibitor
Pmg	*Phytophora megasperma* f.sp. *glycinea*
Pr	red-absorbing phytochrome
PR	pathogenesis related
PR-O	tobacco PR β-1,3-endoglucanase
PRP	proline-rich protein
PS	photosystem
PVX	potato virus X
R	red
RbcS	ribulose bisphosphate carboxylase small subunit
RFLP	restriction fragment length polymorphism
RG	rhamnogalacturonan
RNase	ribonuclease
RSR	race-specific resistance
S-RNase	self-incompatibility associated ribonuclease
SA	salicylic acid
SABP	salicylic acid-binding protein
SAG	salicylic acid glucoside
SAM	*S*-adenosylmethionine
SAM-dc	*S*-adenosylmethionine decarboxylase
SAR	systemic acquired resistance
TBE	Tris-Borate-EDTA
Td	threonin deaminase
TMB-8	8-(*N*,*N*-diethylamino)-octyl 3,4,5-trimethoxybenzoate

TMV	tobacco mosaic virus
$TPMP^+$	triphenylmethylphosphonium ion
TSAG	thio analogue of SAG
VLFR	very low fluence response
XET	xyloglucan endotransglycosylase
XGO	xyloglucan-derived oligosaccharide
YAC	yeast artificial chromosome
ZW 3–12	*N*-dodecyl-*N*,*N*-dimethyl-3-ammonio-1-propane sulphonate

Biochem. Soc. Symp. **60**, 1–4
Printed in Great Britain

Cell wall structural proteins — can we deduce function from sequence?

J.E. Varner

Washington University, Department of Biology, St. Louis, MO 63130, U.S.A.

Proteins that are localized in cell walls, occur widely, are relatively abundant and have highly repetitive sequences have been assumed to have structural functions and have been labelled accordingly. These structural proteins include the extensins, the glycine-rich proteins (GRPs) and the proline-rich proteins (PRPs). These three classes of protein show very different abundances in the walls of different cell types. Thus each can be supposed to have functions specific to its particular cell type. There is little direct evidence as to what these functions might be. Further, we do not have a complete inventory of the components of the walls of any cell type as it occurs in an organized tissue. It is, therefore, difficult to 'put together' a plan of probable arrangements and interactions of cell wall components. One can, however, try to imagine what might happen in any given mix and, perhaps, experiments can be devised to image the real structure.

A recent review by Allan Showalter [1] details the structures, expression and current ideas about the interactions and functions of the extensins, GRPs and PRPs. The review also deals with solanaceous lectins, which I shall not discuss because of their limited occurrence, and the arabinogalactan proteins, which are widespread, readily soluble and for which little sequence data are available. Additionally, Kieliszewski and Lamport have just completed a review that is both factual and speculative [2]. Further, Carpita and Gibeaut [3] have essayed a summary and synthesis of many of the current ideas about how the various cell wall components are assembled to make a wall that is suitable for the cell it surrounds.

For my purposes, in this paper I shall consider a plant cell wall made up of (i) cellulose microfibrils kept evenly spaced by bridging xyloglucan molecules that are tightly hydrogen-bonded to the microfibrillar surfaces; (ii) a pectic matrix made of varying proportions of methylated pectins and randomly demethylated pectins, with occasional blocks of polygalacturonate (fully demethylated); (iii) structural proteins such as extensins, GRPs and PRPs; and (iv) up to 100 different enzyme activities, such as ascorbic acid oxidase, peroxidases, pectin methyl esterases, transglycosylases, invertases, proteases, phosphatases and glycosidases. I have purposefully omitted a number of known cell wall components to focus more tightly on a single speculative question: can we make any useful suggestions about possible functions of cell wall structural proteins by examining what we know of their primary and secondary structure? This approach has allowed Kieliszewski and Lamport [2] to propose probability rules for, first, the hydroxylation of particular proline residues of the proline-rich precursors of the extensin monomers and the PRPs; secondly, the arabinosylation pattern of the extensins and PRP hydroxyproline residues; and, finally, sequence motifs involved in cross-linking extensin monomers *in vitro* and *in vivo*. Val-Tyr-Lys is proposed as one such sequence, because it is present in different extensins that cross-link *in vitro* and is absent from those extensins that do not cross-link *in vitro*. The *in vitro* cross-linking conditions involve hydrogen peroxide and the pI 4.6 extensin peroxidase. The Val-Lys-Pro-Tyr-His-Pro sequence of tomato P1 extensin never appears in P1 peptides isolated from tomato cell wall digests but does appear in a tryptic peptide prepared from isolated P1 monomers; it is, therefore, considered a putative cross-link motif.

If we accept the proposition that Val-Tyr-Lys and Val-Lys-Pro-Tyr-His-Pro are cross link motifs, what is the chemistry of the cross-link? Dityrosine has never been found in cell wall hydrolysates. Iso-dityrosine is present in cell wall hydrolysates but the only iso-dityrosine linkage found in peptides so far has been an intramolecular one. Lysine could form Schiff's base adducts with aldehyde groups. Reduction of this linkage (as occurs in the cytoplasm in the formation of opines) would produce a stable linkage. Oxidation of histidine residues of proteins can open up the imidazole ring and produce an aldehyde group. If this were to form a Schiff's base with the ε-amino group of lysine, and be reduced, a stable protein–protein cross-link could be generated.

Valine and proline do not appear to offer much opportunity for cross-linking; however, proteins can be peroxidized by exposure to reactive oxygen species, such as hydroxyl free radicals produced from hydrogen peroxide and Fe^{2+} ions [4]. The hydroperoxides formed have half-lives of about 1 day at room temperature, but react readily with natural reductants, such as ascorbate or glutathione. Free amino acids are peroxidized by these treatments (Table 1). The hydroperoxides formed involve the amino acid side-chains; they are not peroxyacids, nor do they appear to be peroxides [4]. The formation of these rather stable hydroperoxides raises the possibility that they can initiate further reactions in their vicinity.

An examination of Table 1 shows that the three amino acids most readily peroxidized, Val, Pro and Lys, are abundant in the cross-linking motifs Val-Tyr-Lys and Val-Lys-Pro-Tyr-His-Pro.

Treatment of proteins and free amino acids with hydroxyl free radicals can also generate a reductant. Tyrosine is the most active of the various amino acids after these treatments; therefore, the reductant is thought to be 3,4-dihydroxyphenylalanine produced by the hydroxylation of the tyrosine aromatic ring. Orthodihydric phenols are well known to be strong reducing agents. Although histidine (Table 1) is not so readily peroxidized it can, as already mentioned, be oxidized to open the imidazole ring and generate an aldehyde and an amine group. It is interesting that serine, threonine and hydroxyproline — the amino acid residues most involved in generating the polyproline II structure of extensins — are the least susceptible to peroxidation (Table 1).

Do these considerations allow new proposals about the function of the sequence Ser-Hyp-Hyp-Hyp-Hyp-Thr-Hyp-Val-Tyr-Lys-Tyr-Lys, which is repeated seven times in one carrot extensin, and of the sequence Pro-Hyp-Val-Tyr-Lys, which may occur 29 times in one of the soybean PRPs?

Table 1. Peroxide yields in amino acids oxidized by free radicals. Amino acids (20 mM) in 20 mM phosphate, pH 7.4, were irradiated with λ-rays to a dose of 315 Gy (1 Gy = 1 J/kg). H_2O_2 was removed with catalase and the remaining peroxides were measured. Peroxidation efficiency is the number of peroxide groups formed per $HO^{\cdot}$ radical generated ($\times$ 100). Adapted from Table 3 in [4].

Amino acid	Peroxidation efficiency
V, valine	49
P, proline	44
K, lysine	34
E, glutamate	28
R, arginine	13
A, alanine	11
H, histidine	4
G, glycine	3
Y, tyrosine	3
O, hydroxyproline	2
S, serine	0
T, threonine	0

The soluble PRPs of bean-stem cell walls are apparently insolubilized within a few minutes after wounding the stem [5]. The effect is mediated by hydrogen peroxide because catalase prevents it. The chemistry of this insolubilization is unknown; however, the system seems amenable to further study along the lines suggested above. Olson and Varner [6] have shown by a simple histochemical test (50 mM KI in 4% potato starch is applied to the surface of the freshly cut section to be tested) that hydrogen peroxide is produced abundantly by all cell types in response to crushing, and is produced by xylem elements and phloem fibres in undamaged tissue (P. Olson and J.E. Varner, unpublished work). Thus it is shown simply and directly that hydrogen peroxide is present at sites of lignification and is, therefore, probably involved — as has long been assumed. Because of the tissue co-localization of some of the GRPs, PRPs and lignin [7] it is necessary to consider how these components are arranged in the cell walls. Possible chemistry of the amino acid side-chains of the Pro-Hyp-Val-Tyr-Lys motifs of the PRPs have been discussed above. For the GRPs, notice that bean GRP 1.0 and GRP 1.8 both contain many tyrosine residues [8]. These might easily become linked to the aromatic residues of lignin. GRP 1.8 has several residues each of histidine, valine, glutamine and glutamate, providing opportunities for the kinds of derivatization and linkage already mentioned. Transglutamylation is also a possible source of cross-linking. GRP 1.0 has no glutamine and far fewer residues of histidine, valine and glutamate than GRP 1.8. Perhaps a significant difference between these GRPs is that GRP 1.0 has four Asn-Gly sequences while GRP 1.8 has none. This provides an opportunity for easy selective proteolytic fragmentation of GRP 1.0.

The petunia GRP 1.0 contains no tyrosine residues [9], has only one Asn-Gly sequence (it is near the *N*-terminal end) and is localized in the meristems in expansive tissue, not in lignifying tissue.

Finally, remember that primary walls maintain a uniform thickness of about 50 nm during their extension, and that extensin molecules are of sufficient length, about 80 nm, to be of potential use in determining wall thickness.

References

1. Showalter, A. (1993) Plant Cell **5**, 9–23
2. Kieliszewski, M.J. and Lamport, D.T.A. (1994) Plant J. **5**, 157–172
3. Carpita, N.C. and Gibeaut, D.M. (1993) Plant J. **3**, 1–30
4. Gebicki, S. and Gebicki, J.M. (1993) Biochem. J. **289**, 743–749
5. Bradley, D.J., Kjellbom, P. and Lamb, C.J. (1992) Cell **70**, 21–30
6. Olson, P. and Varner, J.E. (1993) Plant J. **4**, 887–892
7. Ye, Z. and Varner, J.E. (1991) Plant Cell **3**, 23–37
8. Keller, B., Sauer, N. and Lamb, C. J. (1988) EMBO J. **7**, 3625–3633
9. Condit, C.M. and Meagher, R.B. (1986) Nature (London) **323**, 178–181

Biochem. Soc. Symp. **60**, 5–14
Printed in Great Britain

Oligosaccharins as plant growth regulators

Stephen C. Fry

Division of Biological Sciences, Daniel Rutherford Building,
The University of Edinburgh, The King's Buildings, Mayfield Road, Edinburgh
EH9 3JH, U.K.

Synopsis

Oligosaccharides with regulatory effects on living plant tissue have been obtained by partial hydrolysis of xyloglucan, cellulose and pectic polysaccharides. Attention is focused here on xyloglucan-derived oligosaccharides (XGOs), which exert the following two distinct effects on cell growth in pea-stem segments. (i) At approx. 1 nM, the L-fucosylated XGOs, such as XXFG, XFFG and FG (for structure of XXFG, see Fig. 1), antagonize 2,4-dichlorophenoxyacetic acid (2,4-D)-stimulated growth. At approx. 100 nM, XXFG loses this growth-inhibitory effect, probably because it gains a growth-promoting effect [see (ii)]; in contrast, FG retains its growth-inhibitory effect. The growth-inhibitory effect is tentatively attributed to membrane-binding of the active XGOs. (ii) At approx. 1 μM, at least four different cellotetraose-based XGOs (XXXG, XXLG, XXFG and XLLG) mimic auxin in that they induce growth. This effect is thus not L-fucose-dependent and is not exhibited by the cellobiose-based pentasaccharide, FG. Effect (ii) is attributed to the ability of cellotetraose-based XGOs to act as acceptor substrates for xyloglucan endotransglycosylase.

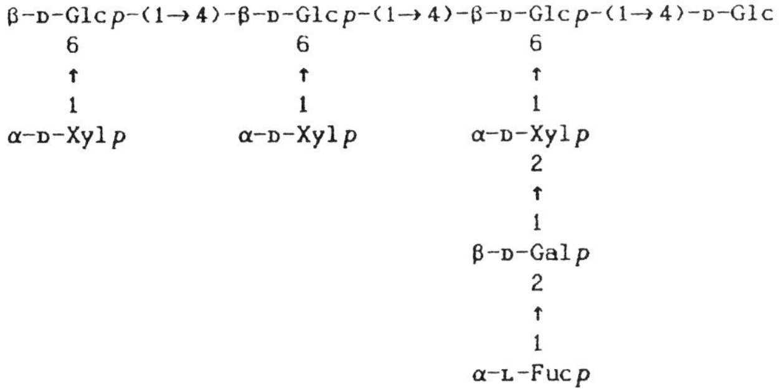

Fig. 1. The structure of XXFG (previously referred to as XG9), a biologically active xyloglucan-derived oligosaccharide. From Valent *et al.* [27] with permission.

The biosynthesis and biodegradation of relevant XGOs has been investigated. By use of labelling with L-[^{3}H]arabinose and L-[^{3}H]fucose *in vivo*, XXFG and O-acetyl derivatives thereof were shown to accumulate extracellularly, in spinach cell cultures, to approx. 0.1 μM. The kinetics of labelling of XXFG showed it to be formed by degradation of pre-formed polysaccharide rather than by *de novo* synthesis of the oligosaccharide. XXFG was remarkably stable *in vivo*, undergoing little hydrolysis in contact with the surfaces of cultured cells; the major metabolic fate of exogenous [^{3}H]XXFG was sequestration into apoplastic polysaccharide by endotransglycosylation.

Oligosaccharins

Plant polysaccharides are immobile within the organism and mainly act as structural building materials (e.g. in cell walls), and as food reserves (e.g. starch). They can, however, be partially degraded by the action of plant and microbial enzymes to yield diffusible oligosaccharides. Although most such oligosaccharides have no significant effect on plant tissues, except as carbon and energy sources, it has recently been established that certain specific oligosaccharides exert hormone-like regulatory effects on living plant tissues. Such oligosaccharides are referred to as oligosaccharins [1–5]. Oligosaccharins are typically effective at very low concentrations (nanomolar to micromolar), and their actions are highly dependent on the oligosaccharide's chemical structure.

Several well-known oligosaccharins are 'elicitors', i.e. they provoke responses (especially the biosynthesis of phytoalexins) which appear to be defence mechanisms against pathogen attack. Other oligosaccharins have been reported to modulate the development and morphogenesis of plants, and the biosynthesis of ethylene. The main theme of the present article is the structure and function of those oligosaccharins that influence the growth (generally taken to mean irreversible cell expansion) of plant cells.

Overview of growth-regulating oligosaccharins

Growth-regulating oligosaccharins have been prepared from three different polysaccharides — xyloglucan, cellulose and pectin. Each of these three classes will be briefly discussed; a more-detailed account of progress on the xyloglucan-derived oligosaccharins will then be given.

Xyloglucan-derived oligosaccharides

Several specific oligosaccharides derived from xyloglucan (XGOs) exert growth-regulating effects on excised plant tissues. Since at least some of these XGOs are formed *in vivo*, considerable interest is focused on these oligosaccharins as possible endogenous plant-growth regulators. Two principal biological effects have been reported: (i) growth inhibition at about 1 nM XGO, and (ii) growth promotion at about 1 μM XGO. These effects are quite distinct, not only in their dose–response curves but also in their structural requirements.

The growth-inhibitory effects have been reported to antagonize endogenous growth (i.e. that which occurs in the absence of any exogenous plant hormone) [6],

2,4-D-stimulated growth [7–13], H^+-stimulated growth [14], gibberellic acid-stimulated growth [6], and fusicoccin-stimulated growth [15]. Indoleacetic acid- and cytokinin-stimulated growth have been reported not to be inhibited by XGOs [13,15]. Thus, under the conditions used for those assays, either the XGO had no effect or endogenous XGO concentration was too high already [13]. The main plant material employed in these studies has been segments (typically 5–10 mm long) cut from the stems of etiolated pea seedlings [7–14]. In addition, however, XGO-bioassays have been carried out with whole seedling shoots (excised just above the cotyledons) [6] and excised pumpkin cotyledons [15]. The growth-inhibiting effect of XGOs cannot be countered by increasing the concentration of 2,4-D. This was shown by the fact that Lineweaver–Burk plots exhibited uncompetitive inhibition by XXFG of 2,4-D-induced growth. Thus XXFG does not act as a classic anti-auxin [13].

The growth-promoting effects of XGOs have been shown to 'mimic' the action of 2,4-D on cell growth in excised stem segments [16] and whole excised shoots [6] of etiolated pea seedlings. The promoting effect on excised shoots is more pronounced when growth is already enhanced by the application of another exogenous plant hormone — gibberellic acid [6].

Cello-oligosaccharides

A mixture of cello-oligosaccharides [oligo-β-(1→4)-D-glucans], enriched in the penta- and hexasaccharides, was shown to antagonize the H^+-stimulated growth of pea-stem segments apparently in a manner similar to that of the XGOs [14]. The cello-oligosaccharides were optimally effective at about 10 nM.

Pectic oligosaccharides

Pectic polysaccharides (pectins) are rich in α-(1→4)-linked galacturonic acid residues. The pectic fraction of the cell wall appears to consist of at least three domains — homogalacturonan and rhamnogalacturonans I and II (RG-I and RG-II) — which, in the intact cell wall, may be glycosidically attached to each other [17]. Homo-oligogalacturonides have recently also been shown to antagonize auxin-induced growth in excised pea-stem segments [18]. The end-effect is similar to the inhibitory effect of XGOs; however, the concentration of oligogalacturonides required for the maximal effect was about 100 μM (cf. 1–10 nM for XGOs), and higher concentrations of oligogalacturonides did not lose their growth-inhibiting activity, in contrast to the behaviour of XGOs. In addition, the effect of the oligogalacturonides appeared to be competitive with the effect of the auxin [18]. The oligogalacturonides specifically interfered with auxin action; cytokinin- and gibberellic acid-induced growth was unaffected.

Apparently unrelated to the growth-inhibiting effect of oligogalacturonides, an unsaturated disaccharide [ΔGalA-(1→2)-Rha], which could possibly have arisen from RG-I, was shown [19] to promote *Amaranthus* hypocotyl elongation. ΔGalA-(1→2)-Rha was active at 3 μM and evoked a 5-fold growth promotion at 1 mM. In addition, recent research at Edinburgh has shown that RG-II inhibits the uptake and incorporation of amino acids into proteins in cultured tomato cells — an indicator of decreased general cell vitality and, therefore, probably of growth (S. Aldington and S.C. Fry, unpublished work).

Structure and nomenclature of xyloglucan oligosaccharides

Xyloglucan has a backbone of β-(1→4)-linked D-glucan. Side-chains (the most common being Xyl, Gal→Xyl and Fuc→Gal→Xyl) are attached to the 6-position of many of the glucose residues [20,21]. Xyloglucan can thus be regarded as a linear polymer mainly composed of the following building blocks (G, X, L and F), whose glucose moieties are β-(1→4)-linked

G	D-Glc*p*-
X	α-D-Xyl*p*-(1→6)-D-Glc*p*-
L	β-D-Gal*p*-(1→2)-α-D-Xyl*p*-(1→6)-D-Glc*p*-
F	α-L-Fuc*p*-(1→2)-β-D-Gal*p*-(1→2)-α-D-Xyl*p*-(1→6)-D-Glc*p*-

In this paper, XGOs are named by listing the appropriate one-letter codes in sequence from non-reducing to reducing terminus. Thus the decasaccharide

$$\begin{array}{llll} \mathrm{Glc} \rightarrow & \mathrm{Glc} \rightarrow & \mathrm{Glc} \rightarrow & \mathrm{Glc} \\ \uparrow & \uparrow & \uparrow & \\ \mathrm{Xyl} & \mathrm{Xyl} & \mathrm{Xyl} & \\ & \uparrow & \uparrow & \\ & \mathrm{Gal} & \mathrm{Gal} & \\ & & \uparrow & \\ & & \mathrm{Fuc} & \end{array}$$

(previously referred to as XG10) is here termed XLFG.

Preparation, purification and structural characterization of XGOs

Oligosaccharides can be produced from xyloglucan by partial digestion with cellulase [β-(1→4)-D-glucanase], which attacks the backbone mainly at unsubstituted glucose residues [22]. The two major oligosaccharides thereby produced from the xyloglucans of dicotyledons are XXXG and XXFG (Fig. 1). Others obtained in smaller amounts include XXLG, XLFG, XFFG and FG. The xyloglucans of grasses have few or no fucose residues.

The XGOs obtained from cellulase digests of xyloglucans are usually initially fractionated by gel-permeation chromatography, e.g. on Bio-Gel P-2 [7,9,22]. The size-classes of oligosaccharides are then further purified by paper chromatography, e.g. in ethyl acetate/acetic acid/water (10:5:6, by vol.) and butan-1-ol/pyridine/water (4:3:4, by vol.) [23], and/or h.p.l.c., e.g. on amino-substituted silica columns [24].

Tests of the homogeneity of XGO preparations include t.l.c. [25], h.p.l.c. on Dionex 'Carbo-Pac PA1' in dilute aqueous NaOH with a gradient of sodium acetate [24], and reversed-phase h.p.l.c. of 2-pyridylamino dervatives of the XGOs [26].

Rigorous determination, from first principles, of the structure of XGOs has been achieved by methylation analysis [27] and by ^{1}H-n.m.r. [28]. In addition, considerable structural information can be obtained by enzymic dissection, especially using a 'xyloglucosidase', i.e. a β-D-glucosidase which acts on 6-*O*-α-D-xylosylated β-D-glucopyranose residues [29]. This enzyme attacks XGOs from the non-reducing terminus, sequentially removing Xyl→Glc groups until the first Gal→Xyl→Glc or Fuc→Gal→Xyl→Glc group is encountered.

Growth-inhibiting effects of XGOs

York *et al.* [7] first reported that XXFG at about 1–10 nM (i.e. about 1.5–15 μg/l) can diminish the growth promotion normally induced in excised stem segments of etiolated pea seedlings by the artificial auxin 2,4-D. XXXG did not bring about the same response. These surprising findings have been confirmed and extended in several other laboratories [6,8–15]. The principal structural requirement for growth inhibition seems to be a terminal α-L-fucose residue: thus XXLG, which differs from XXFG only in lacking the α-L-fucose residue, was inactive [10]. In addition, since L-fucose and methyl α-L-fucopyranoside are not growth inhibitors [11], it seems likely that the β-D-galactose residue, to which the α-L-fucose residue is glycosidically bonded, is also obligatory. It may, therefore, be suggested that the smallest structural unit conferring growth-inhibitory activity is α-L-Fuc*p*-(1→2)-β-D-Gal*p*-, which is present in XFFG, XXFG, FG and 2-fucosyl-lactose — all of which antagonize 2,4-D-induced growth — but not in XXLG or XXXG.

The XXXG core of XXFG does not seem to be essential. For instance, the α-D-xylose residue furthest from the reducing terminus can be removed from XXFG by the action of α-D-xylosidase to generate GXFG, which still retains growth-inhibiting effects [8]. This observation suggests that the growth-inhibiting effect of XGOs is not connected with their ability to act as acceptor substrates for xyloglucan endotransglycosylase (see later), since this enzyme does require the α-D-xylose residue furthest from the reducing terminus [30]. The reducing terminal glucose moiety apparently does not participate either, since XXFGol (i.e. the product of treating XXFG with $NaBH_4$) is active [8]. The fact that 2-fucosyl-lactose is active [11] confirms that the Xyl/Glc-rich backbone of XXFG is not essential.

At 1 nM, XXFG inhibits both 2,4-D- and H^+-promoted growth, suggesting that the effect on 2,4-D-stimulated growth is not mediated via an inhibition of H^+ secretion [14]. Some of the effects of 1 nM XXFG on 2,4-D-treated segments of etiolated pea stems resemble those of added C_2H_4: in particular, 1 nM XXFG antagonizes elongation without significantly affecting the increase in fresh weight (G.J. McDougall and S.C. Fry, unpublished work). It is, therefore, attractive to propose that C_2H_4 may act as a second messenger in the action of nanomolar XXFG. In this context, it may be relevant that some other oligosaccharins have been shown to evoke C_2H_4 biosynthesis [31].

The requirement for an α-L-fucose residue and action at nanomolar concentrations suggests that XGOs may act via a specific receptor. ^{3}H-XGOs are not appreciably taken up by cultured plant cells [32,33], so the most likely location of a receptor would be in the plasma membrane. Membrane-localized binding sites for elicitor-active oligosaccharins derived from fungal β-D-glucans have been detected in plants [34,35], but comparable experiments have not yet been reported for XGOs.

XXFG loses its growth-inhibitory effect at 0.1–1.0 μM [7,9,10,12–14]. This loss of inhibitory effect does appear to be dependent on the Xyl/Glc-rich backbone of the oligosaccharin, unlike the inhibitory effect itself, since the loss is not observed with the pentasaccharide FG or with the trisaccharide 2-fucosyl-lactose [6,11]. It can be proposed that the loss of growth-inhibitory effect at higher concentrations

(0.1–1.0 μM), observed for some XGOs, is connected with the fact that XGOs possessing an XXXG-based backbone acquire a growth-promoting effect at about these concentrations (see below).

Growth-promoting effects of XGOs

Some XGOs (including XXFG) appear to exert an auxin-like effect on growth: at about 1 μM they can promote the elongation of segments excised from etiolated pea stems in the absence of exogenous 2,4-D [6,16]. They can also promote the elongation of whole excised shoots of etiolated pea seedlings, especially in the presence of gibberellic acid [6]. The enhancement of elongation appears to differ in several key features from the growth-inhibitory effects of fucosylated XGOs (see above); in particular, effectiveness is independent of the α-L-fucose residue but dependent on some or all of the XXXG-based backbone.

The growth-promoting effect appears likely to be related to the ability of certain XGOs to act as acceptor substrates for a recently discovered enzyme, xyloglucan endotransglycosylase (XET). Thus, the order of effectiveness of four oligosaccharides (XLLG > XXLG > XXXG > XXFG) was the same for growth promotion [16], for stimulation of the enzymic depolymerization of high-M_r xyloglucan by a crude 'cellulase' preparation from *Phaseolus vulgaris* leaves [16], and for action as acceptor substrate for XET [36]. It now seems probable that the relevant activity of the cellulase preparation was in fact XET, and that the promotion of depolymerization of xyloglucan was due to the oligosaccharides' ability to act as acceptor substrates.

Xyloglucan is a major structural hemicellulose of the primary cell walls of dicotyledons. It hydrogen-bonds strongly to cellulose, as can be demonstrated *in vitro*, and xyloglucan chains have been proposed to tether adjacent cellulosic microfibrils within the plant cell wall [37]. If xyloglucan does indeed occupy this significant position in the cell wall, then its enzymic cleavage would be likely to have a strong impact on wall extensibility [37,38]. Xyloglucan molecules *in vivo* have been well established to be subject to partial degradation [39] — a process that, until recently, was widely assumed to be due to cellulases [i.e. β-(1→4)-D-glucan endohydrolases]. However, it is now clear that XET could also contribute to the cleavage of xyloglucan chains *in vivo* [16,32,33,36,40–42].

XET is proposed to cut an internal β-D-Glc-(1→4)-D-Glc bond in one xyloglucan chain, and to form a transient xyloglucanyl–enzyme conjugate. The xyloglucanyl moiety is then proposed to be transferred from the enzyme molecule on to an appropriate acceptor. The acceptor can be the non-reducing end of either a xyloglucan or an XGO. The XET-catalysed reaction is thus the cleavage of one Glc→Glc bond and the conservation of the energy thereof in the formation of a chemically identical Glc→Glc bond. The reaction may result in no net change in the physical or chemical properties of the reactants:

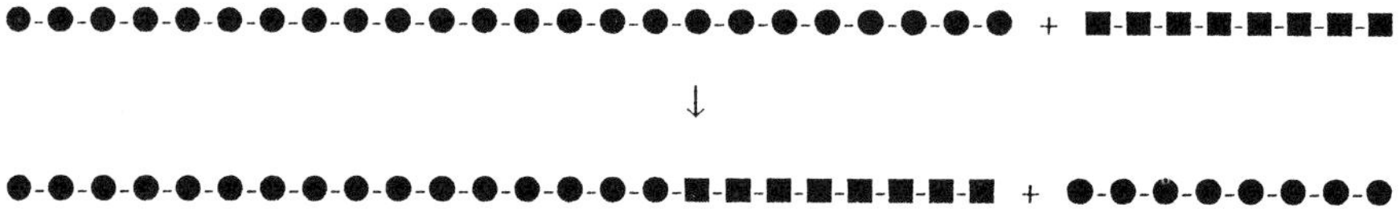

although the breaking and re-making of Glc→Glc linkages may well have caused a highly significant temporary loosening of the cell wall and thereby facilitated cell expansion [43].

XET activity has been found in the growing tissues of all land plants examined, including bryophytes, liliaceous and graminaceous monocotyledons, and dicotyledons.

XET is readily assayed by measurement of its ability to attach a high-M_r xyloglucanyl moiety from a donor polysaccharide on to a ^{3}H-labelled XGO (= acceptor substrate) [36]. After incubation, the reaction mixture is applied to a small square of chromatography paper, which is then washed for 1 h in running water. The product (^{3}H-labelled xyloglucan) hydrogen-bonds firmly to the cellulose of the paper, whereas the unreacted [^{3}H]oligosaccharide is washed off. Bound radioactivity is then assayed by scintillation counting.

It seems plausible that *in vivo*, high-M_r xyloglucan usually acts as acceptor substrate, so that little net change occurs in the M_r of the xyloglucan molecules. The biological significance of this could be to loosen the cell wall transiently, allowing wall creep, followed by the re-formation of intermicrofibrillar xyloglucan tethers. If, however, an exogenous XGO is added, the latter can act as acceptor substrate, in competition with the endogenous polysaccharides [30]. Thus, although xyloglucan chains might continue to be cleaved (loosening the cell wall and favouring cell expansion), the cut ends would frequently be re-annealed to short stubs (XGOs) rather than microfibril-anchored xyloglucan chains. Thus, in the presence of XGOs, the strength of the cell wall is not restored after the cleavage event, and this could, depending on the physiological condition of the cells, lead to an enhancement of cell expansion.

According to this interpretation, at micromolar concentrations XGOs act as growth-regulating oligosaccharins by means of their ability to compete with long-chain xyloglucan molecules as acceptor substrates for the wall-modifying enzyme XET.

Natural occurrence of XGOs

All studies of the biological effects of XGOs have been performed with factitious oligosaccharides, prepared by enzymic hydrolysis of xyloglucan [6–14] or, occasionally, by chemical synthesis [8,15]. However, if a natural growth-regulating role is to be ascribed to XGOs, their natural occurrence at biologically effective concentrations must be established. So far, no studies have been reported of searches for XGOs in intact tissues. Nevertheless, they have been detected in culture filtrates of suspension-cultured spinach cells, and such culture filtrates may reasonably be taken as a model for the apoplast of an intact plant. Oligosaccharides released into the apoplast by cells in an intact plant would be confined to a very small volume — the xylem sap and the water which permeates the cell walls, middle lamella and any extracellular slime etc. In cell cultures, in contrast, apoplastic oligosaccharides would be diluted into a large volume of culture medium, facilitating collection.

When cultured spinach cells were fed L-[1-^{3}H]arabinose or L-[1-^{3}H]fucose, the culture medium accumulated [*xylosyl*-^{3}H]XXFG or [*fucosyl*-^{3}H]XXFG, respectively [44]. These XGOs were mainly present as O-acyl (probably O-acetyl) derivatives. However, these are also of interest because O-acetylated XXFG has been shown to inhibit 2,4-D-induced growth of pea-stem segments almost as well as XXFG [7]. The spinach cells accumulated *O*-acyl-[^{3}H]XXFG to a concentration of about 0.4 μM in the culture medium [44]. Two-dimensional paper chromatography of the culture filtrate revealed at least 16 (*pentosyl*-^{3}H)-labelled oligosaccharides, some of which, when treated with 'Driselase' yielded the xyloglucan-diagnostic disaccharide α-D-Xyl*p*-(1→6)-D-Glc [44].

Soluble extracellular XGOs could arise either by *de novo* synthesis and direct secretion of the oligosaccharides from the protoplast, or, secondarily, by partial degradation of high-M_r polysaccharides. By use of kinetic-labelling experiments, it was demonstrated that the latter occurs. L-[1-^{3}H]Fucose was fed to cultured spinach cells, and the accumulation of (i) soluble extracellular [^{3}H]xyloglucan and (ii) extracellular ^{3}H-XGOs was monitored for up to 6 h [45]. After a short lag period, mainly representing the transit time for the Golgi system, soluble extracellular [^{3}H]xyloglucan accumulated at a constant rate for the whole 6 h duration. In contrast, after the lag period, the ^{3}H-XGOs accumulated in the culture medium with a constant acceleration. It was concluded that the ^{3}H-XGOs were being generated by the action of an enzyme (cellulase?) at a constant rate on a substrate (apoplastic [^{3}H]xyloglucan) that was increasing in specific radioactivity at an approximately constant rate.

It is, however, not certain that the extracellular XGOs arose by the action of cellulase: with the discovery of XET, its possible contribution will have to be evaluated. Indeed, cell cultures contain much higher XET activity than cellulase activity (P.R. Hetherington and S.C. Fry, unpublished work).

Fate of exogenous XGOs *in vivo*

Signalling molecules are generally subject to rapid degradation within the plant so that they do not persist after their message has been 'read'. What is the fate of XGOs *in vivo*? Again, no work has been reported in intact plant tissues, but studies have been made of the fate of ^{3}H-XGOs added to spinach cell-suspension cultures. However, [^{3}H]XXFG underwent surprisingly little hydrolysis [32,33]. Experiments were performed with [*fucosyl*-^{3}H]XXFG, [*xylosyl*-^{3}H]XXFG, [*reducing terminus*-1-^{3}H]XXFG and [*glucitol*-^{3}H]XXFGol. During incubations of up to 72 h, none of these preparations underwent appreciable hydrolysis to lower-M_r products. The major fate of each ^{3}H-labelled XXFG preparation was 'sequestration' by covalent binding to a soluble extracellular polymer [32], later identified as xyloglucan [33].

This is despite the fact that the cell walls of several plants have been shown to contain α-D-xylosidase, α-L-fucosidase, β-D-glucosidase and β-D-galactosidase (which can, in principle, hydrolyse XXFG to monosaccharides). One of the most important of these enzymes would be α-D-xylosidase, which removes the single xylose residue furthest from the reducing terminus of XXFG, to generate GXFG [25,46]. This still possesses growth-inhibitory activity at low concentrations, but is not expected to act as an acceptor substrate for XET [30]. A second important enzyme would be α-L-fucosidase [47,48], whose action would destroy the growth-inhibitory effect of XXFG by converting it into the inactive XXLG [10], but which would not diminish XET acceptor substrate activity [36] and would, therefore, not be expected to prevent the growth-promoting action of micromolar concentrations of XXFG.

The future

XGOs have been shown to have at least two interesting effects on plant growth. These effects have been achieved with factitious oligosaccharides, added exogenously. It will now be important to determine the natural occurrence of XGOs in intact plant organs, the factors regulating their synthesis and degradation, and their possible translocation within the plant. In this way, it will be possible to address the question of their biological significance. Another helpful approach is the use of mutants to dissect oligosaccharin signalling pathways. One interesting study in this respect is that of Reiter *et al.* [49], who have shown that a mutant of *Arabidopsis*, defective in L-fucose formation in shoot tissues, and thus presumably unable to make XXFG in the shoots, is relatively unimpaired in growth control. It will be interesting to determine whether, in this mutant, any XXFG reaches the shoot from the root, and also whether the shoots might synthesize a variant of XXFG in which the L-fucose residue is replaced by its metabolically and structurally close relative, L-galactose [50].

Many questions remain open concerning the mode of action and biological significance of xyloglucan-derived oligosaccharins as growth regulators. Nevertheless, the body of literature which already exists firmly establishes XGOs as interesting signalling molecules which require intensive study.

I am grateful to Miss Joyce Aitken for excellent technical assistance. Our work was supported by a European Community 'BRIDGE' contract.

Manuscript received 13 March 1993.

References

1. Albersheim, P. and Darvill, A.G. (1985) Sci. Am. **253**, 58–64
2. Aldington, S., McDougall, G.J. and Fry, S.C. (1991) Plant Cell Environ. **14**, 625–636
3. Albersheim, P., Darvill, A., Augur, C., Cheong, J-J., Eberhard, S., Hahn, M.G., Marf, V. and Mohnen, D. (1992) Acc. Chem. Res. **25**, 77–83
4. Darvill, A.G., Augur, C., Bergmann, C., Carlson, R.W., Cheong, J-J., Eberhard, S., Hahn, M.G. and Ló, V-M. (1992) Glycobiology **2**, 181–198
5. Aldington, S. and Fry, S.C. (1993) Adv. Bot. Res. **19**, 1–101
6. Warneck, H. and Seitz, H-U. (1993) J. Exp. Bot. **44**, 1105–1109
7. York, W.S., Darvill, A.G. and Albersheim, P. (1984) Plant Physiol. **75**, 295–297
8. Augur, C., Yu, L., Sakai, K., Ogawa, T., Sina, P., Darvill, A.G. and Albersheim, P. (1992) Plant Physiol. **99**, 180–185
9. McDougall, G.J. and Fry, S.C. (1988) Planta **175**, 412–416
10. McDougall, G.J. and Fry, S.C. (1989) Plant Physiol. **89**, 883–887
11. McDougall, G.J. and Fry, S.C. (1989) J. Exp. Bot. **40**, 233–239
12. Emmerling, M. and Seitz, H-U. (1990) Planta **182**, 174–180
13. Hoson, T. and Masuda, Y. (1991) Plant Cell Physiol. **32**, 777–782
14. Lorences, E.P., McDougall, G.J. and Fry, S.C. (1990) Physiol. Plant **80**, 109–113
15. Pavlova, Z.N., Ash, O.A., Vnuchkova, V.A., Babakov, A.V., Torgov, V.I., Nechaev, O.A., Usov, A.I. and Shibaev, V.N. (1992) Plant Sci. **85**, 131–134
16. McDougall, G.J. and Fry, S.C. (1990) Plant Physiol. **93**, 1042–1048
17. O'Neill, M., Albersheim, P. and Darvill, A.G. (1990) Methods Plant Biochem. **2**, 415–441

18. Branca, C.A., De Lorenzo, G. and Cervone, F. (1988) Physiol. Plant **72**, 499–504
19. Hasegawa, K., Mizutani, J., Kosemura, S. and Yamamura, S. (1992) Plant Physiol. **100**, 1059–1061
20. Hayashi, T. (1989) Annu. Rev. Plant Physiol. Plant Mol. Biol. **40**, 139–168
21. Fry, S.C. (1989) J. Exp. Bot. **40**, 1–11
22. Bauer, W.D., Talmadge, K.W., Keegstra, K. and Albersheim, P. (1973) Plant Physiol. **51**, 174–184
23. Fry, S.C. (1988) in The Growing Cell Wall: Chemical and Metabolic Analysis, pp. 262–264, Longman, Harlow
24. McDougall, G.J. and Fry, S.C. (1991) Carbohydr. Res. **219**, 123–132
25. Fanutti, C., Gidley, M.J. and Reid, J.S.G. (1991) Planta **184**, 137–147
26. El-Rassi, Z., Tedford, D., An, J. and Mort, A.J. (1991) Carbohydr. Res. **215**, 25–38
27. Valent, B.S., Darvill, A.G., McNeil, M., Robertsen, B.K. and Albersheim, P. (1980) Carbohydr. Res. **79**, 165–192
28. Hisamatsu, M., York, W.S., Darvill, A.G. and Albersheim, P. (1992) Carbohydr. Res. **227**, 45–71
29. Matsushita, J., Kato, Y. and Matsuda, K. (1985) Agric. Biol. Chem. **49**, 1533–1534
30. Lorences, E.P. and Fry, S.C. (1993) Physiol. Plant **88**, 105–112
31. Campbell, A.D. and Labavitch, J.M. (1991) Plant Physiol. **97**, 699–704
32. Baydoun, E.A-H. and Fry, S.C. (1989) J. Plant Physiol. **134**, 453–459
33. Smith, R.C. and Fry, S.C. (1991) Biochem. J. **279**, 529–535
34. Cosio, E.G., Frey, T. and Ebel, J. (1990) FEBS Lett. **264**, 235–238
35. Cheong, J-J., Birberg, W., Fügedi, P., Pilotti, Å., Garegg, P.J., Hong, N., Ogawa, T. and Hahn, M.G. (1991) Plant Cell **3**, 127–136
36. Fry, S.C., Smith, R.C., Renwick, K.F., Martin, D.J., Hodge, S.K. and Matthews, K.J. (1992) Biochem. J. **282**, 821–828
37. Fry, S.C. (1989) Physiol. Plant **75**, 532–536
38. Taiz, L. (1984) Annu. Rev. Plant Physiol. **35**, 585–657
39. Labavitch, J.M. (1981) Annu. Rev. Plant Physiol. **32**, 385–406
40. Nishitani, K. and Tominaga, R. (1991) Physiol. Plant **82**, 490–497
41. Nishitani, K. and Tominaga, R. (1992) J. Biol. Chem. **267**, 21058–21064
42. Farkaš, V., Sulova, Z., Stratilova, E., Hanna, R. and Maclachlan, G. (1994) Arch. Biochem. Biophys., in the press
43. Fry, S.C., Smith, R.C., Hetherington, P.R., and Potter, I, (1992) Curr. Top. Plant Biochem. Physiol. **11**, 42–62
44. Fry, S.C. (1986) Planta **169**, 443–453
45. McDougall, G.J. and Fry, S.C. (1991) J. Plant Physiol. **137**, 332–336
46. Koyama, T., Hayashi, T., Kato, Y. and Matsuda, K. (1983) Plant Cell Physiol. **24**, 155–162
47. Farkaš, V., Hanna, R. and Maclachlan, G. (1991) Phytochemistry **30**, 3203–3207
48. Augur, C., Benhamou, N., Darvill, A.G. and Albersheim, P. (1994) Plant J., in the press
49. Reiter, W-D., Chapple, C.C.S. and Somerville, C.R. (1993) Science **261**, 1032–1035
50. Baydoun, E.A-H. and Fry, S.C. (1988) J. Plant Physiol **132**, 484–490

Biochem. Soc. Symp. **60**, 15–26
Printed in Great Britain

Molecular and structural features of the pistil of *Nicotiana alata*

Angela H. Atkinson, Jan L. Lind, Adrienne E. Clarke and Marilyn A. Anderson

Plant Cell Biology Research Centre, School of Botany, University of Melbourne, Parkville, Victoria 3052, Australia

Introduction

The structure of the female sexual tissue of flowering plants, the pistil, has been studied for several years as a background for understanding the process of pollination — which leads to the growth of pollen tubes through the stigma and style to the ovary. Over the last 10 years or so, some individual molecules produced by the stigma and style have been identified and partly characterized; the impact of molecular biology has been to clone genes encoding the protein components of these molecules, and to discover genes specific to the female sexual tissues. A new aspect of the study of pistil components has been the realization that some of these, either individually or severally, may be involved in protecting the sexual tissues from potential predators and pathogens. Although the moist secretions which cover the stigma are potentially attractive substrates for fungi, bacteria and insects, it is rare that an infected pistil is observed in the field, even on plants which are infected in other parts. The study of the pistil components now assumes a new significance.

The structure of the pistil of *Nicotiana alata*

The stigma of the pistil of *N. alata* is bi-lobed, with a papillate epidermis, and is covered with exudate at maturity. The upper face of the stigma is formed by a secretory tissue that extends from the stigma through the style, to the placenta of the ovary (Fig. 1). This tissue, referred to as the transmitting tissue or the transmitting tract, forms the pathway that the pollen tubes follow on the way to the ovules. The solid, central core of transmitting tissue in the style is surrounded by a cortex of parenchymatous cells that includes two vascular bundles. The cortex is enveloped in a non-secretory epidermis.

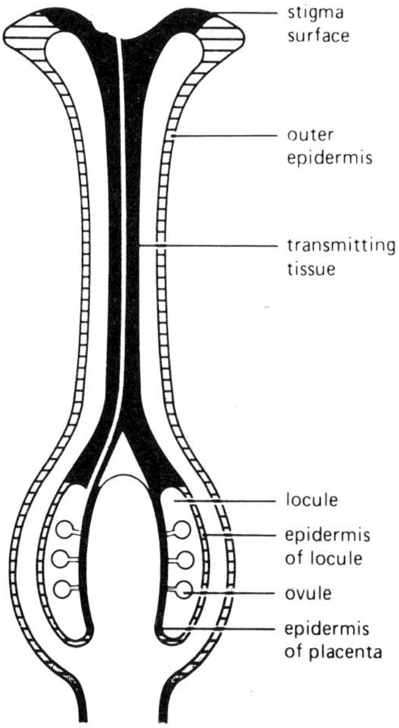

Fig. 1. Schematic representation of the pistil of *N. alata*. Compatible pollen tubes grow extracellularly through the transmitting tissue of the stigma, style and ovary (shaded).

Maturation of the stigma and production of stigma exudate

The development and structure of the stigma of *N. alata* is similar to that of other members of the Solanaceae, including *N. tabacum* [1], *N. sylvestris* [2], *Lycopersicon peruvianum* [3], *L. esculentum* [4] and *Petunia hybrida* [5–8]. The transmitting tissue in the stigma consists of an epidermal layer and the underlying layers of stigmatic secretory cells. At very early stages of development, the papillate surface is covered by a thin cuticle and the surface of the stigma is dry. As the flower matures, exudate accumulates between the cuticle and the cell wall of the papillar cells, causing the cuticle to become disrupted. By maturity, the cuticle is completely broken down, and the stigma surface is bathed in copious secretions. The breakdown of the cuticle as the stigma matures has also been described in *N. tabacum* [1] and *N. sylvestris* [2].

Even at early stages of development, the secretory cells underlying the stigma surface are separated by large spaces, which are filled with exudate. Secretion from these cells at a very early stage was also observed in *N. tabacum* [1]. As the flower matures the exudate increases, the cells separate further, and the exudate may spill out onto the stigma surface [3,5,9]. In *N. tabacum* [1] the most intensive production of exudate occurs 2 to 3 days prior to anthesis. The exudate in *N. alata* stigmas is heterogeneous in appearance, containing lipid-like droplets and electron-dense material (Fig. 2). The stigmatic exudates of other species, both within the Solanaceae and other families, contain lipids [1,2,10], phenolic compounds [10–12], proteins [10,13] and carbohydrate [1,2,4,6,13]. Herrero and Dickinson [8] observed lysis of papillae during maturation of the stigma of *Petunia*, and suggested that the contents of lysed cells also contribute to the stigma exudate.

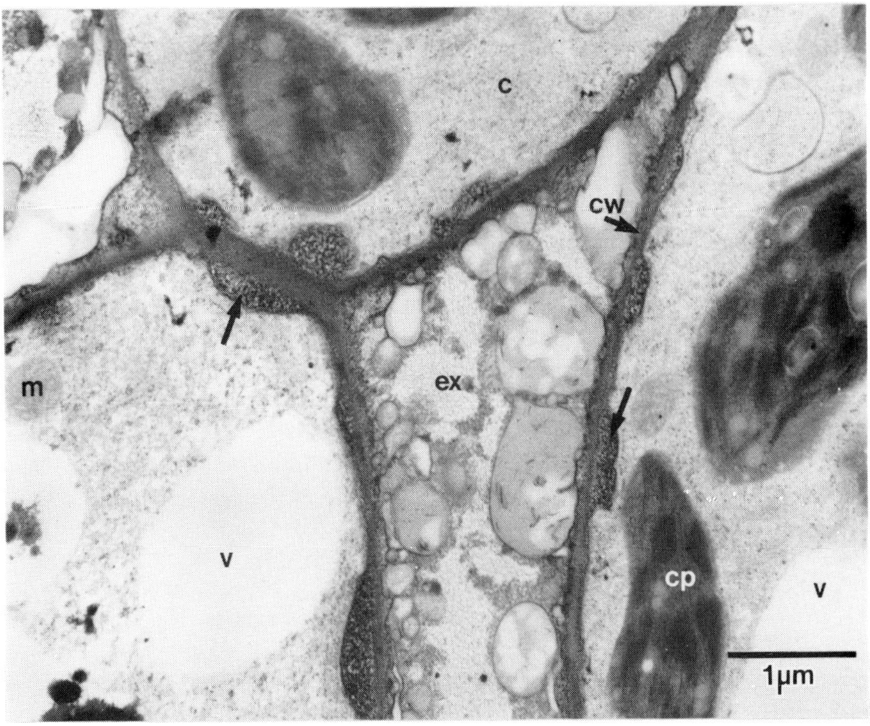

Fig. 2. The stigmatic secretory cells of a mature, receptive flower of *N. alata*. Longitudinal section through the secretory cells below the surface in a mature flower (×20 000). The extracellular spaces are filled with an exudate (ex) containing lipid-like droplets and electron-dense material. Large amounts of electron-dense material are present between the cell wall (cw) and the plasma membrane (arrowed). Abbreviations used: m, mitochondrion; cp, chloroplast; c, cytoplasm; v, vacuole.

Secretion of lipids from the secretory cells in other species coincides with a change from predominantly rough endoplasmic reticulum (ER) at the immature stages, to smooth ER at maturity, and a change in plastid structure [1,2,14,15]. It is not known how the lipid component of the exudate is secreted from the cell. Cresti *et al.* [1] observed that the lipid droplets in the stigma of *N. tabacum*, both in the cytoplasm and extracellular spaces, were not membrane bound, and suggested that the secretion involved passage of lipid molecules through the plasma membrane and cell wall, and re-aggregation in the extracellular spaces (eccrine secretion). In addition, these authors suggested that the same mechanism may function in the secretion of the carbohydrate of the exudate, since no vesicles containing carbohydrate were observed. A similar mode of secretion has been suggested in the stigmas of other species [15–17]. An alternative mechanism has been proposed for the monocotyledon *Gladiolus gandavensis* by Clarke and co-workers, who observed channels in the papillar cell walls that might facilitate the export of lipids and polysaccharides onto the stigma surface [18].

Electron-dense inclusions in the stigma exudate, as seen in *N. alata* (Fig. 2) were reported in *N. sylvestris* [2]. It was suggested that this material may represent terpenes, which have been demonstrated to be a common feature of secretions of leaves of *Nicotiana* species [19]. Terpene synthesis is believed to occur in the plastids [20], and may account for the electron-dense material seen in the plastids of the secretory cells of *N. alata*. In the mature stigma of *N. alata*, large deposits of electron-dense material are evident between the cell wall and the plasma membrane of the secretory cells (Fig. 2) but the nature of this material is unknown.

Development of the transmitting tract of the style

Although the stigma appears to be well developed at the early bud stages of flower development, the transmitting tissue of the style is not fully developed as the cells of the transmitting tract are closely packed. The files of elongated cells of the stylar transmitting tissue separate as the style matures and the region between the cells is filled with mucilage. This mucilage appears to be different from that in the secretory zone of the stigma, because the lipid droplets present in the stigma are not apparent in the transmitting tract of the style (Fig. 3). This is consistent with the results of previous studies which have identified the major components of the transmitting tract matrix as polysaccharides, proteins and glycoproteins [8,21,22]. Histochemical studies have shown that lipids are absent from the stylar exudate in the Solanaceae [22].

Multivesicular bodies are present in the transmitting tract cells of the style of *N. alata*, particularly at the immature stage. These bodies have been described in the stigmas and styles of a number of species [22–24], and have been implicated in the granulocrine secretion of components of the transmitting tract mucilage in *N. alata* [25].

Components of the extracellular matrix of the transmitting tissue

Pollen tubes in *N. alata* grow extracellularly through the matrix of the stigma and style. Some of the components in the matrix have a nutritional role, others function to reject incompatible pollen and there is growing evidence that several molecules function to protect the female reproductive tissues against potential pathogens. The following section describes the molecules from *N. alata* pistils that have been well characterized.

Self-incompatibility-associated ribonucleases

Gametophytic self-incompatibility is a genetically controlled system that prevents fertilization by self-pollen and thus ensures outbreeding and the maintenance of hybrid vigour (see [26] for review). The system is controlled by a single gene locus, the S-locus, which has multiple alleles. If the S-allele in the haploid pollen grain is matched by one of the two S-alleles in the diploid pistil, then pollen tubes are incompatible and growth is terminated in the top 6–7 mm of the style. The S-allele products in the pistil are ribonucleases (RNases) [27] that are produced by the stylar transmitting tissue cells [28] and are secreted into the extracellular matrix as the style matures [29]. The self-incompatibility-associated ribonucleases (S-RNases) are basic proteins, ranging in size from 28–34 kDa [30]. The allelic variation in molecular mass is due mainly to differences in the number of N-linked, complex-type, carbohydrate side-chains [31].

The sequences of seven S-RNases are available for *N. alata* [29,32], and sequences of 13 S-RNases have been obtained from other solanaceous species; *P. hybrida*,

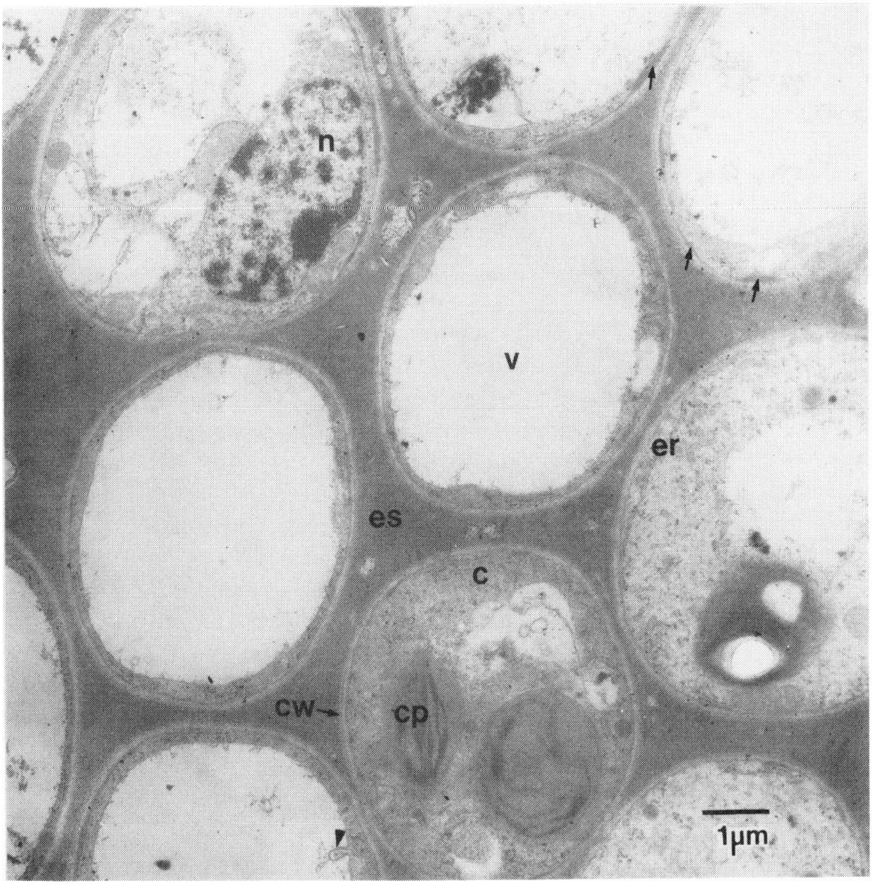

Fig. 3. Transverse section through the stylar transmitting tract of a mature receptive flower of *N. alata*. Transverse section (× 10000) through the transmitting tissue of a mature style. The transmitting tissue cells have large vacuoles (v) and prominent nuclei (n). The extracellular spaces (es) between the cells are larger than in the immature flower and are filled with densely staining mucilage that is homogeneous in appearance. There is some electron-dense material between the cell wall and the plasma membrane (small arrows). In one cell, a multivesicular body (arrowhead) can be seen at the edge of the vacuole. Abbreviations used: c, cytoplasm; cp, chloroplast; er, endoplasmic reticulum; cw, cell wall.

P. inflata, *Solanum tuberosum*, *S. chacoense* and *L. peruvianum* (see [33] for sequence identity). The encoded proteins contain regions that are highly conserved, particularly around the two histidine residues that are believed to be involved in the RNase active site [34]. Highly variable regions, designated hypervariable regions [29,35], are also present, and may be essential for allele specificity.

The S-gene product of the pollen may be involved in selective uptake of the S-RNase from the transmitting tract, when there is a match of S-alleles in the pollen and the style [36]. This is supported by the observation that S_2-RNase from *N. alata* is taken up into S_2-pollen tubes grown *in vitro* [37], although S_2-RNase is also taken up, to a lesser extent, by S_3- and S_6-pollen tubes grown *in vitro* [37]. In addition, the RNA of incompatible pollen tubes growing *in vivo* is degraded, while the RNA of compatible pollen tubes remains intact [36]. Pollen tubes do not synthesize rRNA [38], and rRNA is susceptible to S-RNase digestion [36]. Therefore, the digestion of rRNA would be predicted to cause a slowing of pollen-tube growth and, ultimately, death of the pollen tube [36]. Until the product of the S-locus in pollen is identified, and its relationship to the uptake mechanism established, the precise mechanism of action of the style S-RNase remains a matter for conjecture.

Hydroxyproline-rich molecules in stigmas and styles

Arabinogalactan proteins Arabinogalactan proteins (AGPs) were first observed in pistil exudates by Gleeson and Clarke [39], and subsequently have been detected in the pistils of all angiosperm species surveyed [40]. AGPs are not restricted to pistils; they are present in the extracellular matrix and are associated with the plasma membrane of all plant tissues examined [41]. The AGPs are high-molecular-mass proteoglycans with a negative to neutral overall charge and they are readily soluble in aqueous buffers. A characteristic feature of AGPs is their ability to bind β-glucosyl Yariv reagent, a synthetic compound made by coupling 4-aminophenyl glucoside to phloroglucine [42]. The nature of the binding is not understood but it is likely to involve both the protein and carbohydrate moieties. The majority of AGPs that have been characterized chemically contain less than 10% protein [43]. The carbohydrate side-chains are composed of a branched galactan framework — substituted primarily with arabinose residues but also with rhamnose, mannose, xylose, fucose and uronic acids in some cases [43,44]. The protein backbone is often rich in hydroxyproline, alanine and serine, and a common feature of the limited available sequence is the repeated dipeptide Ala-Hyp [45].

AGPs are present in the extracellular mucilage of both stigmas and styles of *N. alata* [21] and the AGPs in these tissues can be distinguished by their charge properties (see Fig. 4) [46]. Gell and co-workers have observed that pollination with compatible or incompatible pollen, causes a 90% increase in the amount of AGP in the stigma [46]. The increase in AGP was suggested to be a response of the pollinated pistil to the stress generated by the growing pollen tubes [46]. There is no significant change in the amount of AGP in the style after pollination, but the maturation of the fruit after a compatible pollination is associated with an increase in the amount of AGP in the ovary (A. Gane, personal communication), [47]. It has been proposed that AGPs may aid pollen adhesion to the stigma surface on wet stigmas, maintain the gel structure in the extracellular matrix of the transmitting tissue [39] and provide nutrients for uptake by pollen tubes in their heterotrophic phase of growth [39,43,48]. The nutritional role of AGP-like molecules in pistil exudates has been studied in *Lilium longiflorum* by Labarca and Loewus [48,49] who found that radiolabelled sugars from high-molecular-mass glycoconjugates in the pistil exudates were taken up by growing pollen tubes and incorporated into the wall.

It is possible that AGPs have a role in defining tissue identity, as it has been known for some time that AGPs from different tissues have different charge characteristics and subtle differences in their carbohydrate side-chains [43,50–52]. More recently, plasma-membrane-bound AGP-like epitopes have been shown to be under

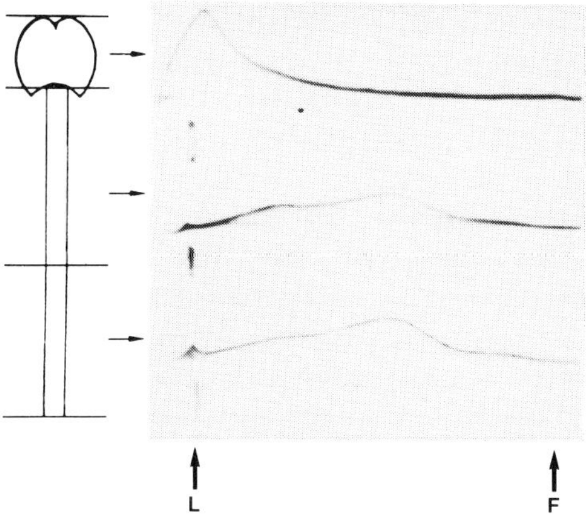

Fig. 4. Differences in the charge properties of the AGPs in the stigma and the upper and lower style of *N. alata*, revealed by crossed electrophoresis of buffer-soluble extracts. L, Loading well; F, dye front. Data compiled from Gell et al. [46].

tight temporal and spatial regulation during development of the male and female gametophytes and the early sporophyte of *Brassica napis* [53,54]. AGP-like epitopes are also developmentally regulated in carrot roots [55].

Extensin-like and proline-rich proteins Chen and co-workers have described a gene that is expressed specifically in the transmitting tissue of the style of *N. alata* and encodes a putative extensin-like protein, called NaPRP3 [56]. Other style-specific genes encoding proline-rich proteins have also been described in *N. tabacum* [57,58], and *Antirrhinum* [59]. The gene from *N. alata* resembles classic extensin in that it contains six copies of the extensin gene-associated repeat unit, Ser(Pro)$_4$, but the gene encodes no tyrosine residues — which are thought to cross-link extensin into the cell wall via intermolecular iso-dityrosine linkages [60]. The polypeptides encoded by the 'extensin-like' genes from *N. tabacum* and *Antirrhinum* have few repeated peptide sequences, and also contain relatively low levels of tyrosine. The protein encoded by the NaPRP3 gene has not been isolated, so the level of hydroxylation of the proline residues is not known. In addition, it is not known whether the protein contains the short tri- and tetra-arabinosyl side-chains linked to hydroxyproline and the single galactose residues linked to serine that are typically found in the extensins [44]. The product of the NaPRP3 gene from *N. alata* has a predicted signal sequence, but it is not known whether the protein is secreted and retained in the cell wall or whether it is a soluble component of the extracellular mucilage.

Defence-related molecules in stigmas and styles

A common feature of many of the characterized proteins from pistils is their relationship to proteins that are induced by wounding or pathogenesis. The extensin-like proteins can be included in this class. The presence of these molecules in pistils is not surprising when one considers the large spaces between the transmitting cells of the stigma and the style (Fig. 2 and Fig. 3) and the absence of a physical barrier, such as a cuticle, to prevent access of potential pathogens. The extracellular exudate, which is rich in proteins, free amino acids, lipids and carbohydrates, is potentially a moist, nutritious environment for pathogen growth, yet infection of pistils is rare [61].

Proteinase inhibitor

A gene encoding a serine proteinase inhibitor (PI), with some sequence identity to the type-II inhibitor family of potato and tomato, is expressed at high levels in the stigmas but not the styles of *N. alata* [62]. In contrast to the type-II PI proteins from potato and tomato that are composed of two domains, one with a chymotrypsin-reactive site and one with a trypsin-reactive site, the precursor PI protein from *N. alata* stigmas contains six repeated domains. Domains 1 and 2 have chymotrypsin-reactive sites and domains 3, 4, 5 and 6 have sites specific for trypsin. The PI protein comprises approximately 20–30% of the total buffer-soluble protein extracted from stigmas, which corresponds to approximately 20–30 μg of PI protein per stigma.

The PI protein was purified from mature stigmas by ammonium sulphate fractionation, gel filtration and affinity chromatography on chymotrypsin–Sepharose. The purified PI protein had an apparent molecular mass of about 6 kDa, rather than 42 kDa as predicted from the cDNA clone. The *N*-terminal sequence of the purified PI, however, corresponded to six sites in the sequence predicted from the cDNA clone, suggesting that the 6 kDa species had been produced by proteolytic processing (Fig. 5). Indeed, the presence of Lys and Thr at position 11 in the *N*-terminal sequence of the purified PI (Fig. 6) indicated that it was a mixture of at least five similar peptides derived from the precursor protein, since two of the peptides were predicted to have Thr at position 11 and three were predicted to have Lys. The presence of both chymotrypsin and trypsin activity in the stigma PI preparation provided further evidence for a mixture of peptides. Processing was confirmed when the specific antibody raised to the purified 6 kDa PI preparation cross-reacted strongly with proteins of 42 kDa and 18 kDa that are present in the green-bud stages of flower development. The 42 kDa component may be the precursor encoded by the cDNA clone, and the 18 kDa component may be a processing intermediate composed of three domains. As the flower matures, the 42 kDa and 18 kDa components decrease in concentration and the 6 kDa species becomes more abundant. The amino acid sequence around the processing site is shown in Fig. 6.

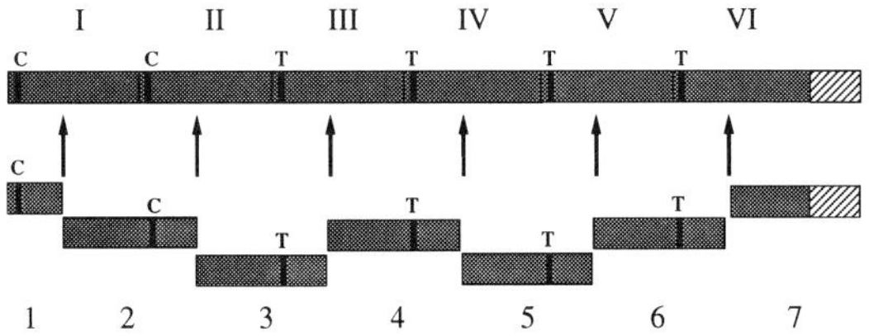

Fig. 5. Processing of the proteinase inhibitor precursor protein from *N. alata* styles. The precursor protein of 42 kDa (minus the signal sequence) contains six repeated domains (I–VI) each with a potential reactive site for chymotrypsin (C) or trypsin (T). Proteolytic processing at a conserved site within each of these domains (Fig. 6) produces five homologous proteinase inhibitors of about 6 kDa (peptides 2–6) and two flanking peptides (1 and 7).

−10 1
ICP(R or L)(S or A)EEKKN**DRICTNCCAG(TorK)KG**

Fig. 6. The amino acid sequence around the processing sites in the precursor PI protein. The sequence in bold is the *N*-terminal sequence obtained from the purified PI protein. The sequence labelled with negative numbers is the flanking sequence predicted from the cDNA clone. The predicted precursor protein contains six repeats of this sequence.

The site of processing has not been determined, but may be located between the aspartate and asparagine residues in the sequence outlined in Fig. 6. Vacuolar proteases with specific requirements for asparagine residues have been isolated from immature soybean seeds [63] and pumpkin cotyledons [64]. In the case of the *N. alata* PI, processing analogous to that of peptide hormones is also possible because each of the possible 6 kDa peptides are flanked by dibasic residues (Lys-Lys, positions −2 and −3 in Fig. 6). However, a system like this has not been described in plants, and it is more likely that the dibasic residues contribute to the predicted hydrophilic loops that present the processing site on the surface of the molecule [62].

The PI has been located in the vacuoles of the papillae and the underlying secretory cells in the stigma of *N. alata* using immunogold electron microscopy. The PI is also found at low levels in the extracellular matrix of the stigma [65].

The PI in stigmas may be involved in deterring insects from feeding on stigmas, or in protecting the stigma from pathogen invasion since the related type-II PIs from potato and tomato are effective against proteases of fungal, bacterial and insect origin (see review by Ryan [66]).

Defence-related molecules identified in the pistils of other members of the Solanaceae

The (1,3)-β-glucanases, identified in the styles of *N. tabacum* [67,68] and *N. alata* [69], have a similar expression pattern to the S-RNases. They are glycosylated polypeptides that are secreted by the transmitting tissue cells into the extracellular matrix during maturation of the style. The (1,3)-β-glucanase polypeptide of *N. tabacum* style is most closely related to the acidic (1,3)-β-glucanases induced in leaves by tobacco mosaic virus infection of *N. tabacum* [67], although the pathogenesis-related glucanase is not glycosylated.

If the stylar (1,3)-β-glucanase functions to protect the style against pathogen invasion, it may be assisted by chitinase. Chitinases and (1,3)-β-glucanases, induced in pea pods by the pathogen *Fusarium solani*, act synergistically to inhibit the *in vitro* growth of several pathogens of pea [70]. A gene that encodes a protein with a high level of sequence identity to a basic chitinase is expressed at high levels in the transmitting tissue of *L. peruvianum* [71] and *Solanum tuberosum* [72]. Chitinase activity has yet to be demonstrated in the style of *L. esculentum*, but there is a recent report of high levels of chitinase activity in the mature stigma of *P. hybrida* [73]. We have not tested *N. alata* styles for the presence of chitinase at this stage.

It is not known whether the stylar (1,3)-β-glucanases are required for pollen-tube growth. Pollen tubes have a large amount of predominantly (1,3)-β-glucan in the inner wall layer [74], but whether it is a substrate for these (1,3)-β-glucanases, and whether the enzyme has access to the inner wall, remains to be determined.

Other defence-related molecules and genes have been detected in stigmas and styles of plants in the Solanaceae by immunological cross-reactivity to characterized antisera, or by nucleic acid hybridization to cDNA probes. Thaumatin-like proteins have been identified in the cortical cells and tracheids of *N. tabacum* styles [75] by their immunological cross-reactivity with antiserum to thaumatin. Recently, thionin-like molecules have also been described in pistils of *N. tabacum* [76]. The role of these stress-related proteins in pistils has not been completely explored. One hypothesis is that they are there to protect the reproductive organs from pathogenic organisms and insects [62,67]. Constitutive expression of these genes may limit the damage to the reproductive organs by potential pathogens, thus increasing the probability of successful fertilization and seed set.

Conclusion

Good progress is being made in establishing information on the molecular components of the pistil transmitting tract. These components are implicated in self rejection during a self-incompatible pollination, in nutrition during the heterotrophic growth phase of pollen tubes through the style and, more recently, as protectants against infection and predation. There are still large gaps in our knowledge, but the molecular tools are now available to establish new knowledge rapidly. Such knowledge will lead to insights into the function of individual molecules in the context of pistil physiology.

References

1. Cresti, M., Keijzer, C.J., Tiezzi, A., Ciampolini, F. and Focardi, S. (1986) Am. J. Bot. **73**, 1713–1722
2. Kandasamy, M.K. and Kristen, U. (1987) Ann. Bot. **60**, 427–437
3. Dumas, C., Rougier, M., Zandonella, P., Ciampolini, F., Cresti, M. and Pacini, E. (1978) Protoplasma **96**, 173–187
4. Kadej, A.T., Wilms, H.J. and Willemse, M.T.M. (1985) Acta Bot. Neerl. **34**, 95–103
5. Konar, R.N. and Linskens, H.F. (1966) Planta **71**, 356–371
6. Konar, R.N. and Linskens, H.F. (1966) Planta **71**, 372–387
7. Kroh, M. (1967) Planta **77**, 250–260
8. Herrero, M. and Dickinson, H.G. (1979) J. Cell Sci. **36**, 1–18
9. Shivanna, K.R. and Sastri, D.C. (1981) Ann. Bot. **47**, 53–64
10. Knox, R.B. (1984) in Cellular Interactions: Encyclopedia of Plant Physiology (Linskens, H.F. and Heslop-Harrison, J., eds.), pp. 197–261, Springer-Verlag, Berlin
11. Martin, F.W. and Brewbaker, J.L. (1971) in Pollen Development and Physiology (Heslop-Harrison, J., ed.), pp. 262–266, Appleton-Century Co., New York
12. Vasil, I.K. (1974) in Fertilization in Higher Plants (Linskens, H.F., ed.), pp. 105–118, North Holland, Amsterdam
13. Kristen, U., Biedermann, M., Liebezeit, G., Dawson, B. and Bohm, L. (1979) Eur. J. Cell Biol. **19**, 281–287
14. Cresti, M., Ciampolini, F., van Went, J.L. and Wilms, H.J. (1982) Planta **156**, 1–9
15. Sedgley, M. and Blesing, M.A. (1983) Bot. Gaz. **144**, 185–190
16. Dumas, C. (1977) Planta **137**, 177–184
17. Heslop-Harrison, J. and Heslop-Harrison, Y. (1983) Ann. Bot. **51**, 571–583

18. Clarke, A.E., Abbot, A., Mandel, T.E. and Pettit, J.M. (1980) J. Ultrastructure Res. **73**, 269–280
19. Heeman, V., Brummer, U., Paulsen, Ch. and Seehofer, F. (1983) Phytochemistry **22**, 133–135
20. Gleizes, M., Pauly, G., Carde, J.P., Marpeau, A. and Bernard-Dagan, C. (1983) Planta **159**, 373–381
21. Sedgley, M., Blesing, M.A., Bonig, I., Anderson, M.A. and Clarke, A.E. (1985) Micron Microsc. Acta **16**, 247–254
22. Cresti, M., van Went, J.L., Pacini, E. and Willemse, M.T.M. (1976) Planta **132**, 305–312
23. Sedgley, M. and Buttrose, M.S. (1978) Aust. J. Bot. **26**, 663–682
24. Heslop-Harrison, J. and Heslop-Harrison, Y. (1980) Acta Bot. Neerl. **29**, 261–276
25. Sedgley, M. and Clarke, A.E. (1986) Nord. J. Bot. **6**, 591–598
26. Clarke, A.E. and Newbigin, E. (1993) Annu. Rev. Genet. **27**, 257–279
27. McClure, B.A., Haring, V., Ebert, P.R., Anderson, M.A., Simpson, R.J., Sakiyama, F. and Clarke, A.E. (1989) Nature (London) **342**, 955–957
28. Cornish, E.C., Pettit, J.M., Bonig, I. and Clarke, A.E. (1987) Nature (London) **326**, 99–102
29. Anderson, M.A., McFadden, G.I., Bernatzky, R., Atkinson, A., Orpin, T., Dedman, H., Tregear, G., Fernley, R. and Clarke, A.E. (1989) Plant Cell **1**, 483–491
30. Jahnen, W., Batterham, M.P., Clarke, A.E., Moritz, R. L. and Simpson, R.J. (1989) Plant Cell **1**, 493–499
31. Woodward, J.R., Craik, D., Dell, A., Khoo, K-H., Munro, S.L.A., Clarke, A.E. and Bacic, A. (1992) Glycobiology **2**, 241–250
32. Kheyr-Pour, A., Bintrim, S.B., Ioerger, T.R., Remy, R., Hammond, S.A. and Kao, T.-H. (1990) Sex. Plant Reprod. **3**, 88–97
33. Tsai, D-S., Lee, H-S., Post, L.C., Kreiling, K.M. and Kao, T-H. (1992) Sex. Plant Reprod. **5**, 256–263
34. Haring, V., Gray, J.E., McClure, B.A., Anderson, M.A. and Clarke, A.E. (1990) Science **250**, 937–941
35. Ioerger, T.R., Gohlke, J.R., Xu, B. and Kao, T.-H. (1991) Sex. Plant Reprod. **4**, 81–87
36. McClure, B.A., Gray, J.E., Anderson, M.A. and Clarke, A.E. (1990) Nature (London) **347**, 757–760
37. Gray, J.E., McClure, B.A., Bonig, I., Anderson, M.A. and Clarke, A.E. (1991) Plant Cell **3**, 271–283
38. Mascarenhas, J.P. (1975) Bot. Ref. **41**, 259–314
39. Gleeson, P.A. and Clarke, A.E. (1979) Biochem. J. **181**, 607–621
40. Hoggart, R.M. and Clarke, A.E. (1984) Phytochemistry. **23**, 1571–1573
41. Clarke, A.E., Anderson, R.L. and Stone, B.A. (1979) Phytochemistry **18**, 521–540
42. Yariv, J., Lis, H. and Katchalski, E. (1967) Biochem. J. **105**, 1C–2C
43. Fincher, G.B., Stone, B.A. and Clarke, A.E. (1983) Annu. Rev. Plant Physiol. **34**, 47–70
44. Showalter, A.M. and Varner, J.E. (1989) Biochem. Plants **15**, 485–519
45. Gleeson, P.A., McNamara, M., Wettenhall, R.E.H., Stone, B.A. and Fincher, G.B. (1989) Biochem. J. **264**, 857–862
46. Gell, A.C., Bacic, A. and Clarke, A.E. (1986) Plant Physiol. **82**, 885–889
47. Webb, M.C. and Williams, E.G. (1988) Ann. Bot. **61**, 415–423
48. Labarca, C. and Loewus, F. (1973) Plant Physiol. **52**, 87–92
49. Labarca, C. and Loewus, F. (1972) Plant Physiol. **50**, 7–14
50. Gleeson, P.A. and Clarke, A.E. (1980) Phytochemistry **19**, 1777–1782
51. Gleeson, P.A. and Clarke, A.E. (1980) Carbohydr. Res. **83**, 187–192
52. van Holst, G-J. and Clarke, A.E. (1986) Plant Physiol. **80**, 786–789
53. Pennell, R.I. and Roberts, K. (1990) Nature (London) **344**, 547–549

54. Pennell, R.I., Janniche, L., Kjellbom, P., Scofield, G.N., Peart, J.M. and Roberts, K. (1991) Plant Cell **3**, 1317–1326
55. Knox, J.P., Linstead, P.J., Peart, J., Cooper, C. and Roberts, K. (1991) Plant J. **1**, 317–326
56. Chen, C-G., Cornish, E.C. and Clarke, A.E. (1992) Plant Cell **4**, 1053–1062
57. Goldman, M.H.D.S., Pezzotti, M., Suerinck, J. and Mariani, C. (1992) Plant Cell **4**, 1041–1051
58. Cheung, A.Y., May, B., Kawata, E.E., Gu, Q. and Wu, H-M. (1993) Plant J. **3**, 151–160
59. Baldwin, T.C., Coen, E.S. and Dickinson, H.G. (1992) Plant J. **5**, 733–739
60. Biggs, K.J. and Fry, S.C. (1990) Plant Physiol. **92**, 197–204
61. Jung, J. (1956) Phytopath. Z. **27**, 405–426
62. Atkinson, A.H., Heath, R.L., Simpson, R.J., Clarke, A.E. and Anderson, M.A. (1993) Plant Cell **5**, 203–213
63. Scott, M.P., Jung, R., Muntz, K. and Nielsin, N.C. (1992) Proc. Natl. Acad. Sci. U.S.A. **89**, 658–662
64. Hara-Nishimura, I., Inoue, K. and Nishimura, M. (1991) FEBS Lett. **294**, 89–93
65. Atkinson, A.H. (1992) in Expression of a Proteinase Inhibitor in the Pistil of an Ornamental Tobacco, *Nicotiana alata*. PhD Thesis, University of Melbourne, Melbourne, Australia
66. Ryan, C.A. (1990) Annu. Rev. Phytopathol. **28**, 425–449
67. Ori, N., Sessa, G., Lotan, T., Himmelhoch, S. and Fluhr, R. (1990) EMBO J. **9**, 3429–3436
68. Lotan, T., Ori, N. and Fluhr, R. (1989) Plant Cell **1**, 881–887
69. Mau, S-L. (1990) in Molecular Studies of Gametophytic Self-Incompatibility. PhD thesis, University of Melbourne, Melbourne, Australia
70. Mauch, F., Mauch-Mani, B. and Boller, T. (1988) Plant Physiol. **88**, 936–942
71. Gassar, C.S., Budlier, K.A., Smith, A.G., Shah, D.M. and Fraley, R.T. (1989) Plant Cell **1**, 15–24
72. Kaufmann, H., Kirch, H., Wemmer, T., Peil, A., Lottspeich, F., Uhrig, H., Salamini, F. and Thompson, P. (1992) in Sexual Plant Reproduction (Cresti, M. and Tiezzi, A., eds.), pp. 115–125, Springer-Verlag, Berlin
73. Leung, D.W.M. (1992) Phytochemistry **31**, 1899–1900
74. Meikle, P.J., Bonig, I., Hoogenraad, N.J., Clarke, A.E. and Stone, B.A. (1991) Planta **185**, 1–8
75. Richard, L., Arro, M., Hoebeke, J., Meeks-Wagner, D.R. and Tran Thanh Van, K. (1992) Plant Physiol. **98**, 337–342
76. Gu, Q., Kawata, E., Morse, M.J., Wu, H.-M. and Cheung, A.Y. (1992) Mol. Gen. Genet. **234**, 89–96

Biochem. Soc. Symp. **60**, 27–33
Printed in Great Britain

Developmental dynamics of plant cell-surface glycoproteins: towards a molecular plant anatomy

J. Paul Knox

Centre for Plant Biochemistry and Biotechnology, University of Leeds, Leeds LS2 9JT, U.K.

Introduction

The internal structure of plants, with a lack of internal organs and a small number of cell types, is relatively simple compared with the highly developed interiors of animal bodies. All vegetative structures consist of a specific arrangement of the three tissue systems — vascular, dermal and ground — that are functionally connected throughout the plant body and serve to integrate the often indeterminate development of an exterior, in both the aerial and soil environments. Although meristems continually form tissue and cell patterns, and the same major cell types occur in all plant structures, there is virtually no understanding of the molecular mechanisms underlying developmental anatomy in plants [1,2]

Plant cells generally only take on precise and distinctive morphologies characteristic of cell types in the context of multicellular structures, i.e. in a developmental context. Plant cells induced to proliferate in culture develop into entire organs rather than specific cell types, although the *Zinnia elegans* system in which isolated mesophyll cells can be induced to differentiate the morphological characteristics of tracheids is clearly an exception [3]. The importance of cell position within a meristem for the determination of cell fate now appears to be well established [4,5]. Major gaps in our understanding now concern the nature of cell-to-cell communication and the form of positional information that underlies the formation of species-specific and organ-specific anatomical patterns.

The cell wall is central to plant anatomy

The internal structure of plants is largely a matter of the cell wall. Two important consequences of the retention of a wall by cells are fixed cell positions and fixed cell shapes. In multicellular plant structures, anatomical patterns are readily identified and characterized by an assessment of wall positions and wall modifications. Indeed, several important functions, and anatomically distinct regions, within the plant body are carried out entirely by modified walls that remain after cell death.

The plant cell wall is both rigid and mutable: rigid in the sense that cells have identifiable and relatively fixed shapes and mutable in that it can stretch to accommodate expansion, and also in that it can be modified in relation to cell specialization. Not only must we consider the shape, form and spatial arrangement of cells, but also how closely connected, or separated, they are. Intercellular space can also be of functional importance and a major anatomical feature of plant structures. (For a consideration of our knowledge of the molecular aspects of cell-to-cell adhesion and tissue cohesion in plants, see [6].)

A significant proportion of what we know about the modulations of the complex architecture of the primary cell wall, in relation to the formation of cell-type specific morphologies or intercellular space, is emerging from the derivation and characterization of monoclonal antibody probes for the plant cell-surface polysaccharides and glycoproteins [7].

The developmental anatomy of the carrot root apex

In recent years, the developmental anatomy of the carrot root apex, and the development of carrot cells in suspension cultures, have been utilized for studies of the developmental regulation of the plant cell-surface components in conjunction with a programme to prepare anti-(cell-surface) monoclonal antibodies. The development of the pattern of the primary tissues at a root apex is ideal for the study of the formation of an anatomical pattern. In addition, cell cultures are highly accessible for studies of cells in the absence of, or in altered, developmental context.

The carrot root has an open apical organization with no clearly identifiable initials [8]. A section through the region of the meristem reveals very little organization in terms of cell shape. Identifiable patterns of cell shape relating to differentiation of the tissue systems are developed and elaborated by the proximal progeny of the meristem. The sequence of events resulting in the formation of the mature primary root anatomy can, perhaps, be thought to occur between two broadly classed developmental phases: the establishment of the spatial arrangement of future cell types and the differentiation of the functional cell types. The formation of the pattern of cell shapes is likely to occur somewhere between the two. However, the molecular basis of these phases and how they are connected or co-ordinated is unknown.

If transverse sections of the carrot root apex are observed in a proximal sequence, the first recognizable cell type is the pericycle [8] which is often the case at root apices [9]. Two different pericycle cell morphologies develop in relation to the future di-arch vascular pattern [8,10]. The first identifiable cellular event is the radial expansion of two groups of 2–3 pericycle cells at opposite sides of the periphery of the emerging stele. These cells can be used to identify not only the boundary between the stele and the cortex, but also the relative positions of the future xylem and phloem within the stele. The two future protoxylem poles will develop adjacent to these first individualized pericycle cells. The number of radially expanded pericycle cells increases and they undergo a series of oblique longitudinal divisions to give rise to two double-layered sectors of the pericycle that will be centred on the future xylem poles [8,10]. Cells of the single-layered regions of the pericycle directly abut the future protophloem sieve elements. (For a schematic representation of the primary root tissue pattern, see Fig. 1).

Two classes of cell-surface glycoprotein epitope have now been identified that display complex and dynamic restricted patterns of expression in the region of the root apex where this pattern is developed, prior to the functional differentiation of the vascular elements.

Modulations of plasma membrane arabinogalactan proteins

Arabinogalactan proteins (AGPs) are a diverse group of macromolecules found in secretions and mucilages and cell culture media [11]. In recent years, it has become apparent that a class of AGP is also associated with the outer face of the plasma membrane of higher plant cells [12–14]. AGPs obtained from seeds and the conditioned medium of cultured carrot cells have recently been implicated in the regulation of carrot somatic embryogenesis [15]. One means of characterization of the plasma membrane class of AGP has been the preparation of monoclonal antibodies recognizing both plasma-membrane-associated glycoproteins and also soluble AGPs obtained from the conditioned media of carrot suspension-cultured cell lines. An extended panel of monoclonal antibodies has now indicated the extensive developmental regulation of plasma membrane AGP epitopes in relation to the acquisition of anatomical patterns. Three of the restricted epitope expression patterns at the carrot root apex are shown schematically in Fig. 1, in relation to the organization of the emerging pattern of tissues and cell types. What they appear to reflect are groups of cells, or collectives, that often occur in the region of emerging cell types, but the boundaries of these collectives are not always precise in anatomical terms. These events are very early and occur well before aspects of functional differentiation (i.e. identifiable modification of walls). They occur at the same developmental time as differences in cell shape allow future positions and patterns of cell types to be identified.

The earliest modulation is that recognized by the monoclonal antibody JIM4 [16,17]. This epitope appears at the cell surface of the two double-layered sectors of the pericycle opposite the future xylem poles (Fig. 1B). During development, these distinct pericycle regions increase in size as the oblique divisions occur in additional pericycle cells, adjacent to the ones that directly abut the protoxylem poles. Recent observations have indicated that the expression of the JIM4 epitope also extends in the pericycle in this way, preceding the oblique divisions (P.J. Casero and J.P. Knox, unpublished work). Subsequent, in developmental terms, to the expression of the JIM4 epitope, cells in the region of the future xylem and the epidermis express the JIM13 epitope (Fig. 1C). At a more proximal distance there is a loss of the JIM15 epitope from cells of the central stele and the epidermis — resulting in the pattern shown schematically in Fig. 1D [18]. Whether these three modulations reflect a series of inducible events is yet to be determined.

Such observations serve to illuminate our ignorance. The restricted epitopes are carried by glycan components of the plasma-membrane glycoproteins, and several other

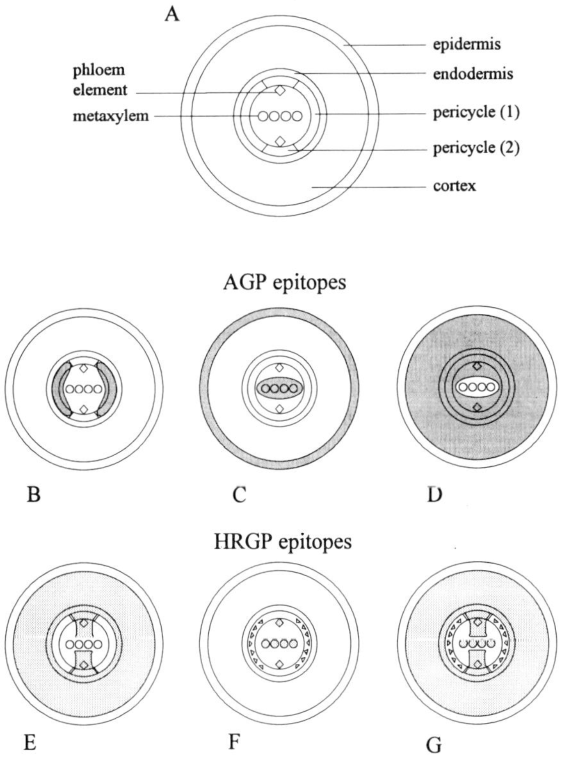

Fig. 1. Molecular anatomy of the carrot root apex. (A) A schematic representation of the pattern of tissues and major cell types seen in a transverse section of the carrot root apex. The pericycle is separated into morphologically distinct regions that reflect the di-arch arrangement of the xylem and phloem. (B–G) Shading indicates examples of some of the restricted epitope expression patterns identified by anti-AGP (B–D) and anti-HRGP (E–G) monoclonal antibodies in relation to the developmental pattern.

epitopes that occur at the surface of all the cells at the carrot root apex have now been identified. The structures of the carbohydrate epitopes are currently being determined, with a view to understanding the biochemical basis of the epitope dynamics. The arabinogalactan epitopes have also been detected to be developmentally regulated in the wall of non-adhered surfaces of plant cells [19,20] and these observations raise questions concerning the relationship of the wall and membrane antigens.

Cell wall glycoproteins and developmental anatomy

In screening hybridoma products for restricted patterns of antibody binding, several further modulations at the surface of carrot root cells have been identified. Recently, the characteristics of three monoclonal antibodies have been reported [21]. These antibodies recognize epitopes of cell wall hydroxyproline-rich glycoproteins

(HRGPs), as indicated by the occurrence of the epitopes in purified samples of carrot tap-root extensin and solanaceous lectins (both HRGPs are known to share common domains rich in hydroxyproline and oligo-arabinoside side-chains) [21]. (For a recent review of cell wall proteins, see [22].) These restricted patterns of HRGP epitopes occur in different cell types and cell collectives from the AGP epitopes, but appear to be complementary in some aspects (see Fig. 1). They also have similar characteristics in that the epitopes are not associated exclusively with single cell types and their expression undergoes dynamic changes in the region of the apex between the meristem and the zone of expansion/vacuolation. The JIM11 epitope is expressed at the surface of inner cortical cells of the emerging cortex, then all cortical cells. Later still, the occurrence of this epitope extends into two regions of the pericycle that directly abut the protophloem element and cells associated by position in the stele. These are cells of the single-layered pericycle — not recognized by the anti-AGP JIM4. As the group of cells develops further, the epitope appears in the epidermal cells.

A second epitope recognized by JIM12 has an even more restricted distribution, occurring initially and specifically at the intercellular spaces arising between the double-layered sectors of the pericycle adjacent to the xylem. These intercellular spaces mature into oil ducts [8]. The same epitope also occurs transiently at the surface of the two sieve elements that abut the pericycle and is finally expressed persistently by several of the large future metaxylem elements (Fig. 1F). Evidence that these two antigens are related is provided by a third epitope that appears to be common to the distinct JIM11 and JIM12 antigens, as revealed by both developmental immunocytochemical and immunoblotting patterns [21].

These observations are similar to those made of AGPs and indicate a further family of cell-surface proteins with developmentally regulated structural elements — the epitopes. In the case of the HRGPs, the loss or creation of epitopes may be related to specific glycosylation patterns or the possible oxidative cross-linking of this class of wall protein.

Prospects for a molecular understanding of developmental plant anatomy

The binding characteristics of the monoclonal antibodies described here appear to indicate that molecular markers for early stages of the formation of anatomical patterns can be detected at the plant cell surface. Cataloguing a series of molecular patterns in relation to morphogenesis form will not, in itself, allow us to unravel the molecular mechanisms that brought that form about. However, the monoclonal antibodies do provide us with molecular probes for biochemical modulations that appear to be correlated with the establishment of anatomical patterns. These probes can now be used to identify the precise nature of the structural elements of these two classes of surface proteins — both oligosaccharides of AGPs and possibly specific configurations of arabinoside side-chains on HRGP peptide domains.

It is not clear how extensive this phenomenon of developmentally regulated epitopes is within one system, such as a root apex. Extensive preparation of monoclonal antibodies for the saturation analysis of possible epitopes has not yet been carried out for either class of antigen.

The functions of the plasma-membrane AGPs and the cell-wall HRGPs are far from clear and, therefore, the functional relationship of the two sets of antigens within one cell or one developing organ is uncertain.

What are the possible functions of these cell-surface glycoproteins? First, they may relate to some aspect of signalling of cell position/cell identity between cells. This has often been speculated upon for AGPs [11] and the recent observations of biological activity of exogenous AGPs in relation to somatic embryo development may support a morphoregulatory role [15]. It is not clear how the added extracellular AGPs influence embryogenesis, or whether they interact with the cell wall or the plasma membrane. The occurrence and relationship of the membrane and wall AGPs need to be understood both for non-adhered cultured cells and also for cells within organized tissues.

Alternatively, these surface antigens may be involved in the construction of specific wall architectures required for the formation of specific anatomical features. Again, this is a possibility for AGP function. The plasma membrane AGPs may be involved in the co-ordination or maintenance of the assembly of cell-specific or cell-process-specific configurations of wall molecules. The recent observations that AGPs may interact with wall pectin [23] and the demonstration of the absence of plasma-membrane AGPs from NaCl-adapted, suspension-cultured tobacco cells, which have a reduced capacity to expand [24], may be important in this respect. Lastly, the glycoproteins may themselves be components of specific wall architectures. This is a likely role for the HRGP antigens with their ability to be insolubilized and possibly cross-linked into the wall. There will clearly be requirements for the strengthening of specific regions of cell walls around cells or intercellular spaces during organ development, and the HRGP epitope developmental patterns may reflect the use of particular wall proteins in this way.

The monoclonal antibodies described here will be invaluble probes for the characterization of the structural elements and interactions of the glycoproteins, as well as their developmental dynamics, and such studies will help to refine our understanding of function.

I am grateful to M. Smallwood for helpful discussions and comments on the manuscript and A. MacGregor for help with figure preparation.

References

1. Becraft, P.W. and Freeling, M. (1992) Curr. Opin. Gen. Dev. **2**, 571–575
2. Jurgens, G. (1992) Curr. Opin. Gen. Dev. **2**, 567–570
3. Fukuda, H. and Komamine, A. (1980) Plant Physiol. **65**, 57–60
4. Poethig, R.S. (1989) Trends Genet. **5**, 273–277
5. Irish, V.F. (1991) Curr. Opin. Cell Biol. **3**, 983–987
6. Knox, J.P. (1992) Plant J. **2**, 137–141
7. Knox, J.P. (1992) Protoplasma **167**, 1–9
8. Esau, K. (1940) Hilgardia **13**, 175–226
9. Steeves, T.A. and Sussex, I.M. (1989) Patterns in Plant Development, 2nd Edn, Cambridge University Press, Cambridge
10. Lloret, P.G., Casero, P.J., Pulgarn, A. and Navascués, J. (1989) Ann. Bot. **63**, 465–475
11. Fincher, G.B., Stone, B.A. and Clarke, A.E. (1983) Annu. Rev. Plant Physiol. **34**, 47–70
12. Pennell, R.I., Knox, J.P., Scofield, G.N., Selvendran, R.R. and Roberts, K. (1989) J. Cell Biol. **108**, 1967–1977

13. Norman, P.M., Kjellbom, P., Bradley, D.J., Hahn, M.G. and Lamb, C.J. (1990) Planta **181**, 365–373
14. Komalavilas, P., Zhu, J. and Nothnagel, E.A. (1991) J. Biol. Chem. **266**, 15956–15965
15. Kreuger, M. and van Holst, G-J. (1993) Planta **189**, 243–248
16. Knox, J.P., Day, S. and Roberts, K. (1989) Development **106**, 47–56
17. Knox, J.P. (1993) Soc. Exp. Biol. Semin. Ser. **53**, 267–283
18. Knox, J.P., Linstead, P.J., Peart, J., Cooper, C. and Roberts, K. (1991) Plant J. **1**, 317–326
19. Pennell. R.I., Janniche, L., Scofield, G.N., Booij, H., de Vries, S.C. and Roberts, K. (1992) J. Cell Biol. **119**, 1371–1380
20. Li, Y., Brun, L., Pierson, E.S. and Cresti, M. (1992) Planta **188**, 532–538
21. Smallwood, M., Beven, A., Donovan, N., Neill, S.J., Peart, J., Shaw, P., Roberts, K. and Knox, J.P. (1994) Plant J. **5**, 237–246
22. Showalter, A.M. (1993) Plant Cell **5**, 9–23
23. Baldwin, T.C., McCann, M. and Roberts, K. (1993) Plant Physiol. **103**, 115–123
24. Zhu, J.K., Bressan, R.A. and Hasegawa, P.M. (1993) Planta **190**, 221–226

Biochem. Soc. Symp. **60**, 35–41
Printed in Great Britain

Glycoprotein domains and their role in gamete recognition in *Fucus*

J.A. Callow*, C.R. Stafford, P.J. Wright and J.R. Green*

School of Biological Sciences, The University of Birmingham, PO Box 363, Birmingham B15 2TT, U.K.

Introduction

The purpose of this paper is to consider the rather specific type of environmental signal presented by gamete surfaces during the process of fertilization, a process which is fundamental to the life-cycle of all organisms that undergo sexual reproduction. We are concerned with the recognition processes which control the mutual association of egg and sperm cells during fertilization and the associated cell responses that occur within a few seconds or minutes of gamete fusion. The exploration of such events in higher plant cells is difficult because the gametes are embedded within tissues, and 'simpler' recognition systems presented by oogamous lower plants, such as *Fucus serratus*, therefore have much to offer. Naked gametes of these algae are released in large enough quantities to permit detailed biochemical studies. The interaction can be followed experimentally, either by measuring fertilization itself, or through a binding assay involving egg and sperm membranes or isolated components derived from them. In either case it is possible to explore the molecular characteristics of fertilization through inclusion of blocking agents, such as antibodies, oligosaccharides, putative receptor fractions and so on.

The general properties of the *Fucus* system have been reviewed several times (for example, see [1–4]) and it is not necessary to give more than brief details. Fertilization in *Fucus* is a species-specific interaction between brown, spherical, apolar, non-motile, naked egg cells, 60–80 μm in diameter, and bi-flagellate sperm cells, some 5 μm long. Sperm and egg cells are liberated naturally at low tide, and the process of fertilization can be considered to involve a number of stages, each involving aspects of recognition.

Chemoattraction

This is mediated by hydrocarbon pheromones, known as octatrienes [5], which are secreted by the egg cells. The polarized motility of sperm cells down the concentration gradient results in 'swarming' of large numbers of sperm around the egg.

*To whom correspondence should be addressed.

Sperm probing and binding to the egg cell

Sperm cells that contact the egg surface appear to engage in active probing of the surface via the anterior flagellum, before fusion occurs [6] and without immediately causing a fertilization potential [7]. Sperm of *Pelvetia canaliculata*, for example, appear to move over the surface of the egg for 10 s or more, attach to one spot then gyrate against the egg for another 10–20 s before evoking a fertilization potential. It has been speculated that this behaviour is evidence for sperm receptors on the egg plasma membrane being present in restricted domains.

Plasmogamy

The earliest perceived response of fucoid eggs to fertilizing sperm is a membrane depolarization from about −60 mV to −25 mV [8]. This activation or fertilization potential serves as a 'fast block' to polyspermy and is well known in various animal egg systems where it is associated with the insertion of sperm-associated ion channels which depolarize the plasma membrane. This is associated with a large increase in intracellular calcium through a positive-feedback loop involving phosphoinositide second messengers. In fucoid algae it appears that there is a transient elevation of cytoplasmic calcium concentration [9]. The plasma membrane of the unfertilized egg contains calcium channels which are opened by depolarization, and at least part of the elevated calcium concentration appears to result from influx.

The fertilization potential lasts a few minutes and is followed by the secretion of β-linked polyuronides from pre-formed cytoplasmic vesicles to form a simple cell wall. This wall can be stained with the fluor Calcofluor White [10] and is formed in *F. serratus* within 3–4 min of adding sperm, providing another, but slower, block to polyspermy.

As a general working hypothesis, we have postulated that species-specific binding and/or fusion of eggs and sperm is mediated by molecular associations involving surface receptors and complementary ligands. However, it remains an open question as to whether specificity lies at the level of the initial binding/adhesion of sperm to egg or at the level of membrane fusion, and it is possible to envisage both specific and non-specific receptor–ligand systems controlling different facets of the overall fertilization process. It is, therefore, pertinent to explore the nature of this cognitive interface, in terms of its complexity and functional properties and we have adopted the following experimental approaches: (i) use of lectin and monoclonal antibody (mAb) probes coupled with high-resolution imaging (confocal microscopy) to explore the spatial distribution of surface proteins/glycoproteins; (ii) indirect identification of receptors and their ligands through functional effects of mAbs and lectins in either the fertilization or the binding assay; and (iii) direct isolation of receptors and ligands from surface membranes or lysates.

Organization of glycoproteins on the egg surface

Jones *et al.* [11,12] raised mAbs to whole sperm cells of *F. serratus*, and it was shown that some sperm surface antigens are highly concentrated in particular regions of the sperm cell. Three of the mAbs raised against *F. serratus* sperm, namely FS2, FS4 and FS5, also bind to egg surfaces through common carbohydrate epitopes of different antigens to those on the sperm surface. It has been shown by confocal laser scanning microscopy that the binding of these mAbs to eggs is not uniform, rather the three antibodies show heterogeneous binding patterns [13]. In particular, the patches or 'domains' recognized by FS2 and FS5 are smaller, and more distinct, than those recognized by FS4. Double-labelling experiments using fluorescence isothiocyanate- and 30 nm gold-conjugated second antibodies [13] clearly show that the domains labelled by FS4 and by FS2/5 are mainly exclusive of each other, although there are some small areas of double-labelling where both antibodies appear to bind. Western blots of egg membrane lysates show that FS2 and FS5 compete for binding to the same set of glycoproteins on the egg surface, between 47 kDa and 185 kDa, and it seems likely that these two antibodies are recognizing closely overlapping epitopes. On the other hand, FS4 binds to a different set of glycoproteins, ranging from 43 kDa to 170 kDa, in egg membrane lysates. Electron microscopy immunogold labelling of egg sections with FS2 showed that this mAb bound in patches to the egg surface in the form of small protruberances.

The egg-surface macromolecules can also be visualized by lectin probes. Concanavalin A (ConA) and FBP (fucose-binding protein) also bind to small, even more discrete domains on the egg surface, and several different procedures have been used to examine the relationship between the glycoproteins identified by FS4, FS2/5 and the two lectins [13]. (i) On Western blots of egg surface-vesicle membrane lysates, ConA binds fewer bands than FS5 but most of them are common. FBP labels one major band at 62 kDa, which is also present on ConA and FS5 blots, plus a few minor bands. (ii) The 62 kDa major band is substantially enriched by pre-absorbing egg vesicle proteins to FBP–agarose beads, and eluting with fucose. Blots of gels probed with FBP show a single band at 62 kDa. The same band is detected by ConA and FS5 but not FS4. (iii) Several bands recognized by FS5 were not labelled after blots of egg vesicle proteins were preincubated with ConA, suggesting that ConA labels a sub-set of glycoproteins recognized by FS5. The same experiment cannot be done with FS4, since this mAb binds to ConA itself. (iv) Since ConA only partially competes with FS5, it is feasible to image the patterns of binding of both probes through confocal laser scanning microscopy. Results show that there are some regions which are $FS5^{+}ConA^{-}$ but none which are $FS5^{-}ConA^{+}$, suggesting that there are no regions outside the $FS5^{+}$ domains that are recognized by ConA.

Overall, these data show that different sets of proteins and glycoproteins on the egg plasma membrane are not distributed homogeneously, but form more-or-less discrete domains. FBP, ConA and FS2/5 appear to recognize progressively larger families of glycoproteins held within small discrete domains on the egg surface and these are, both spatially and in molecular characteristics, largely distinct from a family of glycoproteins recognized by FS4. One interpretation of the spatial distribution of these glycoprotein families is shown in Fig. 1.

The domains are not induced by any form of patching or capping of antigens by the multivalent antibodies or lectins, since they are detectable on both fresh and fixed eggs.

Functional significance of glycoprotein domains

There is, then, a complexity to antigen distribution on the egg surface which is, perhaps, not expected from such an undifferentiated cell. Information on egg plasma

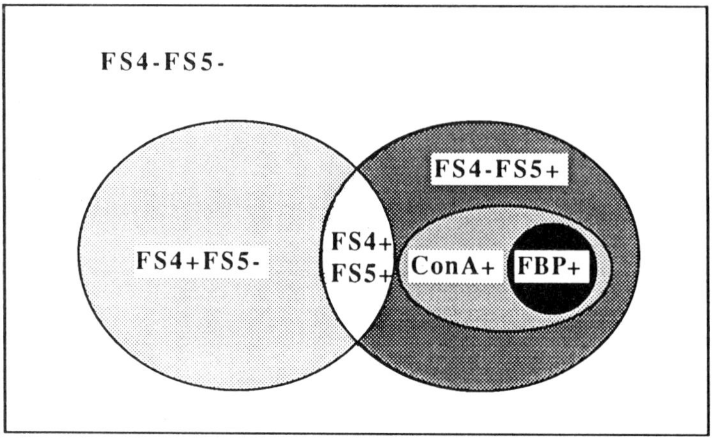

Fig. 1. Interpretive diagram showing the relative distribution of glycoproteins recognized by different mAb and lectin probes on the surfaces of *Fucus serratus* eggs.

membranes is rather limited, but mouse eggs have a microvillus-free area on their cell surface which labels with ConA to a lesser extent [14]. Localization results for the bindin receptor of the sea urchin egg show that it is uniformly distributed [15]. Surface antigen domains are also quite common for highly differentiated cells, e.g. epithelial cells or animal sperm cells [16], and these domains have functional significance. It is possible that the complex antigen and lectin-binding domain structure of the *Fucus* egg surface has some direct relevance to sperm binding and fusion and it was pointed out earlier that observations of the probing movements of the sperm on reaching the egg surface are consistent with non-uniformly distributed sperm receptors.

Evidence which is consistent with this interpretation has recently been obtained from experiments with mAbs [17]. First, fertilization in the standard bioassay can be inhibited by pretreating eggs with as little as 2.5 μg of purified immunoglobulin M (IgM) of FS5, but there is no inhibition by purified IgM of FS4. The inhibition by FS5 is not species-specific, since fertilization in *F. vesiculosus* is also affected. Both mAbs are large, multivalent IgMs, and it is possible that they could bind to egg surface molecules not involved in recognition — thus causing steric hindrance to interacting receptors. However, the Fab (fragment antigen binding) fragments of FS5 also inhibit fertilization, suggesting more direct effects. Control experiments have eliminated the possibility that FS5 and its Fab fragment could be exerting their inhibitory effects through indirect effects on the sperm.

Secondly, egg plasma membrane vesicles are effective in inhibiting fertilization in *F. serratus* by binding to sperm cells [18]. This activity is abolished by pretreating vesicles with periodate (to oxidize carbohydrate residues) but not by trypsin. The inhibition is also species- and genus-specific, vesicles of *F. serratus* eggs having no effect on fertilization in *F. vesiculosus* and vice versa. FS5, but not FS4, pretreatments of egg membranes effectively abolish this inhibitory effect of membranes.

Finally, the 62 kDa glycoprotein bound by FBP, ConA and FS5 (the two lectins also inhibit fertilization by binding to eggs [1,19]) has been isolated from egg vesicle lysates by lectin affinity on FBP–agarose, as previously described. Preliminary experiments show that this FBP-binding protein, which we have called Fse62, inhibits fertilization in the standard bioassay, but with reduced species-specificity compared with the membranes from which it was isolated.

The results clearly point to important functions for one or more glycoproteins in the FS5$^+$ domains, and, since Fse62 is a common band recognized by all three probes, it is likely that the inhibitory effects of these probes are caused by binding to the carbohydrate side-chains of this glycoprotein, which, therefore, assumes the properties of a putative egg receptor ligand. This is consistent with previous evidence suggesting that *Fucus* sperm recognize fucose/mannose-containing glycoproteins on the egg surface during fertilization [2,3,10,19].

One argument against this general interpretation is that the fertilization-blocking effects of FS5, ConA and FBP do not exhibit the species-specificity that might be expected of true receptors. On the other hand, plasma-membrane-enriched egg membrane vesicle preparations do inhibit fertilization species-specifically, and preliminary experiments suggest that Fse62 inhibits 'species-preferentially'. This may suggest that either we have been addressing gamete adhesion which might have a low level of specificity, and that specificity is a function of other, as yet undetected, receptor molecules, or that the lectins and mAbs are binding to conserved parts of the Fse62 rather than regions involved in species recognition. These possibilities will be tested in future work in which a range of mAbs are raised to Fse62 directly.

Sperm surface glycoproteins

These studies on egg surface glycoproteins have been complemented by investigations of sperm receptor proteins employing a new microtitre plate-based quantitative binding assay involving the use of sperm fractions and biotinylated egg membrane vesicles (Fig. 2). (i) Surface proteins, stripped from whole sperm by KCl treatment, inhibit fertilization in a species-preferential manner. (ii) Plate-immobilized KCl extract

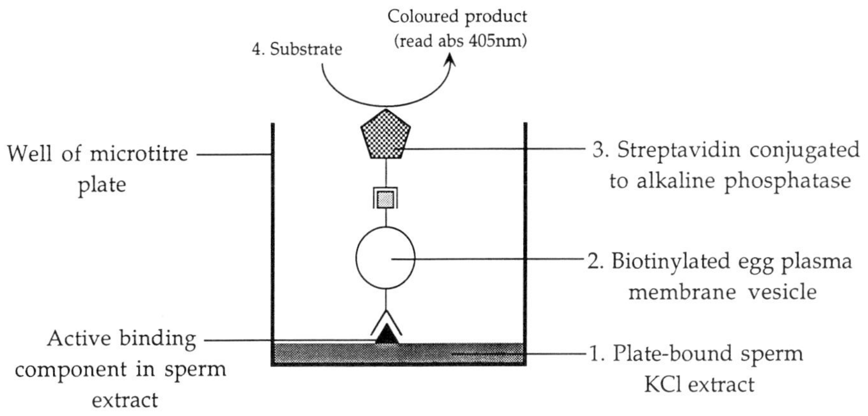

Fig. 2. Components of the plate-binding assay involving sperm extracts and biotinylated egg vesicles of *Fucus serratus*.

binds biotinylated egg plasma membrane vesicles. Binding is saturable, gamete- but not species-specific, destroyed by trypsin/Pronase treatment of the KCl extract but not by periodate, and blocked by certain sulphated polysaccharides (e.g. native but not desulphated fucoidan). In the latter respect, the sperm protein is remarkably similar in character to bindin, a FBP from sea urchin sperm [21,22], and its mammalian analogue proacrosin [23]. For both of these molecules, sulphation of the egg ligand is a critical feature recognized by basic residues within the complementary binding molecule. (iii) KCl extract fractionated by high-performance gel filtration in high ionic strength conditions gives a major peak of binding activity (apparent molecular mass is 60 kDa) which is blocked by fucoidan. The fraction also triggers cell wall release from eggs in the absence of sperm and inhibits fertilization. (iv) This 60 kDa fraction runs on denaturing gels in urea as a 27 kDa 'doublet' to which egg vesicles bind on blots, suggesting that the native form is a dimer or that the apparent molecular mass on high-performance gel filtration is an artefact caused by the high ionic strength promoting hydrophobic interactions. (v) mAb FS17 raised to the 27 kDa component inhibits binding in the plate assay. The full details and characteristics of this binding assay are described elsewhere [20].

We have provisionally termed the 27 kDa component Fss27 and are currently exploring its precise location in or on sperm. Certain of its binding characteristics, notably specificity for sulphated and fucosylated egg ligands, are reminiscent of the selectins, a class of calcium-dependent lectins which recognize carbohydrate-bearing ligands on endothelial cells [24] and we are currently exploring this aspect further.

Conclusion

Two molecules with some of the properties anticipated of receptors and their ligands have been isolated from sperm and egg cells, respectively. In the case of eggs, it has been shown that the putative egg ligand appears to be the fucosylated carbohydrate side-chains of the Fse62 glycoprotein, which is one of the glycoproteins contained within organized spatial domains of other glycoproteins. Fss27, the putative egg-binding protein from sperm, triggers partial cell wall release in the absence of sperm, and its binding to eggs appears to depend on ligands containing sulphated and fucosylated moieties. It must be emphasized, however, that there is no evidence for complementarity of association of Fse62 and Fss27 in egg–sperm recognition, and any substantiation of their role in egg–sperm recognition requires further experimentation.

This work was carried out on programmes supported by AFRC and SERC.

References

1. Callow, J.A. (1985) J. Cell. Sci. (Suppl. 1, 6th John Innes Symposium) 219–232
2. Callow, J.A., Callow, M.E. and Evans, L.V. (1985) Biol. Fert. **2**, 389–407
3. Green J.R., Jones, J.L., Stafford, C.J. and Callow, J.A. (1990) NATO Adv. Sci. Inst. Ser. H **45**, 189–202
4. Callow, J.A., Stafford, C.J. and Green, J.R. (1992) Soc. Exp. Biol. Semin. Ser. **48**, 19–31
5. Muller, D.G. and Seferiades, K. (1977) Z. Pflanzen Physiol. **84**, 85–94
6. Friedmann, I. (1961) Bull. Res. Counc. Isr. Sect. D. **10D**, 73–83

7. Brawley, S.H. (1990) NATO Adv. Sci. Inst. Ser. H **45**, 419–431
8. Brawley, S.H. (1991) Dev. Biol. **144**, 94–106
9. Roberts, S.K., Gillot, I. and Brownlee, C. (1994) Development **120**, 155–163
10. Bolwell, G.P., Callow, J.A., Callow, M.E. and Evans, L.V. (1979) J. Cell Sci. **43**, 209–224
11. Jones, J.L., Callow, J.A. and Green, J.R. (1988) Planta, **176**, 298–306
12. Jones, J.L., Callow, J.A. and Green, J.R. (1990) Planta, **182**, 64–71
13. Stafford, C.J., Green, J.R. and Callow, J.A. (1992) J. Cell Sci. **101**, 437–448
14. Wolf, D.E. and Ziomek, C.A. (1983) J. Cell Biol. **96**, 1786–1790
15. Ruiz-Bravo, N., Janak, D.J. and Lennarz, W.J. (1989) Biol. Reprod. **41**, 323–334
16. Primakoff, P. and Myles, D.G. (1983) Dev. Biol. **98**, 417–428
17. Stafford, C.J., Callow, J.A. and Green, J.R. (1993) J. Phycol. **29**, 325–330
18. Stafford, C.J., Callow, J.A. and Green, J.R. (1992) Br. Phycol. J. **27**, 429–434
19. Catt, J.W., Vithanage, H.I.M.V., Callow, J.A., Callow, M.E. and Evans, L.V. (1983) Exp. Cell. Res. **147**, 127–133
20. Wright, P.J., Green, J.R. and Callow, J.A. (1994) J. Phycol., in the press
21. De Angelis, P.L. and Glabe, C.G. (1987) J. Biol. Chem. **262**, 13946–13952
22. De Angelis, P.L. and Glabe, C.G. (1988) Biochemistry **27**, 8189–8194
23. Jones, R. (1990) J. Reprod. Fert. Suppl. **42**, 89–105
24. Imai, Y., Lasky, L.A. and Rosen, S.D. (1993) Nature (London) **361**, 555–557

Biochem. Soc. Symp. **60**, 43–50
Printed in Great Britain

Formation of embryogenic cells in plant tissue cultures

Sacco C. de Vries*, Anke J. de Jong and Fred A. van Engelen

Agricultural University Wageningen, Department of Molecular Biology, Dreijenlaan 3, 6703 HA Wageningen, The Netherlands

Introduction

Somatic embryogenesis is the process by which somatic cells develop into plants through characteristic developmental stages. For dicotyledons these are globular, heart and torpedo stages. Under *in vitro* conditions, somatic embryos can form directly from protoplasts or microspores, or on the surface of an organized tissue, such as a leaf or stem segment, and indirectly via an intervening step of callus or suspension culture [1].

In *Daucus*, embryogenic cell suspension cultures can be initiated by placing an explant cut from a seedling in culture medium containing 2,4-dichlorophenoxyacetic acid (2,4-D). This treatment induces cell divisions at the wound edges and finally results in an unorganized cell suspension containing both embryogenic and non-embryogenic cells [2]. The embryogenic cells are usually in the form of clusters of small cytoplasmic cells, referred to as pro-embryogenic masses [3]. In almost all embryogenic *Daucus* cultures, the number of cells that are actually embryogenic is quite low, and never amounts to more than about 1–2% [2], so an embryogenic suspension culture of *Daucus* should be seen as a complex mixture of different cell types, of which only a minority has completed the transition from somatic cell into an embryogenic cell [4].

The re-initiation of cell division and subsequent unorganized growth behaviour during the initiation of an embryogenic *Daucus* suspension culture has often been interpreted as dedifferentiation of the cells. However, several observations on the persistence of tissue-specific states in culture clearly show that unorganized growth should not be equated to dedifferentiation. In several cases, unorganized plant cell suspension cultures continue to produce some antigens specific for the organs they were derived from (see [5] and references cited therein). The expression pattern of the *Daucus* EP1 gene [6] (F.A. van Engelen and S.C. de Vries, unpublished work) lends further support to the view that unorganized growth does not necessarily imply a complete loss of cellular differentiation. In seedlings, expression of the EP1 gene is restricted to the epidermis of both the root and the shoot. The fact that EP1 is also expressed in a subset of non-

*To whom correspondence should be addressed.

embryogenic suspension-cultured cells strongly suggests that this cell population is derived from the epidermis of the explant and has retained at least one of the epidermis-specific features. Therefore, embryogenic suspension cultures are probably best described as 'unorganized' rather than 'undifferentiated'.

The most important, but also the least understood, part of somatic embryogenesis is the transition of somatic cells into embryogenic cells. Several recent reports suggest that, whereas the synthetic auxin 2,4-D is widely employed to produce embryogenic cells *in vitro*, other molecules which promote the formation of embryogenic cells in tissue culture, or stabilize embryogenic cells once formed, are now being identified. Here, an overview of these reports will be given and discussed in the light of the cellular mechanisms that may be involved. To facilitate the interpretation, a brief section on what is known about the identity of the cell types involved will precede this discussion.

Identification of embryogenic cells

By using time-lapse photography, Backs-Hüseman and Reinert [7] have described an elongated, single, vacuolated suspension cell that is able to develop into a somatic embryo. Using cell-purification techniques, Nomura and Komamine [8] described a much smaller, almost spherical, cytoplasmic suspension cell, designated a type-1 cell, as being able to develop into a somatic embryo. Both cell types require preculturing in auxin, after which an embryogenic cell cluster is formed. In the case of the type-1 cell, these clusters, which may be equated with pro-embryogenic masses, are designated state-1 cell clusters [9]. Removal of auxin results in the development of a globular stage embryo from such a cluster [9]. Employing automated cell tracking on more than 30 000 single *Daucus* suspension cells, we have recently confirmed the existence of several morphologically different cell types capable of producing embryogenic cell clusters [9a], leading to the conclusion that precursor cells of embryogenic cell clusters cannot be identified solely on the basis of their morphology.

Several molecular markers have been reported that are able to distinguish between embryogenic and non-embryogenic cell cultures [10,11]. One of these is the *Daucus* EP2 gene [12]. Employing *in situ* mRNA localization, the EP2 gene was found to be first expressed in peripheral cells of pro-embryogenic masses that consisted of about 20 cells (E.D.L. Schmidt and S.C. de Vries, unpublished work) and in the protoderm of somatic embryos. In *Daucus* zygotic embryos, EP2 expression was also detected in a protoderm-specific fashion, as early as a 60-celled globular embryo. The EP2 gene encodes a secreted lipid-transfer protein, postulated to function in cutin synthesis [12] (E. Meijer, S.C. de Vries and T. Hendriks, unpublished work).

It has been demonstrated using monoclonal antibodies that a family of plasma membrane arabinogalactan proteins (AGPs) is localized at the outer face of the protoplast and the inner 20–30 nm of the cell wall [13,14]. This observation led to the speculation that AGPs may anchor the primary cell wall to the plasma membrane by binding to wall polysaccharides with their lectin domain. There appear to be subtle differences in AGP structure between tissues or regions of developing organs, as indicated by the differential expression of AGP epitopes recognized by monoclonal antibodies [13,15–18]. AGPs appear to be rapidly turned over by extracellular hydrolytic enzymes [17,19,20]. Another marker for embryogenic cultures consists of a cell wall epitope in *Daucus* suspension cells, that is recognized by the monoclonal antibody JIM8 [10]. The molecule that bears the JIM8 cell wall epitope is probably different from both the plasma membrane AGP epitope described earlier [17] and the epitope present on secreted *Daucus* AGPs [15]. The presence of the cell wall JIM8 epitope in *Daucus* suspension cell cultures is highly correlated with the presence of embryogenic cells. However, immersion immunofluorescence showed that several morphologically different cells react with the JIM8 antibody, but the pro-embryogenic masses do not react with JIM8 [10]. Instead, mainly small single cells, including the type-1 cells, were recognized. The hypothesis put forward by Pennell *et al.* [10] is, therefore, that the JIM8 cell wall epitope marks a transitional state in the formation of embryogenic cells. Because the number of JIM8-reactive single cells exceeds, by far, the number of single cells that are able to develop into an embryo, apparently only a few cells in this transitional state are actually able to reach the status of single cell capable of somatic embryo formation. Although the JIM8 plasma membrane epitope, as observed in *Brassica* flowers, is most probably present on a molecule that is different from the JIM8 cell wall epitope observed in *Daucus* suspension cultures, the observation that both visualize a transient developmental process, which is not restricted to a particular set of morphologically recognizable cells, represents an intriguing parallel.

Formation of embryogenic cells

Although auxins are the best-studied inducers for the formation of embryogenic cells from somatic cells, [21–27], they are certainly not unique in the ability to mediate the transition of somatic cells into embryogenic cells. For example, in *Citrus* a change of the carbon source is sufficient [28]. In *Medicago* the ability of a cell to become embryogenic appears to depend on its sensitivity to auxin, as elegantly illustrated by the totally different response to 2,4-D of leaf protoplasts derived from a genotype that readily forms embryogenic cells *in vitro* and leaf protoplasts from a genotype that does not [29]. In *Daucus*, the ability to modulate plasma-membrane-binding sites for auxin appears to be associated with non-embryogenic cells, either in an embryogenic culture or present in a number of somatic cell variants of *Daucus* [30]. Based on these results, the sensitivity to auxin is apparently important for the formation of embryogenic cells. Although it has often been observed that the ability of plant cells to respond *in vitro* to exogenous growth regulators by the formation of embryogenic cells is genotype dependent, few studies have been aimed at the identification of the genes involved. In *Lycopersicon*, the ability of root explants to undergo shoot organogenesis was recently shown to reside in a single genetically identified locus Rg1 [31]. This study opens the possibility that identification of genes involved in regeneration and the formation of embryogenic cells may, eventually be possible.

Recent evidence suggests that particular purified AGPs, isolated from the culture medium of embryogenic *Daucus* lines and from dry *Daucus* seeds, were able to promote the formation of pro-embryogenic masses — even in previously non-embryogenic *Daucus* cell lines — when added in nanomolar concentrations. Other AGPs, isolated from the medium of a non-embryogenic line, acted negatively on the formation of pro-embryogenic masses [32]. These results show that specific members of the family of AGPs are involved in the formation of embryogenic clusters. Although the underlying mechanisms are unknown, these observations, together with earlier ones employing unfractionated conditioned medium [2,33], suggest that, apart from marking particular events in plant morphogenesis, AGPs are also likely candidates for molecules with the ability to determine developmental processes, perhaps by cell-inductive events [17].

Cellular mechanisms in the formation of embryogenic cells

Cell polarity and asymmetrical cell division

Several observations support the hypothesis that plant growth regulators employed to form embryogenic cells do this by alteration of cell polarity and promotion of subsequent asymmetric divisions. When immature zygotic embryos of *Trifolium* are cultured in the presence of cytokinin, somatic embryos are produced directly from the hypocotyl epidermis. The first sign of the induction of embryogenic cells is a shift from the normal anticlinal division pattern in the epidermis, to irregular periclinal and oblique divisions [34]. The effect of the cytokinin is not entry into mitosis as such, but rather an alteration of the division planes, because regular anticlinal divisions persist for some time in the absence of cytokinin. As pH gradients and low electrical fields can change cell polarity [35], the positive effect on embryo development of pH shifts [36] and low electrical fields [37] may be due to their effect on cell polarity. It is plausible, but unproven, that exogenously applied plant growth regulators directly modify cell polarity, by interference with pH gradients or the electrical field around cells. After stimulation by auxin, asymmetric cell divisions are frequently observed in leaf protoplast cultures derived from an embryogenic *Medicago* cultivar, while in protoplast cultures from a non-embryogenic cultivar cells divide symmetrically [29,38]. The different types of cell division in *Medicago* leaf protoplast cultures appear to be correlated with differences in microtubule organization [39]. In *Daucus*, the first division of single suspension cells capable of forming embryogenic cells is also asymmetric [7,9], and only the smaller daughter cell will ultimately develop into an embryo. As the future root pole of the somatic embryo is always oriented towards the larger cell, the polarity of the entire somatic embryo is already determined prior to the first division of an embryogenic cell. Thus the ability to perform an asymmetric cell division, based on a (possibly auxin-induced) change in cell polarity, seems to be an important and, perhaps, universal mechanism in the formation of embryogenic plant cells from somatic cells.

Control of cell expansion

The plant primary cell wall is faced with the complex task of providing a strong skeleton for the organism as a whole and protecting it from adverse environmental influences. At the same time it has to retain the plasticity to accommodate controlled cell expansion in response to turgor pressure. The latter process is crucial for plant morphogenesis, since, owing to the very presence of the wall, morphogenesis is only determined by regulation of the rate and plane of cell division and the degree and direction of expansion of the cells after division. Therefore, a second mechanism that is of crucial importance in the formation of embryogenic cells *in vitro* is the ability to restrict cell expansion under hypotonic conditions [4,40–42]. The ability to control cell expansion is generally accepted to reside in the cell wall, and is probably mediated by specific sets of cell wall proteins and enzymes [43,44]. In *Daucus*, the glycosylation inhibitor tunicamycin arrests somatic embryogenesis, perhaps by the gradual disruption of pro-embryogenic mass owing to expansion of its outer cell layer. This effect could be counteracted by addition of a single protein, exhibiting peroxidase activity, purified from medium conditioned by a somatic embryo culture [45].

A mechanism that limits cell expansion may also be required at later stages of somatic embryo development, as indicated by the rescue of arrested globular embryos of the temperature-sensitive *Daucus* mutant ts11 with a single secreted acidic endochitinase [46]. Addition of the endochitinase appeared to prevent the formation of an aberrant, irregular protodermal layer, consisting of enlarged, vacuolated cells. A positive effect was also seen on the formation of pro-embryogenic masses and globular embryos from ts11 suspension cells, which implies that more than one stage in the development of embryos is affected in ts11. Although chitinases are generally believed to act only on cell walls of fungal origin, there may be an as yet unidentified substrate for these enzymes in the plant cell wall. Additional clues for the role of the endochitinase in somatic embryogenesis have come from recent findings where the addition of *N*-acetylglucosamine-containing lipo-oligosaccharides of *Rhizobium* were seen to have a rescue effect, similar to that observed for the endochitinase [46a].

Because the culture medium is continuous with the liquid phase of the cell wall, secreted soluble components that are associated with developmental changes diffuse into the medium. This does not only render the spent culture medium a source of biologically interesting molecules, but also allows these molecules to be directly tested for their activity in somatic embryo development by addition to the culture medium. The rationale behind this idea is that cell wall polymers that are involved in, for instance, the control of cell expansion, are accessible to their modifying enzymes from the outside, i.e. from the culture medium. *In planta*, these enzymes are secreted and will most probably reach their substrates from the inside. Further support for this idea comes from a recent study by McQueen-Mason *et al.* [47], who extracted and purified, from *Cucumis* cell walls, two proteins that were able to induce cell wall extension in heat-killed hypocotyl sections, simply by addition to the culture medium surrounding the sections.

In general, young organized structures like embryos and organ primordia are found to consist of small cells [48], suggesting that preventing cell expansion is essential during early stages of morphogenesis. Although the turgor pressure in relation to somatic embryogenesis has not been investigated, a large body of evidence indicates that plant cell expansion is controlled by the extensibility of the cell wall [40,41]. The peroxidase responsible for countering the tunicamycin-induced inhibition of somatic embryogenesis belongs to a well-studied group of plant enzymes that have been proposed to restrict cell expansion by cross-linking of phenolic side-chains of xylans or pectins in the cell wall [40,43,49]. The observed morphological effect of tunicamycin on *Daucus* somatic embryo formation, which is the gradual disruption of the pro-embryogenic mass due to expansion of the surface cells, could then be counteracted by the cross-linking activity of the added peroxidase. Although it remains to be demonstrated that this peroxidase indeed acts on cell wall polymers of the pro-embryogenic mass, the protein appears to fulfil the first criterion for maintenance of embryogenic cells, that is to restrict cell size after division.

Concluding remarks

It appears from the studies that have been carried out so far, that in the formation of embryogenic cells in plants both asymmetric cell division and control of cell expansion are major mechanisms. There is some evidence that cell polarity and a postulated, subsequent, non-random partitioning of cytoplasmic determinants are promoted by auxins and cytokinins, and this might be the cellular basis for their role in somatic embryogenesis. Other molecules, which profoundly influence, for instance, the restriction of cell expansion *in vitro*, can be found with suitable biological assays.

The central role of the plant cell wall in the generation of plant form [44] has been emphasized several times here. The view has been put forward that secreted enzymes act as regulators of morphogenesis by changing the chemical structure of cell wall polymers and, consequently, the mechanical properties of the wall as a whole. When this view is applied to somatic embryogenesis in cell suspension cultures, there is enough reason to be surprised that embryos develop at all in this system. In embryogenic cell suspensions, embryogenic cells are outnumbered, by far, by non-embryogenic cells, which are characterized by extensive expansion and loose attachment to other cells. If these properties of non-embryogenic cells are mediated by secreted enzymes, these enzymes will accumulate in the culture medium, especially in high-density cultures. Thus we might expect from this that conditioned growth medium is essentially hostile towards the development of pro-embryogenic masses, as they are immersed in enzymes promoting cell expansion or loosening cell contacts. However, experience tells us that suspension culture conditions do permit the development of pro-embryogenic masses. This may be explained in two ways. First, the enzymes may simply be diluted too much by the culture medium to reach a critical level. The highest enzyme concentration is clearly found in the walls of the non-embryogenic cells themselves, where they must be active, and probably decreases sharply towards the medium. The effect on the cell wall of an embryogenic cell might then eventually be too small to significantly alter its properties. Secondly, the structure of the wall of an embryogenic cell may *a priori* be such that the enzymes have no access to the primary cell wall polymers themselves. One way to achieve this would be to modify the walls of embryogenic cells in such a way that the primary cell wall is not accessible. The observed total or partial coating of JIM8-reactive single cells with a thick layer of wall material may be an illustration of this phenomenon [10]. Another way to achieve this would be the deposition of a hydrophobic layer on the surface of pro-embryogenic masses and somatic embryos, as suggested by Sterk *et al.* [12], on the basis of the expression pattern of the EP2 gene encoding a secreted lipid-transfer protein thought to be involved in cutin synthesis. The low amounts of protein secreted by pro-embryogenic masses and somatic embryos indicate that this barrier may function both ways, limiting the diffusion of the proteins secreted by these structures. The ability of some cultured cells to protect themselves from adverse influences of culture medium components could, therefore, be an early and crucial step in the formation of embryogenic cells and somatic embryos.

The two secreted proteins identified so far as having promotive effects in *Daucus* somatic embryogenesis have enzyme activities that are also induced when plants are challenged by pathogens or injury [50–52]. The occurrence of these 'defence-related' proteins in plant cell suspension culture media is generally interpreted as a wounding response, since tissue coherency is clearly disrupted in cell suspensions. Yet, there may be alternative explanations for the presence of these proteins in culture media, because some of the genes encoding defence-related proteins have been found to be developmentally regulated at the plant level [53–55]. It will now be of interest to investigate whether analogous cellular mechanisms are employed by plants, both in defence and in development [56].

Primary research in our laboratory is supported by the Foundation for Biological Research, subsidized by the Netherlands Organization for Scientific Research (AJdJ), the Technology Foundation, subsidized by the Netherlands Organization for Scientific Research and the EC-BRIDGE programme.

References

1. Williams, E.G. and Maheswaran, G. (1986) Ann. Bot. **57**, 443–462
2. De Vries, S.C., Booij, H., Meyerink, P., Huisman, G., Wilde, D.H., Thomas, T.L. and Van Kammen, A. (1988) Planta **176**, 196–204
3. Halperin, W. (1966) Am. J. Bot. **53**, 443–453
4. Van Engelen, F.A. and De Vries, S.C. (1993) Morphogenesis in Plants: Molecular Approaches, pp. 181–200, Plenum, New York
5. Meins, F. (1986) Bot. Monogr. **23**, 7–25
6. Van Engelen, F.A., Sterk, P., Booij, H., Cordewener, J.H.G., Rook, W., Van Kammen, A. and De Vries, S.C. (1991) Plant Physiol. **96**, 705–712
7. Backs-Hüsemann, D. and Reinert, J. (1970) Protoplasma **70**, 49–60
8. Nomura, K. and Komamine, A. (1985) Plant Physiol. **79**, 988–991
9. Komamine, A., Matsumoto, M., Tsukahara, M., Fujiwara, A., Kawahara, R., Ito, M., Smith, J., Nomura, K. and Fujimura, T. (1990) in Progress in Plant Cellular and Molecular Biology (Nijkamp, H.J.J., Van der Plas, L.H.W. and Van Aartrijk, A., eds.), pp. 307–313, Kluwer, Dordrecht

9a. Toonen, M.A.J., Hendriks, T., Schmidt, E.D.L., Verhoeven, H.A., vanKammen, A. and de Vries, S.C. (1994) Planta, in the press

10. Pennell, R.I., Janniche, L., Scofield, G.N., Booij, H., De Vries, S.C. and Roberts, K. (1992) J. Cell Biol. **119**, 1371–1380
11. Sterk, P. and De Vries, S.C. (1992) in Synseeds: Applications of Synthetic Seeds to Crop Improvement (Redenbaugh, K., ed.), pp. 115–132, CRC Press, Boca Raton
12. Sterk, P., Booij, H., Schellekens, G.A., Van Kammen, A. and De Vries, S.C. (1991) Plant Cell **3**, 907–921
13. Knox, J.P., Day, S. and Roberts, K. (1989) Development **106**, 47–56
14. Pennell, R.I., Knox, J.P., Scofield, G.N., Selvendran, R.R. and Roberts, K. (1989) J. Cell Biol. **108**, 1967–1977
15. Knox, J.P., Linstead, P.J., Peart, J., Cooper, C. and Roberts, K. (1991) Plant J. **1**, 317–326
16. Pennell, R.I. and Roberts, K. (1990) Nature (London) **344**, 547–549
17. Pennell, R.I., Janniche, L., Kjellbom, P., Scofield, G.N., Peart, J.M. and Roberts, K. (1991) Plant Cell **3**, 1317–1326
18. Stacey, N.J., Roberts, K. and Knox, J.P. (1990) Planta **180**, 285–292
19. Fincher, G.B., Stone, B.A. and Clarke, A.E. (1983) Annu. Rev. Plant Physiol. **34**, 47–70
20. Gibeaut, D.M. and Carpita, N.C. (1991) Plant Physiol. **97**, 551–561
21. Abdullah, R., Cocking, E.C. and Thompson, J.A. (1986) BioTechnology **4**, 1087–1090
22. Halperin, W. (1964) Science **146**, 408–410
23. Jones, T.J. and Rost, T.L. (1989) Bot. Gaz. **150**, 41–49

24. Mórocz, D.J., Donn, G., Németh, J. and Dudits, D. (1990) Theor. Appl. Genet. **80**, 721–726
25. Song, J., Sorensen, E.L. and Liang, G.H. (1990) Plant Cell Rep. **9**, 21–25
26. Vasil, V., Redway, F. and Vasil, I.K. (1990) BioTechnology **8**, 429–434
27. Wernicke, W. and Brettell, R. (1980) Nature (London) **287**, 138–139
28. Gavish, H., Vardi, A. and Fluhr, R. (1991) Physiol. Plant **82**, 606–616
29. Bögre, L., Stefanov, I., Abrahám, M., Somogyi, I. and Dudits, D. (1990) in Progress in Plant Cellular and Molecular Biology (Nijkamp, H.J.J., Van der Plas, L.H.W. and Van Aartrijk, J., eds.), pp. 427–436, Kluwer, Dordrecht
30. Filippini, F., Terzi, M., Cozzani, F., Vallone, D. and Lo Schiavo, F. (1992) Theor. Appl. Genet. **84**, 430–434
31. Koornneef, M., Bade, J., Hanhart, C., Horsman, K., Schel, J., Soppe, W., Verkerk, R. and Zabel, P. (1993) Plant J. **3**, 131–141
32. Kreuger, M. and Van Holst, G.J. (1993) Planta **189**, 243–248
33. De Vries, S.C., Booij, H., Janssens, R., Vogels, R., Saris, L., Lo Schiavo, F., Terzi, M. and Van Kammen, A. (1988) Genes Dev. **2**, 462–476
34. Maheswaran, G. and Williams, E.G. (1985) Ann. Bot. **56**, 619–630
35. Quatrano, R.S. (1978) Annu. Rev. Plant Physiol. **29**, 487–510
36. Smith, D.L. and Krikorian, A.D. (1990) Plant Cell Rep. **9**, 468–470
37. Dijak, M., Smith, D.L., Wilson, T.J. and Brown, D.C.W. (1986) Plant Cell Rep. **5**, 468–470
38. Dudits, D., Bögre, L. and Györgyey, J. (1991) J. Cell Sci. **99**, 475–484
39. Dijak, M. and Simmonds, D.H. (1988) Plant Sci. **58**, 183–191
40. Fry, S.C. (1990) in Progress in Plant Cellular and Molecular Biology, (Nijkamp, H.J.J., Van der Plas, L.H.W. and Van Aartrijk, J., eds.), pp. 504–513, Kluwer, Dordrecht
41. Taiz, L. (1984) Annu. Rev. Plant Physiol. **35**, 585–657
42. Van Engelen, F.A. and De Vries, S.C. (1992) Trends Genet. **8**, 66–70
43. Fry, S.C. (1986) Annu. Rev. Plant Physiol. **37**, 165–186
44. Carpita, N.C. and Gibeaut, D.M. (1993) Plant J. **3**, 1–30
45. Cordewener, J., Booij, H., Van der Zandt, H., Van Engelen, F., Van Kammen, A. and De Vries, S. (1991) Planta **184**, 478–486
46. De Jong, A.J., Cordewener, J., Lo Schiavo, F., Terzi, M., Vandekerckhove, J., Van Kammen, A. and De Vries, S.C. (1992) Plant Cell **4**, 425–433
46a. De Jong, A.J., Heidstra, R., Spaink, H. *et al* (1993) Plant Cell **5**, 615–620
47. McQueen-Mason, S., Durachko, D.M. and Cosgrove, D.J. (1992) Plant Cell **4**, 1425–1433
48. Steeves, T.A. and Sussex, I. (1989) Patterns in Plant Development, Cambridge University Press, Cambridge
49. Greppin, H., Penel, C. and Gaspar, T. (eds.) (1986) Molecular and Physiological Aspects of Plant Peroxidases, University of Geneva Press, Geneva
50. Bowles, D.J. (1990) Annu. Rev. Biochem **59**, 873–907
51. Chibbar, R.N., Cella, R., Albani, D. and Van Huystee, R.B. (1984) J. Exp. Bot. **35**, 1846–1852
52. Kragh, K.M., Jacobsen, S., Mikkelsen, J.D. and Nielsen, K.A. (1991) Plant Sci. **76**, 65–78
53. Lotan, T., Ori, N. and Fluhr, R. (1989) Plant Cell **1**, 881–887
54. Neale, A.D., Wahleithner, J.A., Lund, M., Bonnett, H.T., Kelly, A., Meeks-Wagner, D.R., Peacock, W.J. and Dennis, E.S. (1990) Plant Cell **2**, 673–684
55. Ye, Z.-H. and Varner, J.E. (1991) Plant Cell **3**, 23–37
56. Collinge, D.B., Kragh, K.M., Mikkelsen, J.D., Nielsen, K.K., Rasmussen, U. and Vad, K. (1993) Plant J. **3**, 31–40

Biochem. Soc. Symp. **60**, 51–60
Printed in Great Britain

Lipo-oligosaccharide signalling: the mediation of recognition and nodule organogenesis induction in the legume–*Rhizobium* symbiosis

Jean Dénarié*†, Georges Truchet† and Jean-Claude Promé‡

†Laboratoire de Biologie Moléculaire des Relations Plantes-Microorganismes, CNRS-INRA, BP 27, 31326 Castanet-Tolosan Cedex, France and ‡Laboratoire de Pharmacologie et Toxicologie Fondamentales, CNRS, 205 Route de Narbonne, 31077 Toulouse Cedex, France

Introduction

Soil bacteria belonging to the genera *Rhizobium*, *Bradyrhizobium* and *Azorhizobium*, collectively referred to as rhizobia, elicit the formation on legume roots (and stems of some species) of specific organs, the nodules, in which they fix nitrogen. An important feature of these symbiotic associations is their specificity: every rhizobial strain has a definable host range [1,2]. Some rhizobia have a narrow host range, whereas others have a broad host range. For example, *Rhizobium meliloti* elicits the formation of nitrogen-fixing nodules on species of only three genera — *Medicago*, *Melilotus* and *Trigonella* [2] whereas *Rhizobium* sp. NGR234 nodulates more than 70 legume genera, as well as the non-legume *Parasponia* [3].

The mechanisms by which rhizobia infect legumes are varied. In the *R. meliloti*/alfalfa symbiosis, the bacteria induce a marked curling at the tip of root hairs, followed by the formation of tubular structures derived from plant cell wall material, the infection threads, through which bacteria pass on their way across the root hairs and the root cortex [4]. At some distance from the advancing infection thread, the induction of cell divisions in the root cortex leads to the formation of a nodule primordium and a nodule meristem. This meristem gives rise to a differentiated nodule.

A number of *Rhizobium* genes which control host specificity, infection and nodulation, the *nod* (and *nol*) genes, have been identified and characterized. During the

*To whom correspondence should be addressed.

last three years, it has become clear that a major function of *nod* genes is to determine the production of extracellular lipo-oligosaccharidic signals, the Nod factors, which are capable of eliciting many of the early plant responses which are observed in the course of infection and nodule initiation [5–7]. In this chapter we will summarize recent results on the rhizobial production of Nod factors and the responses that these signal molecules elicit on legumes.

The common *nodABC* genes and the production of lipo-oligosaccharides

The structural *nod* genes are not expressed in pure culture. Their expression requires the presence of plant signals secreted in root exudates, generally flavonoids, and the presence of regulatory NodD proteins which are transcription activators. This basic *nodD* regulatory system is present in all *Rhizobium*, *Bradyrhizobium* and *Azorhizobium* strains studied so far. Each NodD protein is activated by a given set of specific inducers [5] and the transfer of *nodD* genes between rhizobia has been reported to be associated with changes in the host range [8,9]. Thus *nodD* genes represent a first class of determinants of host range.

The structural *nod* genes are divided into common and species-specific genes. Common *nodABC* genes are structurally conserved among all rhizobia, and are absolutely required for both root hair curling and infection, for eliciting mitosis in the root cortex, and for nodule formation. The species-specific *nod* genes, such as *nodPQ* and *nodH* in *R. meliloti* and *nodFE* in *R. leguminosarum* and *R. meliloti*, are major determinants of the specificity of infection and nodulation [5] (see Fig. 1).

When *nod* gene expression is induced with appropriate flavonoids, the sterile supernatants of cultures from *R. meliloti* and *R. leguminosarum* bv. *viciae* elicit root-hair deformations of host plants, and this activity is abolished by mutations in the *nodABC* genes, indicating that *nod* genes are required for the production of specific

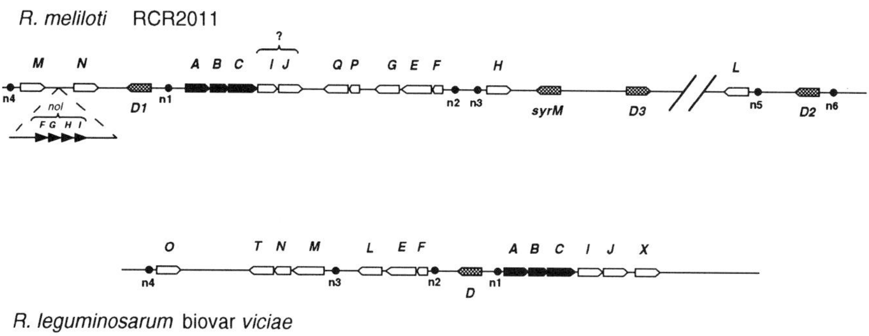

Fig. 1. Organization of nodulation (*nod* and *nol*) genes in *R. meliloti* and *R. leguminosarum* bv. *viciae*. Reproduced from [5] with permission. The *nod*-boxes preceding the *nod* operons are represented by the letter n. Letters above the maps represent the structural *nod* genes, and letters below the maps represent the regulatory genes.

extracellular Nod factors [10–12]. Root-hair-deformation assays were used in an attempt to purify these Nod factors in *R. meliloti*, but it rapidly became apparent that the quantities of secreted Nod factors were too small to allow determination of their structures. Increasing the copy number of the two transcription activators *syrM* and *nodD3* increased the transcription level of the structural *nod* genes, and the amount of secreted Nod factors increased approximately 1000-fold [13,14]. This made possible the purification of milligram quantities of factors for chemical analysis.

Nod factors secreted by *R. meliloti* (NodRm factors) were found to be lipo-oligosaccharides. They are tetra- and pentamers of *N*-acetylglucosamine, β-1,4-linked (chitin oligomers), O-acetylated and N-acylated at the non-reducing end and O-sulphated at the reducing end (see Fig. 2) [13,15,16]. The major acyl chain is $C_{16:2}$, unsaturated in positions 2 and 9 [13,15,16]; however, factors having an acyl chain of $C_{16:3}$ have also been found [14–17]. Nod factors from *R. leguminosarum* bv. *viciae* (NodRlv factors) have the same overall structure: they are also mono-N-acylated chitin

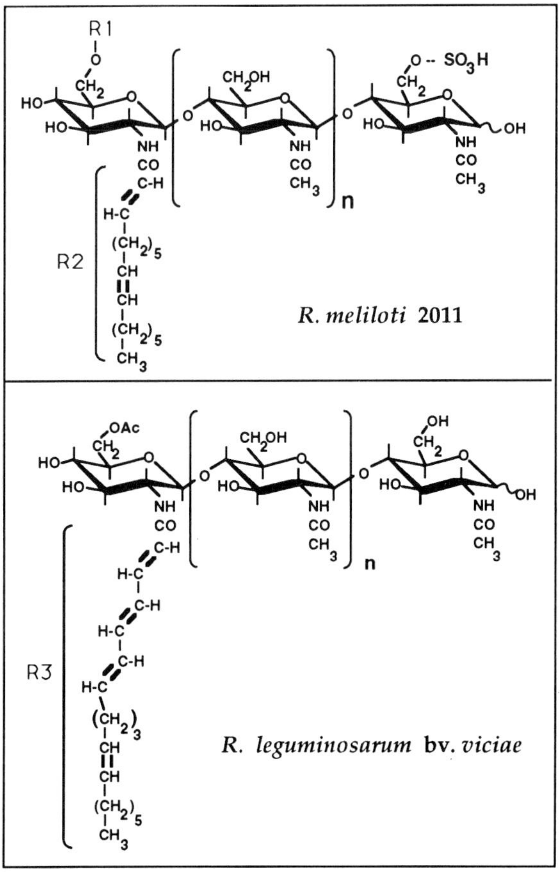

Fig. 2. Structures of Nod factors of (a) *R. meliloti* [13–17] and (b) *R. leguminosarum* bv. *viciae* [18]. n = 2 or 3; R1 = H or $COCH_3$; R2 = $C_{16:2(\Delta 2,9)}$, $C_{16:1(\Delta 9)}$, $C_{16:3(\Delta 2,4,9)}$; R3 = $C_{18:4(\Delta 2,4,6,11)}$, $C_{18:1(\Delta 11)}$.

oligomers but they are not sulphated and the N-acyl chains are C_{18}, either vaccenic acid ($C_{18:1}$) or a highly unsaturated ($C_{18:4}$) fatty acid [18].

In both *R. meliloti* and *R. leguminosarum* bv. *viciae*, mutations in the *nodABC* genes abolish the production of Nod factors. Moreover, it was shown, in an *R. leguminosarum* bv. *viciae* strain cured of the symbiotic plasmid, that the introduction of a clone carrying the *nodABC* genes is sufficient to specify the production of mono-N-acylated chitin oligomers with vaccenic acid. These results indicate that the *nodABC* genes are involved in the synthesis of the sugar backbone and in its N-acylation. NodC, which shows sequence similarity with fungal chitin synthases, is likely to specify the β-1-4-polymerization of *N*-acetylglucosamine building blocks, ensuring the synthesis of the chito-oligosaccharidic backbone [19,20]. NodB has recently been shown to deacetylate the terminal, non-reducing *N*-acetylglucosamine residue [21], an activity that may be a necessary prerequisite for the N-attachment of the acyl chain, which in turn might be specified by NodA. NodM shares sequence similarity with glucosamine synthases and exhibits glucosamine synthase activity [22,23]; it is, therefore, likely to contribute to the production of building blocks for the synthesis of the glucosamine Nod factor backbone.

Nod factors are signal molecules

Nod factors elicit various responses on legumes in a specific manner. Purified NodRm and NodRlv factors are active on epidermal cells and root hairs of alfalfa and vetch, respectively. At low concentrations, in the range 10^{-8}–10^{-12} M, they elicit root-hair induction (Hai response), i.e. differentiation of epidermal cells into root hairs, and also root hair deformations and branching (Had and Hab responses) [13,16,18]. Van Brussel *et al.* [24] have shown that the NodRlv factors with the $C_{18:4}$ polyunsaturated acyl moiety elicit the formation, in the root cortex of vetch, of radial cytoplasmic bridges which are proposed to be pre-infection threads. NodRm and NodRlv factors, with conjugated double bonds on the N-acyl chain, induce on alfalfa and vetch, respectively, cell divisions in the inner cortex and the formation of nodule primordia [18,25]. In the case of alfalfa, NodRm factors elicit the development of root-derived structures which have the shape, the ontogeny and the anatomy of bacteria-induced nodules, and their formation is repressed by an excess of combined nitrogen [25]. Thus the Nod factors elicit legume reactions which mimic plant responses observed during bacterial infection: deformation of root hairs, induction of cortical cell divisions and formation of nodule primordia.

Molecular basis of Nod factor specificity

Rhizobial genetic analysis has revealed the presence of species-specific *nod* genes which determine the host-range of infection and nodulation. It has been shown recently that these genes are involved in the synthesis of Nod factors, mediating the decoration of the chito-oligosaccharidic core molecules. The genetic mechanisms that ensure *nod* variation between rhizobia species or biovars are both non-allelic and allelic. Some of these genes are present in some species and not in others (non-allelic variation): for example, the *nodPQ* and *nodH* genes are present in *R. meliloti* and not in *R. leguminosarum* bv. *viciae*. In contrast, genes like *nodFE* are present in *R. meliloti* and *R. leguminosarum* bv. *viciae* but mediate different host specificities (allelic variation) [2]. We will examine the types of Nod factor decoration that these two categories of genes specify.

An example of non-allelic variation: the *nodH* and *nodPQ* genes

R. meliloti nodulates alfalfa and not vetch. However, *R. meliloti nodH* mutants infect and nodulate vetch and fail to nodulate alfalfa, exhibiting a shift in the host range of nodulation [11]. *nodQ* mutants have an extended infection host range and are able to infect both alfalfa and vetch [12]. Initially, a correlation between the specificity of the symbiotic behaviour of a given mutant and the specificity of the Had activity of its culture filtrate was reported [11,12,26]. These data suggested that one function of the host range *nod* genes might be to specify the chemical modification of Nod factors to make them plant-specific. Indeed, determination of the structure of Nod factors produced by *nodH* mutants has revealed that they differ (from those produced by *nodH*$^{+}$ strains) by the absence of the O-sulphate group.The *nodQ* mutants produce a mixture of sulphated and non-sulphated factors [16]. Two functional copies of *nodPQ* are present in *R. meliloti* [27] and mutant strains altered in both *nodPQ* copies do not produce sulphated factors [16]. Thus *nodH* and *nodPQ*, major host-range genes of *R. meliloti*, are involved in the sulphation of the lipo-oligosaccharidic Nod factors. NodPQ proteins share sequence similarity with ATP sulphurylase and APS kinase and are involved in the synthesis of adenosine-5′-phosphosulphate (APS) and 3′-phosphoadenosine-5′-phosphosulphate (PAPS), the ATP-derived activated forms of sulphate [27]. NodH shares identity with sulphotransferases, and its function is probably to transfer a sulphate group to the Nod factor chito-oligosaccharidic precursors [16].

Purified sulphated factors are active on alfalfa, and *R. meliloti* strains producing sulphated molecules infect alfalfa. Unsulphated factors are active on vetch, and *R. meliloti* strains producing such factors infect vetch. It is, thus, reasonable to hypothesize that it is by encoding the sulphation of Nod factors that the *nodPQ* and *nodH* genes of *R. meliloti* determine alfalfa specificity. The introduction of the *nodPQ* and *nodH* genes into *R. leguminosarum* bv. *viciae* and bv. *trifolii* — symbionts of pea and clover, respectively — results in the transfer of the ability to infect and nodulate alfalfa [12,28].

Another case of non-allelic variation is illustrated by the *nodL* gene which is present in *R. meliloti* and *R. leguminosarum* but absent in other species. NodL is similar to bacterial acetyltransferases [29] and specifies, in *R. leguminosarum* bv. *viciae*, the O-acetylation of the C_6 of the non-reducing, terminal glucosamine residue [18]. The same function has been attributed to the *R. meliloti* NodL (M. Ardourel and C. Rosenberg, unpublished work). *nodL* has not been found in *Bradyrhizobium japonicum* 110, and the Nod factors secreted by this strain are not acetylated on the C_6 of the non-reducing, terminal glucosamine residue [30].

An example of allelic variation: the case of *nodFE* genes

In *R. leguminosarum* and *R. meliloti*, the *nodFE* genes are host-range genes [31]. Protein sequence similarity and biochemical studies strongly indicate that NodF is an acyl-carrier protein [32,33], whereas NodE is similar to β-ketoacylsynthases [31]. In *R.*

leguminosarum bv. *viciae*, NodE and NodF are required for the synthesis of the highly unsaturated $C_{18:4}$ acyl chain [18]. *R. leguminosarum* bv. *viciae nodE* mutants produce NodRlv factors which, because of the absence of synthesis of the polyunsaturated $C_{18:4}$ fatty acid, are acylated by default with vaccenic acid ($C_{18:1}$), the most common fatty acid in Gram-negative bacteria [7,18]. In *R. meliloti*, the *nodFE* genes also determine the synthesis of the N-acyl chain but, in this case, a C_{16} chain with two (or three) double bonds [17]. *R. meliloti* strains, in which *nodFE* genes have been replaced by the *nodFE* genes of *R. leguminosarum* bv. *viciae*, produce Nod factors acylated with polyunsaturated C_{18} fatty acids. Thus the molecular basis of allelic variation for the *nodFE* genes relies on the production of NodFE enzyme variants which determine the synthesis of fatty acids differing in the carbon length and the number of unsaturations. The observed variations in the length of the acyl chain and the number of unsaturations might be due to a difference in the number of the chain-elongation cycles in *R. leguminosarum* and *R. meliloti* [7]. Therefore, one function of species-specific *nod* genes, allelic or not, is to decorate the core Nod factor molecules and to make them plant-specific.

Rhizobial Nod factors: a variety of decorations on the same core molecules

Rhizobium, *Bradyrhizobium* and *Azorhizobium* are very distant genetically. However, the Nod factors from strains of these three genera were recently shown to belong to the same chemical family: they all are chito-oligosaccharides, mono-N-acylated on the terminal, non-reducing glucosamine residue. The substitutions on the chito-oligosaccharide core, however, vary with the species or biovars. *B. japonicum* 110, a symbiont of soybean, produces Nod factors in which the reducing sugar is substituted on C_6 with methyl fucose [30]. At the non-reducing end, the fatty acid is $C_{18:1}$ (see Fig. 3). Interestingly, *R. fredii*, another symbiont of soybean, produces similar Nod factors which are also substituted with a fucose group on the C_6 of the reducing glucosamine residue (M.P. Bec and J.C. Promé, unpublished work).

Rhizobium NGR234, a tropical strain with a very broad host range, nodulates more than 70 different legume genera and even the non-legume *Parasponia*. This strain produces a family of Nod factors which are mono-N-acylated chitopentasaccharides with a variety of possible substitutions [34] (see Figure 3). The non-reducing end may be O-substituted with one or two carbamoyl groups and the N atom is acylated with palmitic or vaccenic acids and is also methylated. The reducing sugar is substituted with methyl fucose which, in turn, may be either O-acetylated or O-sulphated, in a mutually exclusive manner. Thus, in contrast to *R. meliloti*, which produces only sulphated factors, and to *R. leguminosarum* and *B. japonicum*, which produce only non-sulphated factors, NGR234 secretes a mixture of sulphated (charged) and non-sulphated factors, and this property may be part of the basis of its promiscuity [34].

In *Azorhizobium caulinodans*, the reducing glucosamine is substituted by a different sugar, arabinose [35]. It thus appears that all rhizobia produce Nod factors which belong to the same family of molecules — mono-N-acylated chito-oligosaccharides. The common core is probably synthesized by enzymes encoded by the common *nodABC* genes, which are present in *Rhizobium*, *Bradyrhizobium* and

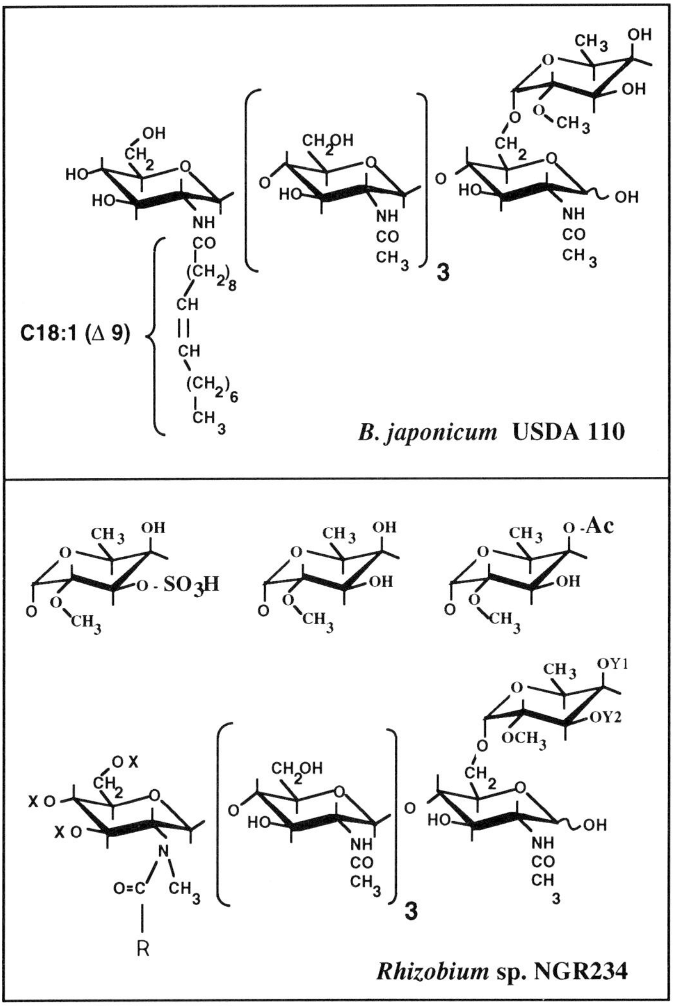

Fig. 3. Structures of Nod factors of (a) *B. japonicum* [30] and (b) *Rhizobium* sp. NGR234 [34]. For NodNGR factors: x = carbamoyl; Y1 = H or COCH3; Y2 = H or SO_3H; R = vaccenic acid [$C_{18:1\ (\Delta 11)}$] or palmitic acid ($C_{16:0}$). Above the general formula, are the mutually exclusive substitutions of O-acetyl and O-sulphate on the O-fucosyl residue [34].

Azorhizobium. The function of the species-specific *nod* genes is to code for enzymes which decorate the core molecules, and make them plant-specific.

During the last two years, the genetic and biochemical dissection of Nod-factor biosynthesis in a few rhizobial species has led to the identification of a number of *nod* genes that code for transferases catalysing the transfer of various groups (for example, sulphate, methyl, acetyl, carbamoyl, fucosyl, arabinosyl, and so on) at specific positions on the glucosamine backbone. A very large number of rhizobial isolates, exhibiting an extremely large variety of host ranges, are now available and are likely to infect and nodulate hundreds of legume species. Future studies will tell which molecular strategies these bacteria have evolved to synthesize Nod factors differing in the decoration of the chito-oligosaccharide backbone.

Carbohydrates are frequently involved in recognition between cells; this is probably due to the large possibility of O- and N-substitutions offered by these molecules, resulting in a great structural diversity. The genetic and biochemical dissection of oligosaccharide synthesis and decoration is much easier to perform in prokaryotes, such as rhizobia, which are amenable to simple genetic analysis, than in eukaryotes. We can thus expect that rhizobia will serve as a source of genes and enzymes specifying the transfer of different groups on oligosaccharides and facilitate the search for homologous genes and proteins in eukaryotes.

A model for signal perception and transduction in plants

The study of Nod factors as signalling molecules is interesting for a number of reasons. First, in contrast to most of the plant–microbe interaction systems studied so far, the biological significance of these molecules is supported by a great deal of genetic data. Loss-of-function as well as gain-of-function experiments have clearly established that the *nod* genes determine both host-specificity of infection and nodulation, and the production of Nod factors. Moreover, the exogenous supply of Nod factors elicits plant responses similar to those induced by rhizobia. In contrast, the physiological significance of numerous fungal elicitors remains to be established. Secondly, Nod factors are highly specific signalling molecules and the study of their recognition by plant cells will contribute to the understanding of mechanisms of cell recognition in plants. Thirdly, Nod factors elicit plant responses at very low concentrations, down to 10^{-12}–10^{-14} M, which suggests the existence of very efficient mechanisms of recognition and signal transduction by the plant cells. Finally, Nod factors elicit different types of developmental responses: stimulation of plant cell wall growth for infection thread formation, and induction in cortical cells of mitosis and of a major developmental switch which will lead to the formation of a new organ.

A number of laboratories are now trying to understand how these bacterial signals are recognized by the plants. Since these molecules are active at low concentrations and determine host specificity, it seems reasonable to hypothesize that receptors are involved in their recognition and amplification. These receptors would have to recognize both ends of the sugar backbone as well as the lipid moiety, and they should be located on epidermal and root-hair cells, and possibly in cortical cells. To analyse the perception and transduction of these signals, experimental tools allowing the detection of early plant responses are needed. Sharon Long and co-workers have developed a system to study very early reactions of root-hair cells to Nod factors. Using microelectrodes impaled in root hairs, they have observed that the addition of *R. meliloti* culture filtrates or purified NodRm factors induces a rapid (few minutes) depolarization of root-hair membrane potential [36]. Such an approach will be useful to study the effect on plant responses of inhibitors specific for some classes of receptors and of signal transduction pathways.

Another strategy, the search for plant genes which are specifically expressed during the early steps of symbiosis, has been followed for many years by Ton Bisseling and co-workers. They have identified a number of pea early nodulin genes, e.g. ENOD5 and ENOD12, which are associated with *Rhizobium* infection [37,38]. ENOD12 and ENOD5 probes were used to detect the expression of these genes in root hairs after amplification of mRNA using the polymerase chain reaction. Infection of pea seedlings by a compatible *R. leguminosarum* strain results in the induction of transcription of both genes [38]. Similarly, the addition of *R. leguminosarum* bv. *viciae*-purified Nod factors triggers the transcription of both genes in pea root hairs [39]. Another method to study the pattern of expression of early nodulin genes during infection and nodulation has been developed using transgenic alfalfa transformed with fusions between the *Medicago truncatula* ENOD12 promoter and the GUS reporter gene [40]. This approach allows a clear visualization of ENOD12 gene expression in different plant tissues after either bacterial infection or addition of Nod factors.

These different cellular and molecular approaches will, in the near future, allow the study of structure–function relationships for the activity and specificity of Nod factors in the induction of plant responses at different steps of the symbiosis, and contribute to the analysis of the perception and transduction of these signals. Nicolaou and co-workers [41] have recently achieved the complete chemical synthesis of sulphated and non-sulphated factors of *R. meliloti*, starting from glucosamine derivatives as building blocks. This breakthrough opens the way to the synthesis of radioactive and photoactivatable derivatives which will be valuable tools for the search of Nod factor receptors.

Part of the work described in this overview was supported by grants from the Conseil Régional Midi-Pyrénées and from the Human Frontier Science Program.

References

1. Young, J.P.W. and Johnston, A.W.B. (1989) Trends Ecol. Evol. **4**, 331–349
2. Dénarié, J. and Roche, P. (1991) in Molecular Signals in Plant–Microbe Communication (Verma, D.P.S., ed.), pp. 295–324, CRC Press, Boca Raton
3. Lewin, A., Cervants, E., Wong, C.H. and Broughton, W.J. (1990) Mol. Plant-Microbe Interact. **3**, 317–326
4. Brewin, N.J. (1991) Annu. Rev. Cell Biol. **7**, 191–226
5. Dénarié, J., Debellé, F. and Rosenberg, C. (1992) Annu. Rev. Microbiol. **46**, 497–531
6. Fisher, R.F. and Long, S.R. (1992) Nature (London) **357**, 655–660
7. Spaink, H.P. (1992) Plant Mol. Biol. **20**, 977–986
8. Horvath, B., Bachem, C.W.B., Schell, J. and Kondorosi, A. (1987) EMBO J. **6**, 841–848
9. Spaink, H.P., Wijffelman, C.A., Pees, E., Okker, R.J.H. and Lugtenberg, B.J.J. (1987) Nature (London) **328**, 337–340
10. van Brussel, A.A.N., Zaat, S.A.J., Canter Cremers, H.C.J., Wijfellman, C.A., Pees, E., Tak, T. and Lugtenberg, B.J.J. (1986) J. Bacteriol. **165**, 517–522
11. Faucher, C., Maillet, F., Vasse, J., Rosenberg, C., van Brussel, A.A.N., Truchet, G. and Dénarié, J. (1988) J. Bacteriol. **170**, 5489–5499
12. Faucher, C., Camut, S., Dénarié, J. and Truchet, G. (1989) Mol. Plant-Microbe Interact. **2**, 291–300
13. Lerouge, P., Roche, P., Faucher, C., Maillet, F., Truchet, G., Promé, J.C. and Dénarié, J. (1990) Nature (London) **344**, 781–784
14. Schultze, M., Quiclet-Sire, B., Kondorosi, E., Virelizier, H., Glushka, J.N., Endre, G., Géro, S.D. and Kondorosi, A. (1992) Proc. Natl. Acad. Sci. U.S.A. **89**, 192–196

15. Roche, P., Lerouge, P., Ponthus, C. and Promé, J.C. (1991) J. Biol. Chem. **266**, 10933–10940
16. Roche, P., Debellé, F., Maillet, F., Lerouge, P., Faucher, C., Truchet, G., Dénarié, J. and Promé, J.C. (1991) Cell **67**, 1131–1143
17. Demont, N., Debellé, F., Aurelle, H., Dénarié, J. and Promé, J.C. (1993) J. Biol. Chem. **268**, 20134–20142
18. Spaink, H.P., Sheeley, D.M., van Brussel, A.A.N., Glushka, J., York, W.S., Tak, T., Geiger, O., Kennedy, E.P., Reinhold, V.N. and Lugtenberg, B.J.J. (1991) Nature (London) **354**, 125–130
19. Atkinson, E.M. and Long, S.R. (1992) Mol. Plant-Microbe Interact. **5**, 439–442
20. Debellé, F., Rosenberg, C. and Dénarié, J. (1992) Mol. Plant-Microbe Interact., **5**, 443–446
21. John, M., Röhrig, H., Schmidt, J., Wieneke, U. and Schell, J. (1993) Proc. Natl. Acad. Sci. U.S.A. **90**, 625–629
22. Baev, N., Endre, G., Petrovics, G., Banfalvi, Z. and Kondorosi, A. (1991) Mol. Gen. Genet. **228**, 113–124
23. Marie, C., Barny, M.A. and Downic, J.A. (1992) Mol. Microbiol. **6**, 843 852
24. van Brussel, A.A.N., Bakhuisen, R., van Spronsen, P.C., Spaink, H.P., Tak, T., Lugtenberg, B.J.J. and Kijne, J.W. (1992) Science **257**, 70–72
25. Truchet, G., Roche, P., Lerouge, P., Vasse, J., Camut, S., de Billy, F., Promé, J.C. and Dénarié, J. (1991) Nature (London) **351**, 670–673
26. Banfalvi, Z. and Kondorosi, A. (1989) Plant Mol. Biol. **13**, 1–12
27. Schwedock, J. and Long, S.R. (1990) Nature (London) **348**, 644–647
28. Debellé, F., Maillet, F., Vasse, J., Rosenberg, C., de Billy, F., Truchet, G., Dénarié, J. and Ausubel, F. (1988) J. Bacteriol. **170**, 5718–5727
29. Downie, J.A. (1989) Mol. Microbiol. **3**, 1649–1651
30. Sanjuan, J., Carlson, R.W., Spaink, H.P., Bhat, U.R., Barbour, W.M., Glushka, J. and Stacey, G. (1992) Proc. Natl. Acad. Sci. U.S.A. **89**, 8789–8793
31. Spaink, H.P., Weinman, J., Djordjevic, M.A., Wijffelman, C.A., Okker, J.H. and Lugtenberg, B.J.J. (1989) EMBO J. **8**, 2811–2818
32. Shearman, C.A., Rossen, L., Johnston, A.W.B. and Downie, J.A. (1986) EMBO J. **5**, 647–652
33. Geiger, O., Spaink, H.P. and Kennedy, E.P. (1991) J. Bacteriol. **173**, 2872–2878
34. Price, N.P.J., Relic, B., Talmont, F., Lewin, A., Promé, D., Pueppke, S.G., Maillet, F., Dénarié, J., Promé, J.C. and Broughton, W.J. (1992) Mol. Microbiol. **6**, 3575–3584
35. Mergaert, P., Van Montagu, M., Promé, J.C. and Holsters, M. (1993) Proc. Natl. Acad. Sci. U.S.A. **90**, 1551–1555
36. Ehrhardt, D.W., Atkinson, E.M. and Long, S.R. (1992) Science **256**, 998–1000
37. Nap, J.P. and Bisseling, T. (1990) Science **250**, 948–954
38. Scheres, B., Van De Wiel, C., Zalensky, A., Horvath, B., Spaink, H., Van Eck, H., Zwartkruis, F., Wolters, A.M., Gloudemans, T., Van Kammen, A. and Bisseling, T. (1990) Cell **60**, 281–294
39. Horvath, B., Heidstra, R., Lados, M., Moerman, M., Spaink, H.P., Promé, J.C., Van Kammen, A. and Bisseling, T. (1993) Plant J. **4**, 727–733
40. Pichon, M., Journet, P.E., Dedieu, A., de Billy, F., Truchet, G. and Barker, D.G. (1992) Plant Cell **4**, 1199–1211
41. Nicolaou, K.C., Bockovich, N.J., Carcanague, D.R., Hummel, C.W. and Even, L.F. (1992) J. Am. Chem. Soc. **114**, 8701–8702

Biochem. Soc. Symp. **60**, 61–73
Printed in Great Britain

Bacterial and plant glycoconjugates at the *Rhizobium*–legume interface

N.J. Brewin*, A.L. Rae, S. Perotto, E.L. Kannenberg, E.A. Rathbun, M.M. Lucas, A. Gunder, L. Bolaños, I.V. Kardailsky, K.E. Wilson, J.L. Firmin and J.A. Downie

John Innes Institute, Colney Lane, Norwich NR4 7UH, U.K.

Synopsis

Many classes of bacterial and plant glycoconjugate have been shown to be involved in establishing the *Rhizobium* root nodule symbiosis with peas (*Pisum sativum*). It was demonstrated, using techniques of molecular genetics, that a group of *Rhizobium* nodulation genes (*nod* genes) co-operate to synthesize a lipo-oligosaccharide signal molecule that specifically initiates nodule development on legume hosts. An additional gene function, encoded by *nodX*, has been found to extend the host range of *Rhizobium leguminosarum* bv. *viciae* to include nodulation of a pea mutant, cultivar Afghanistan; the *nodX* gene product specifies the addition of an acetyl group to the terminal *N*-acetylglucosamine residue at the reducing end of the pentasaccharide core of this signal molecule. Several other classes of bacterial glycoconjugate have also been shown by genetic analysis to be essential for normal nodule development and function: these include a capsular extracellular polysaccharide; lipopolysaccharide in the outer membrane; and cyclic glucans present in the periplasmic space. Potential functions for these glycoconjugates are discussed in the context of tissue and cell invasion by *Rhizobium*. Some plant components involved in symbiotic interactions have been identified by the analysis of nodule-specific gene expression (early nodulins). Several of the cDNA clones encoding these early nodulins specify proline-rich proteins that presumably correspond to cell wall glycoproteins or membrane arabinogalactan proteins. Other plant glycoconjugates have been identified using monoclonal antibodies as probes. A plant glycoprotein present in intercellular spaces has been identified as a component of the luminal matrix of infection threads. Because it attaches to the surface of bacteria and is itself susceptible to oxidative cross-linking, this glycoprotein may be involved in limiting the progress of microbial infections. Endocytosis of bacteria into the plant cytoplasm is apparently driven by direct interactions between the bacterial

*To whom correspondence should be addressed.

surface and the plasma membrane that is exposed within an unwalled infection droplet; glycoprotein and glycolipid components of the plant membrane glycocalyx have been defined using monoclonal antibodies. Differentiation of endosymbiotic bacteroids is preceded by differentiation of the plant-derived peribacteroid membrane which encloses the symbiosome compartment. Using a monoclonal antibody that identifies a group of plant membrane-associated, inositol-containing glycolipids, we have identified a very early marker for the differentiation of peribacteroid membrane from plasma membrane.

Introduction

The *Rhizobium*–legume symbiosis is characterized by the development of a new plant organ, the root nodule, with a specialized central infected tissue that provides a suitable micro-environment for biological nitrogen fixation [1]. Within each infected host cell, endosymbiosis is characterized by the development of a new form of quasi-organelle structure termed the 'symbiosome'. This comprises a differentiated form of *Rhizobium*, the 'bacteroid', enclosed by a plant-derived peribacteroid membrane, somewhat similar in structure to the host plasma membrane, but lacking the capacity to synthesize an associated cell wall and acquiring instead specialized transport functions that regulate the nutritional, osmotic and physiological status of the endosymbiont.

In terms of cell-to-cell signalling in nodule development (Fig. 1), there are several important issues which merit investigation, and in which the *Rhizobium*–legume symbiosis may serve as a model for the discovery of more general truths concerning the nature of plant development and of plant–microbe interactions. How is the nodule initiated? What sustains the process of nodule development and the accompanying

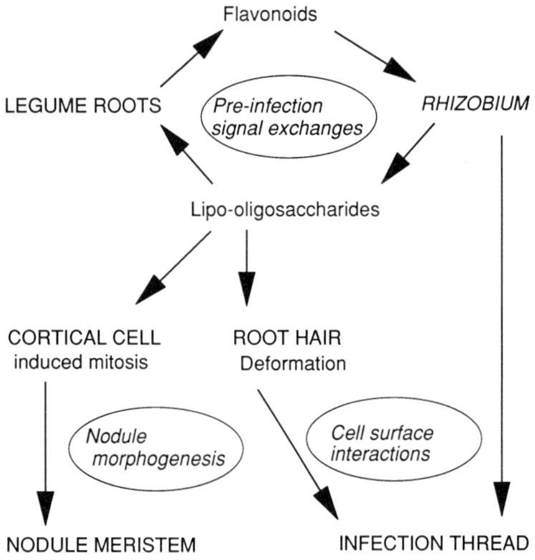

Fig. 1. Diagram showing three phases of cell-to-cell signalling involved in legume nodule development. (A) Pre-infection signal exchanges; (B) nodule organogenesis; (C) tissue and cell invasion by *Rhizobium*.

process of cell differentiation? What forces and intercellular signals drive the process of tissue and cell invasion by *Rhizobium*? What distinguishes a symbiotic form from a pathogenic form of host–microbial interaction?

One way to answer these questions has been to use genetics, because the analysis of malfunctions in plant and bacterial mutants that are symbiotically defective can tell us much about signal exchanges that are involved in normal nodule development [2,3]. Another approach has been to use the techniques of cell biology, either to investigate the phenomenon of tissue-specific gene expression [4,5], or to use monoclonal antibodies as probes to build up a picture of the antigens on the differentiating plant and bacterial cell surfaces at successive stages of nodule development. This last approach has illuminated the particular role of cell surface glycoconjugates in nodule development.

Our experimental system involves *R. leguminosarum* bv. *viciae*, which has been very well characterized genetically [2,6]. Among its hosts are the small-seeded, hairy vetch *Vicia hirsuta* (convenient for cytological analysis) and the garden pea *Pisum sativum* (convenient for biochemical analysis). Ever since the days of Mendel, the pea has been a classic plant for genetic analysis, and at least 30 genes affecting symbiotic development have been described [7]. In the present chapter, we will briefly review what has been learned about plant–microbial signal exchange from an analysis of pea mutants, and then we will examine the role of cell surface glycoconjugates in the process of tissue and cell invasion by *Rhizobium*.

Host plant genetics

The symbiotically defective mutants that have been isolated in pea have either been derived by screening for natural variation in field isolates or, more commonly, they are the product of mutagenesis with either chemicals or irradiation [7,8]. Examples of the five major phenotypic classes of symbiotically defective mutant are given in Table 1. Each of these classes provides some interesting clues concerning the nature of cell-to-cell signalling during nodule development.

The non-nodulating mutants either fail to respond to the bacterial lipo-oligosaccharide signalling molecule, or the response of root-hair cells is abnormal, resulting in stunted or distorted root hairs or the early abortion of the *Rhizobium*-induced infection thread [9]. Such mutants may either lack the receptor system for bacterial lipo-oligosaccharide signal molecules, or lack the mechanisms for transducing these signals into an effective host-cell response. The occurrence of several mutants with deformed and distorted root hairs indicates that reorganization of cell wall growth, presumably involving reorganization of the cytoskeleton, is an essential precondition for infection thread development. A fascinating subclass of these non-nodulating mutants is a group (involving several different loci) where the phenotype is pleiotropic, preventing establishment of the vesicular arbuscular mycorrhizal symbiosis as well as of the symbiosis with *Rhizobium* [10]. Two possible inferences can be drawn from this phenotype: either that both symbionts have a common system for suppressing the normal host defence system which prevents invasion by micro-organisms (and that this suppression system is inoperative in the mutant phenotype) or, alternatively, that the two symbionts make use of a common, non-disruptive system of cell and tissue invasion. Such a system might involve the formation of transcellular tunnels sheathed by plant

Table 1. Examples of symbiotically defective pea mutants.

Non-nodulating
Most mutants fail to respond to lipo-oligosaccharide signal molecules
Some mutants show abnormal root hair growth
Many mutants are also Myc^- i.e. non-mycorrhizal
Strain-specific nodulation
Pea cultivar Afghanistan is *sym2 sym2* (homozygous recessive), i.e. only responds to acetylated form of lipo-oligosaccharide
Low frequency of nodulation
sym5 — mutant is ethylene hypersensitive
Non-fixing mutants (Fix^-)
sym13 — no nitrogenase induction by bacteroids, premature senescence
Sprint-2 Fix^- — no nitrogenase, multiple bacteroids per symbiosome
Supernodulating mutants (nitrate-tolerant symbiosis)
nod3 — phenotype is root controlled
sym28 — phenotype is shoot controlled

References to each of these classes of mutant are given in the text.

cell wall and plant cell membrane, whose organized deposition is, in some way, controlled by the plant cytoskeleton.

A second group of pea mutants (derived both from natural variants and from induced mutagenesis) are mutants showing strain specificity for nodulation. The best-studied example is that of pea cultivar Afghanistan, which is resistant to nodulation by most strains of *R. leguminosarum* bv. *viciae*. The resistance character by which cultivar Afghanistan differs from normal cultivated garden peas corresponds to a single recessive Mendelian allele [11,12]. In the case of pea species, the evolutionary origin and centre of divergence is thought to be the eastern Mediterranean region, and it was found that certain field isolates of *R. leguminosarum* derived from soils in the Middle East were capable of nodulating effectively with cultivar Afghanistan, as well as with normal garden peas. One of these strains (designated TOM, a field isolate from Tomask in Turkey) was subjected to genetic analysis to investigate the basis for its extended host-range character (Table 2). It was found that the particular nodulation characteristics of TOM could be co-transferred to other strains of *R. leguminosarum* by transfer of the symbiotic plasmid (P*sym*) which also carries all the normal nodulation genes [13]. It was subsequently established, both by mutagenesis and by gene-cloning experiments, that the gene conferring the extension of host-range character in strain TOM was an extra gene (termed *nodX*) with no counterpart in a normal strain of *R. leguminosarum* [14–16]. When the *nodX* gene was sequenced, it was deduced, on the basis of motifs found in its predicted amino acid sequence, that the protein was likely to function as an acetyltransferase. Furthermore, in collaboration with Dr R.W. Carlson (University of Georgia, U.S.A.), we have recently demonstrated that strains of *R. leguminosarum* bv. *viciae* carrying the *nodX* gene synthesize a chemically modified form of lipo-oligosaccharide which carries an additional acetyl substitution on the terminal reducing sugar of

Table 2. Nodulation of peas by strains of *R. leguminosarum*.

R. leguminosarum strain or genotype	Pea cultivar Afghanistan *sym2 sym2*	Cultivated garden pea *Sym-2 Sym-2*
TOM field isolate (contains *nodX*)	+	+
300 field isolate (lacks *nodX*)	−	+
300 P*sym*⁻	−	−
300 P*sym*⁻ + P*sym* (TOM)	+	+
300 + *nodX*	+	+

nodX is an extra gene from strain TOM with no counterpart in strain 300 or other strains of *R. leguminosarum* bv. *viciae*. The DNA sequence of the *nodX* gene indicates homology with acetyltransferase. Lipo-oligosaccharide isolated from TOM or 300 *nodX* carried an additional acetyl group on the reducing sugar of the penta-glucosamine core.

the penta-glucosamine backbone (Fig. 2). Thus it appears that acetylation of the bacterial signal molecule (by the NodX gene product) is required to overcome the host-resistance character associated with the homozygous recessive genotype of cultivar Afghanistan. Now it will be interesting to discover the function of the dominant allele, Sym2, that is present in normal garden peas. Could it be that this plant gene product also serves to convert the unacetylated bacterial oligosaccharide (synthesized by strains of *R. leguminosarum* lacking *nodX*) into a form that is active in plant tissue? Perhaps the *Sym2* gene also encodes an acetyltransferase equivalent in function to the bacterial NodX gene product — but such possibilities are still only speculations at present.

A third group of symbiotically defective pea mutants corresponds to a phenotypic class that gives poor nodulation and frequently a temperature-sensitive nodulation phenotype. One such mutant is *sym5* [17,18]. This mutant has recently been shown to be hypersensitive to ethylene, reminding us that the process of nodule initiation and development is peculiarly sensitive to ethylene, far more so than other aspects of plant development, such as root growth [19]. It is interesting to note that ethylene is often synthesized by plant cells in response to stress, and hence its involvement in the control of nodule development is particularly intriguing [20].

A fourth general class of symbiotically defective phenotype, includes all mutants that create nodules in which nitrogen fixation does not take place. Some of these mutants show abnormalities in the process of nodule development. For example, in Sprint-2 Fix⁻, surface contacts between intracellular bacteroids and the peribacteroid membrane are apparently altered so that the bacteroids are enclosed in groups of a dozen rather than singly, as in the wild-type symbiosis [21]. In another Fix⁻ mutant, *sym13*, morphological development of bacteroids is normal but nitrogenase is not induced, presumably as a result of the failure of an essential plant-derived signal or metabolic transport function [22]. Interestingly, in such Fix⁻ mutants, the bacteroids senesce prematurely, implying that the host plant has a surveillance system that can act to close down unproductive nodules that have been colonized by rhizobia, but where

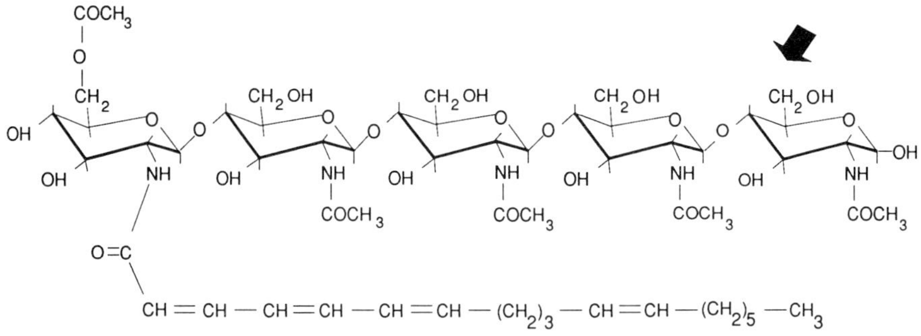

Fig. 2. Chemical structure of the lipo-oligosaccharide signalling molecule encoded by the nodulation genes of *R. leguminosarum* bv. *viciae*. This compound specifically initiates the development of nodules in pea roots by causing cortical cell divisions. The molecule has an oligosaccharide core, comprising four or five sugar residues of *N*-acetylglucosamine; this is a common feature of lipo-oligosaccharides synthesized by all species of *Rhizobium*. In each species, a different polyunsaturated fatty acid is attached to the amino group of glucosamine at the non-reducing end of the oligosaccharide chain; this confers one element of host specificity on the molecule. Other components of host specificity are an acetyl substitution on the terminal non-reducing sugar and a variety of possible substitutions on the terminal reducing sugar (but not shown on the molecular form illustrated here). In *R. leguminosarum* strain TOM, the action of the *nodX* gene product confers an additional acetyl substitution to the reducing sugar of the pentasaccharide core, as indicated by the arrowhead, thereby extending host range.

no nitrogen fixation results: such bacteria would be more accurately described as pathogens rather than as symbionts.

The fifth and final class of pea nodulation mutants are the so-called supernodulating mutants. These mutants were originally identified because their nodule development was not suppressed by growth of plants in high levels of fixed nitrogen (nitrate-tolerant symbiosis). In addition, these mutants have far more nodules on their roots than is normal [9,23]. They appear to be defective in a feedback control system that acts systemically to control nodule number (or nodule mass) on the whole plant. Furthermore, it has been ascertained from root–shoot grafting experiments, that in some of these lines the mutant phenotype is expressed in root tissue, whereas in others the phenotype is expressed by the shoot tissue, implying that both roots and shoots are involved in the mediation of this whole-plant control system.

Bacterial cell surface glycoconjugates

Recent genetic and biochemical analysis has shown that at least four kinds of *Rhizobium*-derived glycoconjugate molecule are essential for tissue and cell invasion by *Rhizobium*. These are (i) the diffusible lipo-oligosaccharide 'Nod-factor', which alone can stimulate root-hair curling and cortical cell division [6]; (ii) acidic extracellular polysaccharide (EPS), which is a Ca^{2+}-gelling succinoglycan that encapsulates rhizobial cells and is apparently essential for infection thread initiation and development [24]; (iii) lipopolysaccharide (LPS), the major component of the bacterial outer membrane, which is essential for proper infection thread development, bacterial release into host cells and the proper development and functioning of nitrogen-fixing bacteroids [25,26]; and (iv) periplasmic cyclic glucan polymers, which may function in osmotic adaptation to the endophytic environment [27]. The initiation of an infection thread in pea root-hair cells involves the participation of live bacteria carrying appropriate EPS and LPS components in their cell wall.

Our work has focused on the role of bacterial lipopolysaccharide and its possible interaction with plant cell surface components at successive stages of pea nodule development, both in the early stages of infection thread development and subsequently during the differentiation of nitrogen-fixing bacteroids within the symbiosome compartment. The importance of LPS is suggested by the fact that mutants with a variety of modifications in the structure and biosynthesis of their LPS macromolecules are unable to establish a normal nitrogen-fixing symbiosis [26]; these mutants induce the development of abnormal root nodules on peas and other legumes [28–30]. We have analysed the development of these nodules and the fate of the LPS-defective mutant bacteria within them by using monoclonal antibodies and cytochemical techniques. The mutants fell into three general classes: severe mutants, inducing an 'empty nodule' phenotype; moderately severe mutants, inducing a delayed nodule development and reducing nitrogen fixation to less than 5% of the rate for wild-type nodules; and mildly disabled mutants, which only slightly impaired the normal processes of nodule development and nitrogen fixation.

The most severe LPS-defective mutant failed to invade nodule tissue. Inoculation of pea seedling roots with strain B659 induced the development of empty nodule-like structures with peripheral vasculature, a rudimentary endodermis and a central uninfected tissue which secreted quantities of an extracellular matrix glycoprotein that accumulated in the intercellular spaces [31]. However, in the absence of invading bacteria, no infection thread structures were seen.

The second group of LPS-defective mutants induced nodules in which only a small proportion of the central nodule tissue was colonized by bacteria. Consequently, much of the central tissue was occupied by uninfected parenchyma, particularly underneath the nodule endodermis. Despite the abnormal development of infection threads and the relatively low number of infected host cells in these nodules, the infected host cells still induced leghaemoglobin production and the endosymbiotic bacteria enclosed by peribacteroid membranes also induced the synthesis of nitrogenase. The presence of both of these proteins was detected by immunostaining with specific antisera. The appearance of nodule cells invaded by LPS-defective mutants was found to be highly heterogeneous. Adjacent to viable infected cells with nitrogenase-containing bacteroids, other cells were found with clear signs of cytoplasmic disorganization and collapse. This suggests that the signal eliciting the host defence response was very localized.

The ultrastructural analysis of infected nodule cells also revealed clear differences between wild-type and LPS-defective bacteroids. The mutant bacteria were always released into the plant cytoplasm surrounded by a peribacteroid membrane; but, in contrast to wild-type bacteroids which are normally Y-shaped, the bacteroids formed by mutant strains were usually highly branched and much larger in size. Moreover, in nodules formed by LPS-defective mutants, several bacteroids were commonly seen inside the same peribacteroid membrane envelope, whereas in wild-type nodules bacteroids were always individually enclosed. Mutant bacteroids also showed premature senescence and induced the formation of apparently lytic vesicles in the host-cell cytoplasm. From these observations of abnormal nodule development, it seems that LPS-defective mutants are a little closer to the borderline between symbiotic and pathogenic interactions with the host plant. We conclude that the correct LPS structure is essential for the avoidance of host defence responses and for the physiological adaptation of rhizobia to the endophytic environment.

Glycoconjugates in the plant intercellular matrix

Using the monoclonal antibody MAC 265 as a probe, we have identified a 95–110 kDa plant glycoprotein secreted into the lumen of infection threads as an early response to *Rhizobium* infection [32,33]. The glycoprotein (or family of glycoproteins) recognized by MAC 265 has recently been purified from pea nodules, using a procedure based on immunoaffinity chromatography, but the *N*-terminal sequence that has been obtained does not conform to any of the known classes of plant extracellular glycoproteins. Some indications are now emerging about the physical properties of the plant matrix glycoprotein and how it might interact with invading rhizobia in the lumen of infection threads.

We have recently demonstrated that the matrix glycoprotein is capable of attachment through ionic binding to the surface of rhizobial cells, derived either from free-living culture or isolated from nodules. This attachment occurs also with *Rhizobium* mutants lacking the outer (O-antigen) components of LPS, and also with mutants lacking the extracellular capsular polysaccharide (Fig. 3). Moreover, the binding of matrix glycoprotein to the bacterial surface was always stronger after the capsular polysaccharide had been removed by prior washing in a high-salt buffer. Thus, it seems probable that the capsular polysaccharide (and perhaps also the O-antigen components of LPS) may actually inhibit binding of matrix glycoprotein to the bacterial cell surface by masking the highly charged groups that lie closer to the surface of the bacterial outer membrane. It should also be noted that the attachment of matrix glycoprotein to the bacterial surface is not specific to *Rhizobium*: it occurs equally well with cultured cells of *Escherichia coli*. Therefore, it is conceivable that the matrix glycoprotein functions as part of a general antimicrobial defence system that entraps invading micro-organisms, preventing their further penetration into host tissues. In this case, the essential feature of a successful invasion might be the ability to escape entrapment by the intercellular plant matrix glycoprotein. On this model, the role of *Rhizobium* extracellular polysaccharide in infection thread growth in peas might be to provide a capsular sheath that masks the charged components on the bacterial surface from interaction with plant matrix glycoprotein: the outer (O-antigen) components of LPS might also have an enhancing role in this function, since mutants lacking these components show some abnormalities in infection thread growth (Fig. 4).

Support for the model that matrix glycoprotein functions in a microbial entrapment system comes from observation of the physical behaviour of the glyco-

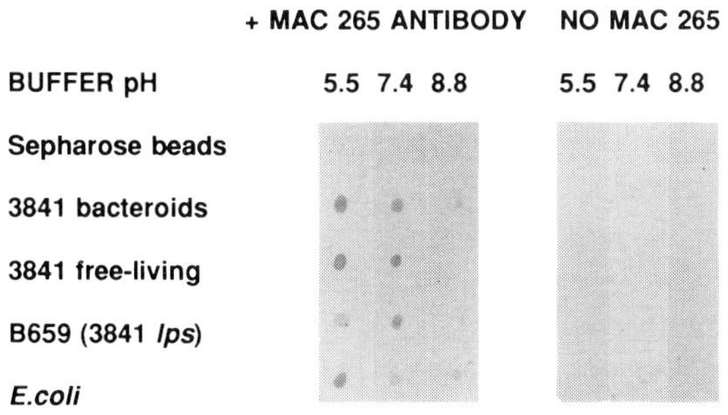

Fig. 3. Binding of plant matrix glycoprotein to the surface of bacteria, as revealed by immunostaining with MAC 265, a rat monoclonal antibody recognizing the glycoprotein. Bacterial cell suspensions (10^9 cells per ml) were incubated with crude matrix glycoprotein (derived from nodule homogenates) and diluted in buffer solutions (50 mM) containing dithiothreitol (5 mM) and adjusted to the pH values indicated. After incubation for 1 h, bacteria were re-isolated by centrifugation and washed extensively. Aliquots (1 μl) were immobilized on nitrocellulose sheets and probed with MAC 265 (left-hand sheet) or, as a negative control, with an irrelevant antibody (right-hand sheet), followed by peroxidase anti-(rat immunoglobulin) and chromogenic substrates.

protein under conditions of oxidative stress. It was found that the glycoprotein could only be purified from nodule homogenates in the presence of antioxidants (ascorbic acid and dithiothreitol). In high oxygen concentrations or in low protein concentrations the glycoprotein had a tendency to become insoluble. This process was enhanced by the introduction of peroxidase enzyme and hydrogen peroxide into the medium. It is well known that plant cells, particularly those that have been subjected to microbial attack, often respond to stress by inducing an oxidative burst [34]. If these oxidative conditions resulted in the intercellular glycoprotein becoming cross-linked and insolubilized, it is clear that this could also result in the entrapment of invading bacteria, thus limiting the progress of invasion. What is less clear, however, is the nature of the signals from the bacterium that might induce the host plant cells to respond by the generation of the oxidative burst. Perhaps ethylene is also an important component of the host defence response, since this gas is known to enhance the activity of peroxidases [35].

Plant glycoconjugates associated with the peribacteroid membrane

There is a very interesting transition which occurs during nodule development when rhizobia change from being in the extracellular (apoplastic) environment to being intracellular, having been engulfed by plasma membrane that is exposed in an unwalled infection droplet structure. The engulfment, or endocytosis, of rhizobia seems to involve a close surface interaction between unencapsulated rhizobia and the surface of the plasma membrane, which still carries an apparently typical glycocalyx composed of glycolipid and glycoprotein components [36]. Endocytosis is accompanied by the

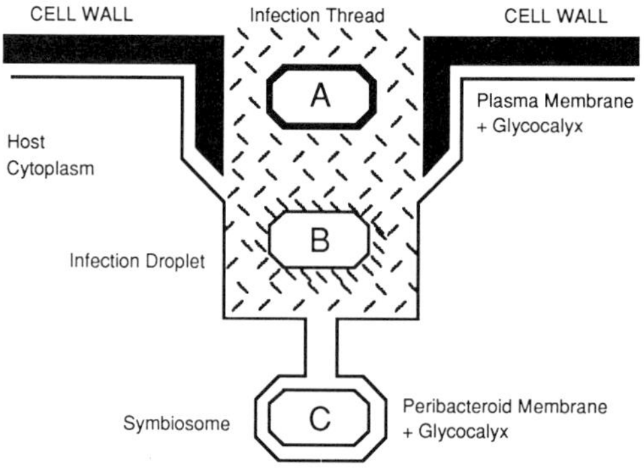

Fig. 4. Model for cell-surface interactions during tissue and cell invasion by *Rhizobium*. (A) The infection thread: bacteria are enclosed within a tubular ingrowth of the plant cell wall. A capsular sheath of extracellular polysaccharide cocoons the bacteria, which are embedded in a matrix of plant glycoproteins similar in composition to that found in intercellular spaces. (This is represented as hatched material, which is recognized by antibody MAC 265.) Close contact between the bacterial outer membrane and the plant cell membrane is prevented by the presence of bacterial capsule and the plant cell wall. (B) The infection droplet: synthesis of plant cell wall and bacterial capsule is reduced or inhibited. Plant matrix glycoprotein becomes attached to the bacterial cell surface. Subsequently, endocytosis occurs as a result of surface interactions between plant and bacterial membranes. (C) The symbiosome: after engulfment of bacteria, the plant matrix glycoprotein is apparently excluded from the symbiosome compartment. The peribacteroid membrane retains a glycocalyx similar to that of the plasma membrane (recognized, for example, by antibody MAC 206). Subsequent division and differentiation of intracellular rhizobia leads to the development of nitrogen-fixing bacteroids. The ensheathing plant membrane, which acquires new material by fusion of vesicles from the Golgi and endoplasmic reticulum, becomes functionally specialized as the peribacteroid membrane. In pea nodules, the peribacteroid membrane divides in synchrony with dividing bacteroids so that only a single bacteroid is enclosed within each symbiosome unit.

apparent loss of the matrix glycoprotein from the bacterial surface and thus it is possible that the interaction between the bacterial surface and the matrix glyoprotein is replaced by a similar interaction with a component of the plant membrane glycocalyx [33]. Furthermore, as intracellular bacteria continue to divide, the accompanying peribacteroid membrane also divides concomitantly, implying continued surface interactions between plant and bacterial membrane surfaces. It is interesting to note that the function of the plasma membrane glycocalyx of plant cells has been postulated to involve some form of physical interaction with the plant cell wall [37]. However, although the peribacteroid membrane has a glycocalyx, there is no associated plant cell wall, but only a bacterial cell wall with which the plant membrane might possibly interact.

As the infected cells of the nodule gradually mature, the peribacteroid membrane becomes differentiated by acquiring nodule-specific proteins (nodulins) and transport functions associated with the specialized metabolism of this nitrogen-fixing 'organelle'. These developmental changes proceed in phase with the progressive differentiation of the intracellular rhizobia into nitrogen-fixing bacteroid forms. A monoclonal antibody has recently been isolated that recognizes a new class of plant glycolipid membrane antigen. *In situ* immunostaining of pea nodule sections with this antibody reveals that the corresponding antigen is always present on the plasma membrane but it disappears from the peribacteroid membrane at a precise point in nodule differentiation which slightly precedes the induction of leghaemoglobin. Thus, loss of the glycolipid antigen from the peribacteroid membrane coincides with differentiation of this membrane and the enclosed bacteria into nitrogen-fixing organelles. This glycolipid seems to be a marker for a very early developmental switch from a 'juvenile' to a 'mature' form of peribacteroid membrane, which may be very relevant to the subsequent course of symbiosome differentiation.

After extraction with various organic solvents, the glycolipid antigen from pea nodule membranes has been examined by thin-layer chromatography. Because a similar glycolipid component can be identified in extracts of carrot cell membranes, carrot cell suspension cultures have been used for incorporation studies with radiolabelled sugar precursors. It has thus been demonstrated that the glycolipid incorporates label from inositol and glucosamine. Moreover, the presence of phosphate is indicated by the fact that the antibody reacts in dot immunoassays with phosphatidylinositol monophosphate (PIP). Although further work is needed to characterize the chemical structure of this membrane glycolipid from pea nodules, the current experimental evidence suggests that it belongs to one of the groups of inositol-containing glycolipids, most probably the glycosyl-phosphatidylinositols or the glycophosphosphingolipids. The properties of these groups of glycolipids have not been well characterized in plants. In animal systems, the glycosyl-phosphatidylinositols can act either as membrane anchors for Golgi-derived extracellular proteins or they can be cleaved as part of a signal transduction pathway, releasing diacylglycerol and inositol phosphoglycan as intracellular messengers [38]. Very little is known about the intracellular signals controlling differentiation of the peribacteroid membrane and of the consequent onset of nitrogen fixation. However, using a range of bacterial mutants in conjunction with molecular probes for *in situ* cytological analysis, it is now possible to embark on this investigation by analysing the synthesis and metabolism of this family of inositol-containing membrane glycolipids.

Conclusions

Rhizobium invades host cells and tissues as a result of a reorganization of plant cell wall growth which is initiated by a host-specific signal, the lipo-oligosaccharide, that is synthesized and secreted by the appropriate *Rhizobium* strain. This leads to the development of an intracellular tunnel, the infection thread, that traverses the host plant cell from one side to the other and provides a channel for the entry of rhizobial cells embedded in an extracellular matrix material secreted by the plant. Some of the components involved in cell surface interactions have been identified using monoclonal antibodies as molecular probes. The role of LPS in cell and tissue invasion was investigated by examining pea nodules induced by mutants of *R. leguminosarum* with defects in LPS structure and biosynthesis. We conclude that the correct LPS structure is essential for invasion of plant cells and tissues, for avoidance of host defence responses and for physiological adaptation to the endophytic microenvironment. Differentiation of endosymbiotic bacteroids is preceded by differentiation of the plant-derived peribacteroid membrane which encloses the symbiosome compartment. Using monoclonal antibody probes, we have identified a number of glycolipid and glycoprotein components of the peribacteroid membrane that may be involved in surface interactions with bacteroids. In addition, we have identified a group of plant membrane-associated, inositol-containing glycolipids, which serve as a very early marker for the differentiation of peribacteroid membrane from plasma membrane. Further analysis of plant–microbe signal exchanges during nodule development is likely to depend heavily on the analysis of plant mutants as well as bacterial mutants that are symbiotically defective.

We thank Andrea Davies for help with bacterial strains and for assistance with preparing the figures. We also thank B. Drobak and N. Donovan for help with the identification of plant membrane glycolipids, and R.W. Carlson for assistance with chemical analysis of lipo-oligosaccharides and lipopolysaccharides. This work was funded by the U.K. Agricultural and Food Research Service.

References

1. Brewin, N.J. (1991) Annu. Rev. Cell Biol. **7**, 191–226
2. Dénarié, J., Debellé, F. and Rosenberg, C. (1992) Annu. Rev. Microbiol. **46**, 497–531
3. Caetano-Anolles, G. and Gresshoff, P.M. (1991) Annu. Rev. Microbiol. **45**, 345–382
4. Franssen, H.J., Vijn, I., Yang, W.C. and Bisseling, T. (1992) Plant Mol. Biol. **19**, 89–107
5. Verma, D.P.S., Hu, C.-A. and Zhang, M. (1992) Physiol. Plant **85**, 253–265
6. Spaink, H.P. (1992) Plant Mol. Biol. **20**, 977–986
7. LaRue, T.A. and Weeden, N.F. (1992) Pisum Genet. **24**, 5–12
8. Weeden, N.F., Kneen, B.E. and LaRue, T.A. (1990) in Nitrogen Fixation: Achievements and Objectives (Gresshoff, P.M., Roth, J., Stacey, G. and Newton, W.E., eds.), pp. 323–330, Chapman and Hall, New York
9. Duc, G. and Messager, A. (1989) Plant Sci. **60**, 207–213
10. Duc, G., Trouvelot, A., Gianinazzi-Pearson, V. and Gianinazzi, S. (1989) Plant Sci. **60**, 215–222
11. Young, J.P.W. and Matthews, P. (1982) Heredity **48**, 203–210
12. Kneen, B.E. and LaRue, T.A. (1984) Heredity **52**, 383–389
13. Brewin, N.J., Beringer, J.E. and Johnston, A.W.B. (1980) J. Gen. Microbiol. **120**, 413–420
14. Hombrecher, G., Götz, R., Dibb, N.J., Downie, J.A., Johnston, A.W.B. and Brewin, N.J. (1984) Mol. Gen. Genet. **184**, 293–298
15. Götz, R., Evans, I.J., Downie, J.A. and Johnston, A.W.B. (1985) Mol. Gen. Genet. **201**, 296–300
16. Davies, E.O., Evans, I.J. and Johnston, A.W.B. (1988) Mol. Gen. Genet. **212**, 531–535
17. Fearn, J.C. and LaRue, T.A. (1991) Plant Cell Env. **14**, 221–227
18. Guinel, F.C. and LaRue, T.A. (1991) Plant Physiol. **97**, 1206–1211

19. Lee, K.H. and LaRue, T.A. (1992) Plant Physiol. **100**, 1759–1763
20. Boller, T. (1991) in The Plant Hormone Ethylene (Mattoo, A.K. and Suttle, J.C., eds.), pp. 293–314, CRC Press, Boca Raton
21. Borisov, A.Y., Morzina, E.V., Kulikova, O.A., Tchetkova, S.A., Lebsky, V.K. and Tikhonovich, I.A. (1993) Symbiosis, **14**, 297–313
22. Kneen, B.E., LaRue, T.A., Hirsch, A.M., Smith, C.A. and Weeden, N.F. (1990) Plant Physiol. **94**, 899–905
23. Postma, J.G., Jacobsen, E. and Feenstra, W.J. (1988) J. Plant Physiol. **132**, 424–430
24. Leigh, J.A. and Coplin, D.L. (1992) Annu. Rev. Microbiol. **46**, 307–346
25. Kannenberg, E.L. and Brewin, N.J. (1994) Trends Microbiol., **2**, 277–283
26. Carlson, R.W., Bhat, U.R. and Reuhs, B. (1992) in Plant Biotechnology and Development (Gresshoff, P.M., ed.), pp. 33–44, CRC Press, Boca Raton
27. Ielpi, L., Dylan, T., Ditta, G., Helinski, D.R. and Stanfield, S.W. (1990) J. Biol. Chem. **265**, 2843–2851
28. Priefer, U.B. (1989) J. Bacteriol. **171**, 6161–6168
29. de Maagd, R.A., Rao, A.S., Mulders, I.H.M., Goosen-de Roo, L., van Loosdrecht, M.C.M., Wijffelman, C.A. and Lugtenberg, B.J.J. (1989) J. Bacteriol. **171**, 1143–1150
30. Perotto, S., Brewin, N.J. and Kannenberg, E.L. (1994) Mol. Plant–Microbe Interact. **7**, 99–112
31. Rae, A.E., Perotto, S., Knox, J.P., Kannenberg, E.L. and Brewin, N.J. (1991) Mol. Plant–Microbe Interact. **4**, 563–570
32. VandenBosch, K.A., Bradley, D.J., Knox, J.P., Perotto, S., Butcher, G.W. and Brewin, N.J. (1989) EMBO J. **8**, 335–342
33. Rae, A.L., Bonfante-Fasolo, P. and Brewin, N.J. (1992) Plant J. **2**, 385–395
34. Devlin, W.S. and Gustine, D.L. (1992) Plant Physiol. **100**, 1189–1195
35. Cassab, G.I., Lin, J-J., Lin, L-S. and Varner, J.E. (1988) Plant Physiol. **88**, 522–524
36. Perotto, S., VandenBosch, K.A., Butcher, G.W. and Brewin, N.J. (1991) Development **112**, 763–773
37. Knox, J.P., Linstead, P.J., Peart, J., Cooper, C. and Roberts, K. (1991) Plant J. **1**, 317–326
38. Merida, I. (1992) New Biologist **4**, 207–211

Biochem. Soc. Symp. 60, 75–88
Printed in Great Britain

Barley–fungal interactions: signals and the environment of the host–pathogen interface

Sarah Gurr*, Elena Titarenko*, Zümrüt Ögel*, Conrad Stevens*, Tim Carver†, Molly Dewey* and John Hargreaves‡

*Department of Plant Sciences, University of Oxford, South Parks Road, Oxford OX1 3RA, U.K., † Institute of Grasslands and Environmental Research, Aberystwyth SY23 3EB and ‡Long Ashton Research Station, Bristol BS18 9AF, U.K.

Introduction

Plants are subjected to attack by a diverse array of pathogenic micro-organisms during their life-time, yet disease is a relatively rare event. Two factors account for this. First, most pathogens are highly specialized and are pathogenic on a limited range of plant species with which they are 'basically compatible' [1,2]; if they attack a plant of an inappropriate species, they lack this basic compatibility and the plant is said to exhibit 'non-host resistance'. Secondly, if basic compatibility between host and pathogen exists, then resistance can operate at a level governed by the genotype of host and pathogen which either prevents or limits disease development. In this case, the outcome of a plant–parasite interaction may be determined by the interaction of simple genetic factors in the host and pathogen or by more complex, polygenetically controlled characteristics. Understanding the molecular basis of these various mechanisms of resistance should facilitate the development of effective and durable resistance in crop plants.

For many crop species and their pathogens which show host-specific resistance, the 'gene-for-gene' hypothesis [3] appears to be satisfied. This hypothesis implies that resistance is functionally dominant and results from the molecular recognition of complementary plant and pathogen gene products. The rapid selection of virulent pathogen phenotypes which can overcome single-gene resistance may lead to disease epidemics; this is particularly serious where a single genotype of host plant occupies a large acreage, as is often the case with the small grain cereals. Under suitable environmental conditions, huge losses may be suffered as disease rampages through the crop.

Communication between a plant and an invading pathogen during the initial stages of infection and during the subsequent colonization of host cells is a critical, but poorly understood, aspect of pathogenesis. A two-way exchange of information can take place very early in the infection process, even before the fungus has penetrated the surface of the plant cell. This may occur either through the release of specific diffusable products, such as enzymes and low-molecular-mass metabolites or by contact sensing of topographical features on the surface of the plant [4]. In addition, these early signalling processes act in concert with other channels of communication that are established later in the infection process, and result in either the failure of the pathogen to become ensconced within the plant (resistance) or unhindered growth and colonization by the pathogen (susceptibility).

To gain a better understanding of the nature of host–pathogen communication during these early stages of infection, we are studying two interactions of barley (*Hordeum vulgare*) with fungal pathogens. The first is a non-host-resistance reaction of barley epidermal cells to infection by a wheat isolate of *Septoria*, *S. nodorum*, and the second is a more specialized interaction with the biotrophic powdery mildew pathogen *Erysiphe graminis* f. sp. *hordei* on different barley cultivars. The non-host-resistance aspect has focused on the events that trigger plant defence gene expression, whereas the work with *E. graminis* has two main objectives — namely, to follow changes in gene expression in the germinating mildew conidia and to monitor cytological changes in host epidermal cells during both incompatible and compatible interactions with resistant and susceptible barley cultivars.

S. nodorum–barley interaction

Active mechanisms, collectively termed non-host resistance, prevent the pathogen from gaining access to the plant tissue. These mechanisms often operate early during the infection process, i.e. soon after attempted penetration, and are triggered when plants are challenged by fungi that are not normally pathogens of that species. Such resistance is extremely effective and, in the case of barley, is associated with modifications to the outer wall of epidermal cells beneath the sites of penetration. This type of resistance has been shown to occur in many Graminaceous plants [5].

Cell wall modifications

The interaction between a wheat isolate of the fungus *S. nodorum* and barley coleoptiles offers a useful system for studying this type of resistance. Soon after invasion the infection hyphae penetrate the cuticle, and structural modifications occur to the plant cell wall beneath the penetration site. These changes are associated with a reduction in the growth of the invading pathogen and the penetrating hyphae become restricted to a region beneath the cuticle that is within the outer layers of the cell wall. Such cell wall modifications are often observed only after the cuticle is breached and this implies that, in this particular interaction, recognition and generation of defence signals occurs before the cell wall is penetrated.

Cell wall modifications associated with this type of resistance are the result of the deposition of material directly beneath the site of penetration, as a cell wall apposition or a papilla, and an alteration to the cell wall surrounding the penetration site to form a halo or disc, as illustrated in Fig. 1.

The material deposited within these modified walls is heterogeneous, confers rigidity and is resistant to degradation by fungal cell-wall-degrading enzymes. Such

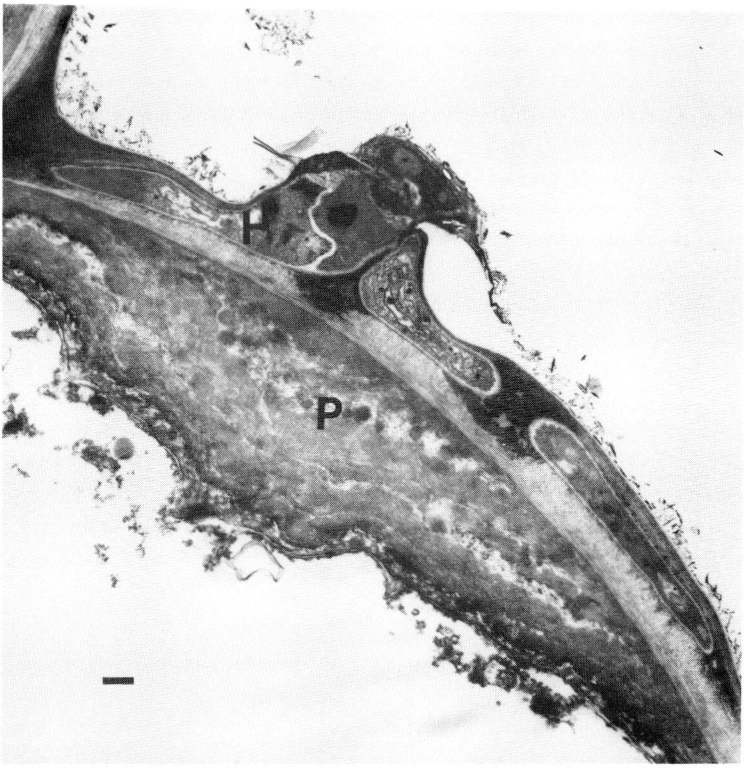

Fig. 1. A transmission electron micrograph through an attempted penetration of a barley epidermal cell by the wheat isolate of *S. nodorum* [7,43]. The growth of infection hyphae is inhibited within the outer layers of the cell wall. Abbreviations used: H infecting hypha; P, papilla. Bar represents 4 μm.

changes to the cell wall may hinder the exchange of metabolites between the pathogen and the host and may, thereby, prevent the movement of nutrients, toxins and signalling molecules associated with pathogenicity. Furthermore, Ride [6] proposes that hyphal extension is prevented because the invading hyphae become impregnated with some of this host-derived material.

A wide variety of cellular components may contribute to these wall modifications [7], including the synthesis of antifungal proteins, such as thionins, callose and glycoproteins, and cross-linking metabolites, such as phenolic compounds including lignin. Rapid responses, such as the accumulation of callose and the oxidative cross-linking of proteins, are unlikely to involve gene activation, while longer-term processes, such as the accumulation of thionins and the synthesis of phenolic precursors, may require the transcriptional activation of specific genes. Indeed, the oxidative gelation of glycoproteins within both cell walls and papillae could explain the mechanism of cell wall modification in barley in response to *Septoria* infection. The rapid *in vitro* development of highly visco-elastic gels by wheat flour glycoprotein in the presence of calcium and peroxidase has been demonstrated where intermolecular cross-linking occurs via pentose-esterified ferulic acid residues in the moiety of the glycoproteins [8]. Pentosans are ubiquitous in the cell walls of higher plants [9] and this, coupled with a rapid increase in peroxidase-encoding mRNA species seen in *Septoria*-infected barley coleoptiles (Fig. 2A), makes the *in vivo* oxidative gelation of glycoproteins at sites of attempted fungal penetration a distinct possibility. This mechanism of resistance may also occur in other plants, e.g. Lamb *et al.* [10] report the oxidative cross-linking of a proline-rich plant cell wall protein in bean or soybean cell suspension cultures and intact plants of resistant cultivars after treatment with fungal elicitor or, indeed, with glutathione. Such treatments lead to the insolubilization of pre-existing proline- or hydroxyproline-rich structural proteins in the cell wall, which is so rapid that it may precede the expression of transcription-dependent defence responses.

Although the rapid, H_2O_2-mediated oxidative cross-linking bears similarities with the rapid oxidative gelation proposed to occur in cereals, there are several constraints to prevent too close a comparison with this non-host interaction; elicitors of the reaction have yet to be described, although Hiramoto *et al.* [11] have demonstrated the presence of endogenous elicitors in barley seeds which could induce local and systemic resistance (in seedlings) and Kristensen [12] reported an exogenous elicitor in germinating mildew spores. Moreover, there is very little evidence for the involvement

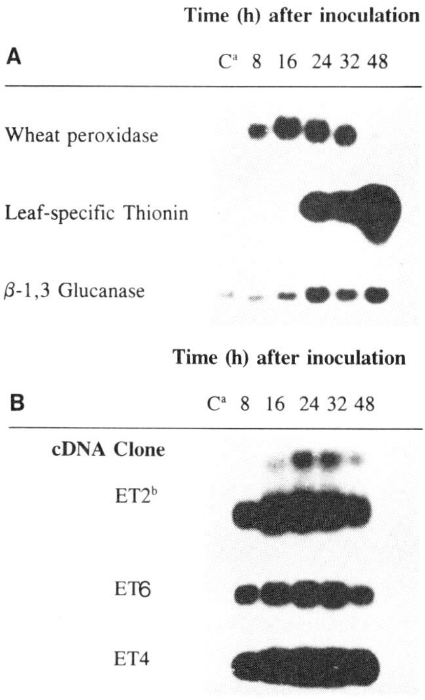

Fig. 2. Transcriptional activation of known barley defence-related genes (A) and of three clones identified by differential screening of a cDNA library from *S. nodorum*-infected coleoptiles (B). C[a], Uninfected coleoptile tissue collected 16 h after incubation. [b] Sequence of clone ET2 is identical to the barley β-1,3-1,4-glucanase gene.

of hydroxyproline-rich glycoproteins (HRGPs) in plant defence responses in cereals [13] as HRGPs are found sparsely in the Gramineae, as compared with their abundance in dicotyledon primary cell walls. However, Munk-Kruse and Meldgaard-Madsen [14] and Showalter and Varner [15] have shown that dicotyledon HRGP-encoding gene sequences cross-hybridize with several monocotyledon nucleic acids. Clearly, the function of cell wall cross-linking in disease resistance merits further investigation.

E. graminis–barley interaction

The interaction between races of barley powdery mildew *E. graminis* f. sp. *hordei* and cultivars of *H. vulgare* L. offers three main advantages as compared with many other host–pathogen systems. First, there is a well-defined genetic basis for pathogenicity, with resistance and susceptibility genes in the host matched by avirulent and virulent genes in different fungal pathotypes. Secondly, differentiation of the pathogen through distinct morphological stages occurs at specific stages during infection (see Fig. 3). Finally, early events associated with the development of infection structures can be synchronized.

Race-specific resistance

Several resistance alleles which conform to the gene-for-gene hypothesis have been identified and used to control *E. graminis* f. sp. *hordei* infection. Race-specific resistance to barley powdery mildew, conferred by the *Ml-a*, *Ml-c*, *Ml-g*, *Ml-h*, *Ml-at*, *Ml-k*, *Ml-La* and *Ml-p* alleles, can be manifested at two different stages in the interaction, i.e. by hypersensitive cell-death response (HR) or else by a papilla response at

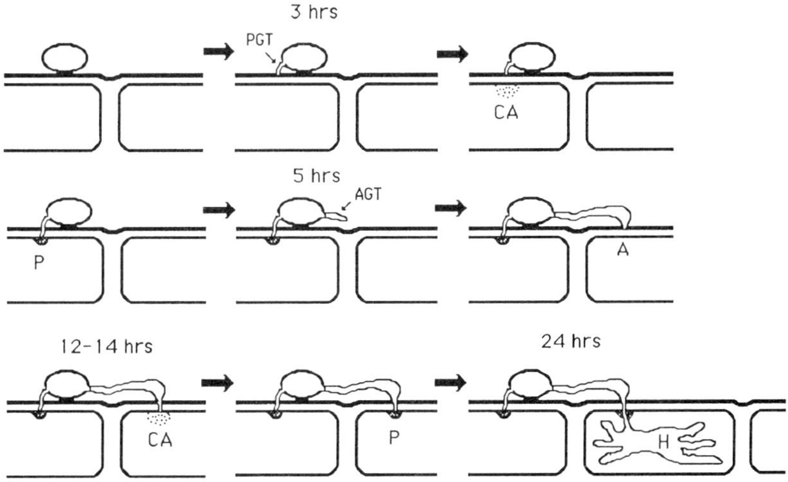

Fig. 3. The sequence of events leading from the arrival of a conidium to the formation of a haustorium in barley infected with *E. graminis*. Abbreviations used: PGT, primary germ tube; CA, cytoplasmic aggregation; AGT, appressorium germ tube; A, appressorium; P, papilla; H, haustorium.

time of penetration. Both mechanisms effectively delimit fungal growth, either by compartmentalization of the pathogen through the death of invaded epidermal cells (HR) or by barring pathogen entry by papilla deposition. However, the effectiveness of resistance determined by a single gene has eventually been lost through the emergence of corresponding virulence in the pathogen population [16,17]. Current pragmatic strategies to combat disease are, thus, to breed two or more race-specific resistance (RSR) genes from different genetic backgrounds into desirable cultivars to give combined resistance. The longer-term aim is to increase the durability of resistance mechanisms by understanding the factors that underly both the papilla response and hypersensitive cell death. A recent survey of cereal pathogen virulence [18] concludes that 'the barley mildew population has continued to increase in complexity in response to wide-spread use of specific resistances in winter and spring barley cultivars. Consequently, most combinations of specific resistance factors are largely ineffective.' This is well illustrated by the introduction, in 1986, of a spring barley cultivar, Klaxon, with combined resistance that carries the mildew resistance alleles *Mla-7, Ml-k* and *Ml-La* [19]. This cultivar constituted 10% of all barley seed sold for the 1986 spring crop in the U.K. [20], but in the space of just 5 months, from June to October 1986, a race of *E. graminis* f. sp. *hordei* virulent on cv. Klaxon increased in frequency from <1% to 36% [17]. In the light of such rapid breakdown in disease resistance the need to breed barley cultivars with more durable resistance is obvious.

Ml-o resistance

Resistance conferred by the recessive ml-o allele is not race-specific, but is durable and effective against mildew worldwide [21]. Although originally identified in mutagen-treated barley, an allele was also discovered in an Ethopian landrace of barley which has been used in breeding programmes. Several spring barley cultivars popular in the U.K. and Europe now carry ml-o resistance.

Structural changes induced in the host cell carrying the ml-o gene

Although cell wall appositions or papillae form during both incompatible and compatible interactions between *E. graminis* and barley, such structures have been assigned a role in resistance because penetrating infection hyphae are often inhibited at the papillae stage [22]. A rapid papilla response is also initiated in ml-o-resistant interactions with larger amounts of callose deposited on the host cell wall beneath the germ tube, as compared with ml-o-mother varieties [23].

Furthermore, in the ml-o resistant barley isolines, the papillae contain a light-absorbing component (LAC), which is rich in phenylpropanoids, together with basic staining material (BSM). These are not present in papillae induced in the susceptible isoline. On the basis of these observations, Aist *et al.* [24] proposed that the LAC and BSM in papillae are the components influenced by the ml-o resistance gene. Recently, however, Carver and Zeyer [25] have shown that inhibitors of phenylalanine ammonia-lyase (PAL) applied to an ml-o genotype do not affect resistance to penetration. This contrasts with other barley genotypes (and the papillae of other cereals) where inhibition of PAL decreases the accumulation of phenolic autofluorogens in the papillae, and this is associated with increased success in penetration through the cell wall. Bayles *et al.* [26] showed inhibition of callose deposition dramatically increased penetration success in ml-o barley coleoptiles, and suggested that calcium, known to activate

1,3-β-glucan synthesis in other systems, may be important in modulating resistance. Indeed, Bayles and Aist [27] showed that the addition of calcium enhanced ml-o resistance, while the addition of calcium chelators had the opposite effect. Thus it is proposed that the ml-o mutation may alter calcium regulation in the cell, and that this, in turn, might affect the rate and quality of callose deposition in papillae.

Development of the powdery mildew pathogen

The early stages in the differentiation of the barley powdery mildew fungus have been described in detail [28]. Contact with a barley leaf, or with a non-host surface such as cellophane, induces the powdery mildew conidia to undergo a fairly dramatic change in morphology [29]. Within a few minutes of contact, the reticulate surface of the conidium disappears beneath a film of liquid exuded from the spore surface. This conidial exudate comprises a mixture of proteins, three with esterase activity, of which one has recently been ascribed to the serine esterase, cutinase [30]. The secretion of esterases by conidia on the surface of barley leaves takes place in two stages: namely a rapid release, which occurs within 5 min of the spore making contact, and a second rise between 10 and 15 min later, which remains constant for a further 45 min [29]. After 30 min of contact, the surface morphology of the conidia changes; the surface texture becomes granular and is interspersed with globose bodies [29].

The conidia germinate with the emergence of the primary germ tube (PGT) about 1 h after surface contact and, some 2 h later, the appressorial germ tube (AGT) appears (Fig. 2 and Fig. 4). The functions of the PGT remain unclear; they quickly attach conidia to the leaf surface, induce localized host responses [31] and form a short infection peg which breaches the host epidermal wall [28,32]. The PGT may gain access to host water, under dry incubation conditions, which supports the development of the larger AGT [33]. The AGT is septate at the base and becomes elongated and thicker in diameter as it forms a button- or hook-shaped structure directly above the site of penetration (Fig. 4 and Fig. 5). The mode of penetration is believed to be by enzymic means as penetration holes have round and smooth edges [29]. The penetration peg breaches the cell wall, invaginates the host plasma membrane and differentiates to form a haustorium. This specialized feeding cell transfers nutrients from the host to the growing hyphae, which ramify across the leaf surface, penetrate neighbouring cells and develop new haustoria, and, in turn, hyphae are formed [34].

Role of cutinase in penetration

During the course of evolution, different groups of fungi have discovered different strategies for attacking their host. It is not surprising, therefore, that the role of cutinase in relation to fungal pathogenicity appears to be variable and controversial [35]. In some plant–pathogen interactions, cutinase may be required for the efficient penetration of the host cuticle [29,32]. In others it may be involved in surface recognition [4] and for adhesion to the leaf surface. Some reports suggest that the secretion of cutinase is not essential for pathogenicity [36], while in others there is good evidence as to the importance of cutinases in pathogenicity [37]. In the case of *E. graminis* f. sp. *hordei*, the release of cutinase and non-specific esterases has been shown to be associated with adhesion [29]. Release of cutinase by this fungus appears to be

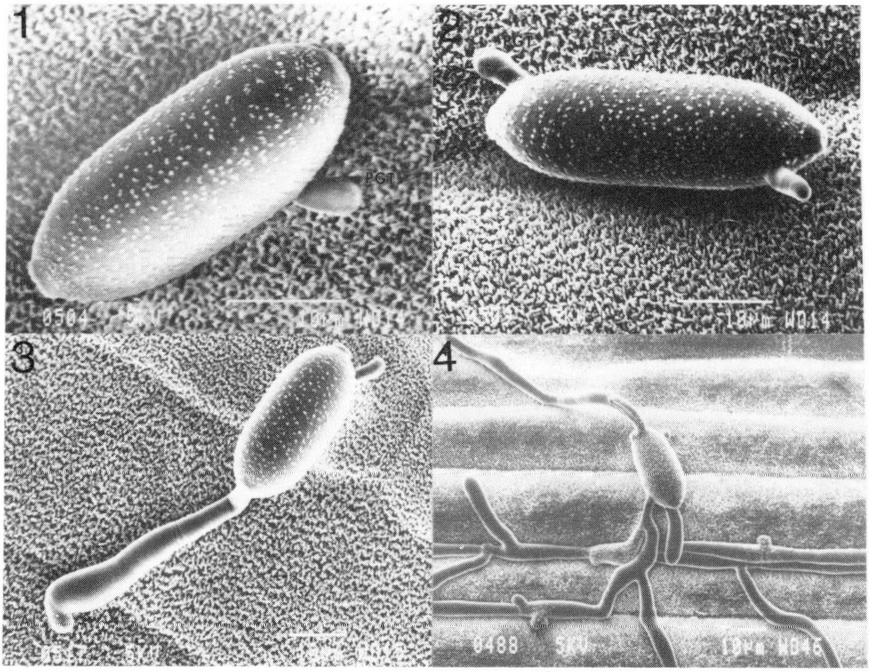

Fig. 4. Cold-stage scanning electron micrographs of cryo-fixed barley leaves attacked by *E. graminis* f. sp. *hordei* (T. Carver and B. Thomas). (1) At 2 h after inoculation: the PGT has emerged from the conidium and attached to the host cell surface. Note the crystalline epicuticular plate wax on the leaf surface. (2) At 4 h after inoculation: the second germ tube destined to form the AGT has emerged from the conidium. (3) At 12 h after inoculation: the fungal appressorium matures at the tip of the elongated AGT. (4) At 48 h after inoculation: secondary hyphae ramify across the host surface and hyphal appressoria have differentiated from lateral hyphal buds.

quite different from that of the necrotrophic fungus *Fusarium solani* f. sp. *pisi* [38] where cutinase is not present in spores but can be induced in fungal hyphae following germination on a medium containing cutin as the sole source of carbon [37,39] (Fig. 6). In *E. graminis* the secretion of the cutinase-containing exudate from the spores takes place on both barley and cellophane [29], suggesting the release is from constitutively produced cutinase, and is a response to surface contact rather than a molecular induction process. Using cutinase-specific polyclonal antiserum and monoclonal antibodies raised to cutinase purified from *F. solani* f. sp. *pisi*, we have shown by immunofluorescence that cutinase is present on the surface of ungerminated spores of *E. graminis* but that cutinase could not be detected in ungerminated spores of *F. solani* f. sp. *pisi*.

To study the involvement of cutinase in the adhesion of conidia to barley leaves we have isolated the cutinase gene from *E. graminis*. This forms part of a line of research towards the goal of isolating genes that are expressed by *E. graminis* at the

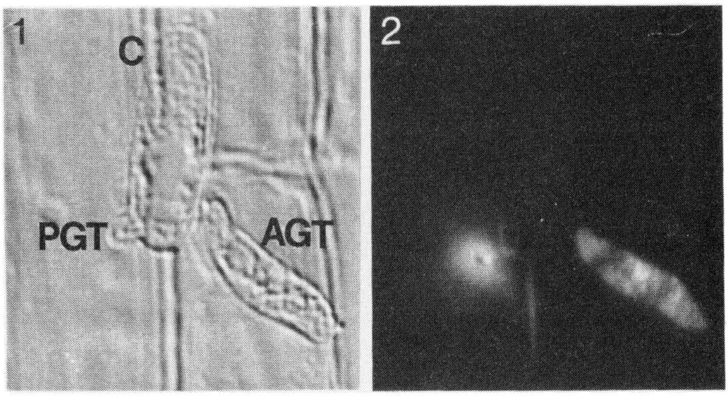

Fig. 5. Confocal microscopy image analysis (argon ion laser, 488 nm) of an incompatible interaction between *E. graminis* f. sp. *hordei* and barley carrying ml-o resistance, 18 h after inoculation. (1) Conventional image showing the PGT and AGT. (2) Confocal image shows a 'halo' around the PGT. (J. Patrick, M. Fricker, T. Carver and S. Gurr).

initial stages of infection, and also part of a series of comparative studies on cutinase gene expression at various stages of barley–fungal communication. In this respect, degenerate nucleotide primers were designed by aligning two conserved regions of the cutinase genes from *F. solani* f. sp. *pisi* [40], *Colletotrichum gloeosporoides*, *C. capsici* and *Magnaporthe grisea* [40,41], and were used to amplify part of the *Erysiphe* cutinase genes by genomic polymerase chain reaction (PCR) amplification [42]. The identity of the amplified fragment was confirmed by Southern-blot analysis and heterologous hybridization, to a region of the *F. solani cut*A gene. The DNA fragment amplified from the *Erysiphe* DNA has been used as a homologous hybridization probe to isolate clones containing the *cut* gene from an *E. graminis* genomic library.

Changes in gene expression during resistance

While specific resistance genes have been used extensively to breed for disease resistance in barley, little is known about the defence reactions influenced by these genes; considerably less is known at the molecular level about the non-host reaction of barley to fungal infection [43]. However, Gregersen *et al.* [44] have characterized nine different cDNA clones induced in barley by *E. graminis*. These were isolated by subtractive hybridization to a cDNA library prepared from barley leaves 6 h after inoculation. The clones included transcript encoding a peroxidase, sucrose synthetase, various pathogenesis-related proteins and five cDNA species, which represent genes that had not previously been implicated in host defence responses: these included a putative signal transduction regulatory gene, '14-3-3', and an endoplasmic reticulum-resident member of the heat shock family, GRP94, which encodes a glucose-regulated protein. Transcript analysis of these clones showed a twin-peak accumulation with all but one of the pathogenesis-related genes; mRNA profiles showed maximum levels at times coincident with the induction of the papilla response by PGTs and AGTs. Similar

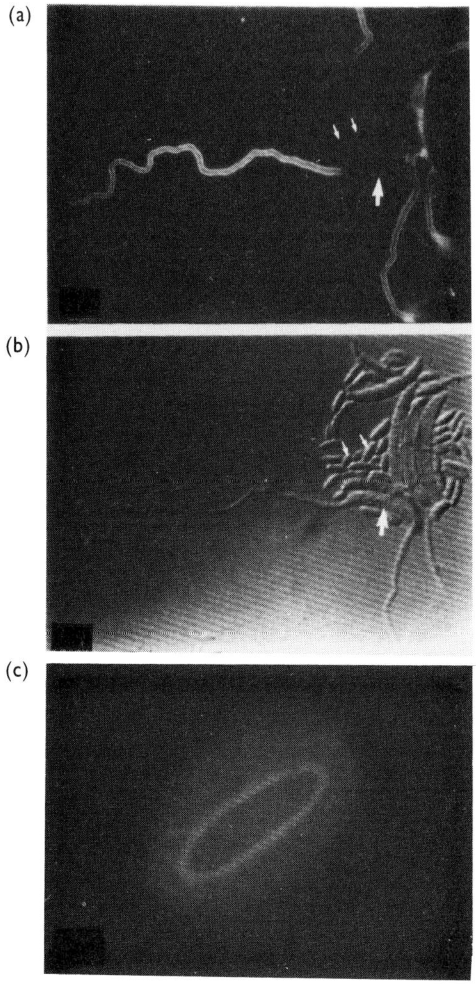

Fig. 6. Immunolocalization of cutinase in hyphae of *F. solani* f. sp. *pisi* spores germinated on cutin water agar (*a*) using polyclonal cutinase antiserum [39]. Note the micro- and macronidia in the DIC light micrograph (*b*) and the immunofluorescent hyphae but no fluorescent conidia in (*a*). (*c*) Shows the immunofluorescence of ungerminated *E. graminis* conidia on cutin water agar.

twin-peak transcript accumulation had previously been shown in barley by Clark *et al.* [45] and, more recently, by J. Brown (personal communication), who also analysed the expression of a peroxidase, chitinase and a PAL-like gene in response to barley powdery mildew infection.

We have also investigated the transcriptional activation of several defence-related genes in barley challenged by the wheat pathotype of *S. nodorum* (Fig. 2a). Unlike the barley–powdery mildew interaction, no common pattern of mRNA accumulation is seen in response to infection. However, unlike *E. graminis,* it is not possible to synchronize infection by *S. nodorum*. Transcripts homologous to a wheat peroxidase clone [46,47] are elevated 8 h after infection with *S. nodorum*, but decline 24 h later (Fig. 6a). In contrast to the peroxidase pattern of accumulation transcripts encoding leaf-specific thionins [48] appear later in the reaction, accumulating 24 h after inoculation. Transcripts encoding a β-1,3-glucanase are present in control tissue and exhibit only a slight increase following infection (Fig. 2a). No hybridization was recorded to the wheat sucrose synthetase gene [49]. A barley annexin, generated by PCR amplification with highly degenerate primers, i.e. a part of a gene which encodes a calcium-dependent phospholipid-binding protein [50], also shows elevated transcript levels 8 h after infection — with peak activity sustained through the subsequent 40 h (Fig. 7). This data, taken in conjuction with '14-3-3' data [51], provides an interesting forum for speculation as to their apparent involvement in signal transduction pathways.

A cDNA library prepared from barley coleoptile tissue 16 h after inoculation with *S. nodorum* [43] has been screened with cDNA probes with RNA from 'healthy' or 'infected' coleoptiles. Several differentially expressed clones have been isolated (Fig. 2b), and include, among others, a β-1,3-1,4-glucanase (ET2). Two clones have been more extensively characterized; a cDNA clone (ET4) which shows some similarity to an unidentified wheat gene WHTWAL15a (GenBank), induced following treatment of root tips with aluminium, and low identity with a wound-induced protease inhibitor from maize, ZMWIP1-1, and ET6, a gene bearing considerable identity to the PR-1 proteins (Fig. 2b). ET4 transcript is induced not only after infection with *S. nodorum*, but also during the inbibition of water by dry seed barley embryos (Fig. 8). This supports the dual role of certain genes expressed both in development and defence [52].

To study early expression events in infected coleoptiles, a 'cold-plaque' screen was undertaken, i.e. recombinant library plaques were picked which hybridize to neither the healthy nor the infected first-strand cDNA probes and used in dot-blot analysis of healthy versus infected RNA. A time-course profile of expression of one such cDNA clone, which bears no sequence identity with any sequences lodged within current data-bases, is shown in Fig. 9. This transcript is maximally expressed 3 h post inoculation and decreases after 9 h. This merits further analysis.

The molecular characterization of genes involved in host–pathogen communication during the early stages of infection is in its infancy. Towards the goal of under-

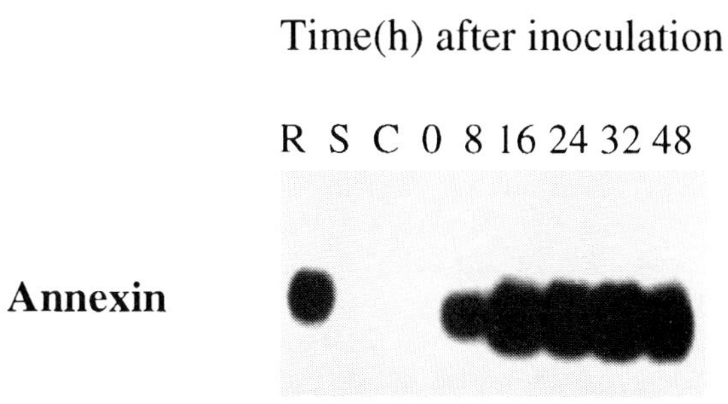

Fig. 7. Transcriptional activation of a barley annexin gene [42,50] after infection by *S. nodorum*. Abbreviations used: R, root; S, shoot; C, coleoptile, uninfected tissue collected 16 h after incubation.

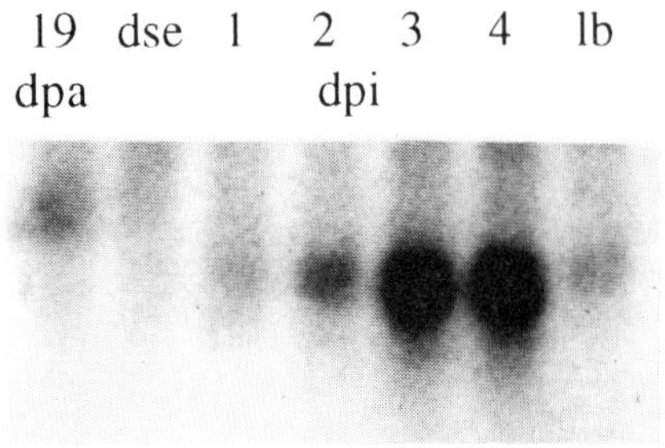

Fig. 8. Transcriptional activation of ET4 in developing barley embryo tissue. Abbreviations used: dpa, days post anthesis; dse, dry seed embryo; dpi, days post imbibition; lb, leaf base. Northern blot courtesy of Li Yi and D.J. Bowles.

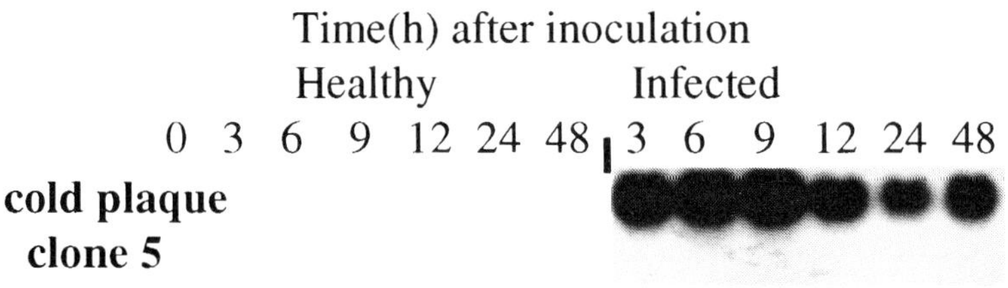

Fig. 9. Transcriptional activation of cold plaque clone 5, comparing an absence of activity in healthy coleoptile tissue with elevated levels after infection by *S. nodorum*.

standing the molecular and cellular interplay between plants and fungi, we are studying the non-host interaction between barley and *Septoria* and the more specialized biotrophic interaction between *E. graminis* f. sp. *hordei* and barley.

We are grateful to The Royal Society and AFRC for their support and to Oxford University for a 'pump-priming' grant.

References

1. Heath, M.C. (1981) Phytopathology **71**, 1121–1123
2. Heath, M.C. (1982) Phytopathology **72**, 967
3. Flor, H.H. (1971) Annu. Rev. Phytopath. **9**, 275–296
4. Hoch, H.C., Staples, R.C., Whitehead, B., Comeau, J. and Wolf, E.D. (1987) Science **235**, 1659–1662
5. Sherwood, R.T. and Vance, C.P. (1976) Phytopathology **6**, 503–510
6. Ride, J.P. (1975) Physiol. Plant Path. **5**, 125–134
7. Hargreaves, J.A. and Keon, J.P.R. (1986) in Biology and Molecular Biology of Plant Pathogen Interactions (Bailey, J., ed.), pp. 133–140, Springer-Verlag, Berlin
8. Painter, T.J. and Neukom, H. (1968) Biochim. Biophys. Acta **58**, 363
9. Baker, J.C., Parker, H.K. and Mize, M.D., (1943) Cereal Chem. **20**, 267–271

10. Lamb, C.J., Brisson, D.J., Bradley, D.J. and Kjelbom, P. (1992) in Mechanisms of Plant Defence Responses (Fritig, B. and Legrand, M., eds.), pp. 250–256, Kluwer, The Netherlands
11. Hiramoto, T., Tobimatsu, R., Shiraishi, T., Yamada, T., Ichimose, Y. and Oku, H. (1992) J. Phytopath. **35**, 167–176
12. Kristensen, H.J. (1989) Master's Thesis. The Royal Veterinary and Agricultural University, Copenhagen
13. Clarke, J.A, Lisker, N., Ellingboe, A.H. and Lamport, D.T. (1983) Plant Sci. Lett. **30**, 339–346
14. Munk-Kruse, A.G. and Meldgaard-Madsen, K.M. (1990) Master's Thesis. The Royal Veterinary and Agricultural University, Copenhagen
15. Showalter, A. and Varner, J. (1989) Biochem. Plants **15**, 485–520
16. Brown, J.K.M. and Wolfe, M.S. (1990) Plant Path. **9**, 391–401
17. Brown, J.K.M., Simpson, C.G. and Wolfe, M.S. (1993) Plant Path. **42**, 108–115
18. Michell, A.G. and Slater, S.E. (1992) in UK Cereal Pathogen Virulence Survey. UKCPVS Committee, Cambridge
19. Brown, J.K.M. and Jørgensen, J.H. (1991) in Integrated Control of Cereal Mildews: Virulence Patterns and their Change (Jørgensen, J.H., ed.), pp. 268–286, Riso National Laboratory, Denmark
20. Anon (1987) in Cereals Statistics, Home-grown Cereals Authority, London
21. Jørgensen, H.J. (1992) Euphytica **63**, 141–152
22. Koga, H., Zeyen, R.J., Bushnell, W.R. and Ahlstrand, G.G. (1988) Physiol. Mol. Plant Path. **32**, 395–409
23. Skou, J.P., Jørgensen, J.H. and Lilholt, U. (1984) Phytopathol. Z. **109**, 147–168
24. Aist, J.R., Gold, R.A., Bayles, C.J., Morrison, G. H., Chandra, S. and Israel, H.W. (1988) Physiol. Mol. Plant Pathol. **33**, 17–32
25. Carver, T.L.W. and Zeyen, R.J. (1992) in Mechanisms of Plant Defence Responses (Fritig, B. and Legrand, M., eds.), pp. 324–327, Kluwer, The Netherlands
26. Bayles, C.J., Ghemawat, M.S. and Aist, J.R. (1990) Physiol. Mol. Plant Pathol. **33**, 17–32
27. Bayles, C.J. and Aist, J.R. (1987) Physiol. Mol. Plant Path. **30**, 337–345
28. Kunoh, H. (1981) in Microbial Ecology of the Phylloplane (Blakeman, J.P., ed.), pp. 85–101, Academic Press, London
29. Nicholson, R.L., Yoshioka, H., Yamaoka, N. and Kunoh, H. (1988) Exp. Mycol. **12**, 336–349
30. Pascholati, S.F., Yoshioka, H., Kunoh, H. and Nicholson, R.L. (1992) Physiol. Mol. Plant Path. **41**, 53–59
31. Woolacott, B. and Archer, S.A. (1984) Plant Pathol. **33**, 225–231
32. Kunoh, H., Nicholson, R.L., Yoshioka, H., Yamaoka, N. and Kobayashi, I. (1990) Physiol. Mol. Plant Pathol. **36**, 397–407
33. Carver, T.C. and Bushnell, W.R. (1983) Physiol. Plant Pathol. **23**, 229–240
34. Jenkyn, J.F. and Bainbridge, A. (1978) in Powdery Mildews (Spencer, D.M., ed.), pp. 283–321, Academic Press, London
35. Chasan, R. (1992) Plant Cell **4**, 617–618
36. Stahl, D.J. and Schäfer, W. (1992) Plant Cell **4**, 621–629
37. Kolattakudy, P.E. (1985) Annu. Rev. Phytopath. **23**, 223–250
38. Dickman, M.B., Podila, G.K. and Kolattakudy, P.E. (1989) Nature (London) **342**, 446–448
39. Hiscock, S.J., Dewey, F.M., Coleman, J.O.D. and Dickinson, H.G. (1994) Planta **193**, 377–384
40. Soliday, C.L., Dickman, M.B. and Kolattukudy, P.E. (1989) J. Bacteriol. **171**, 1942–1951

41. Ettinger, W.F., Thurkal, S.K. and Kolattukudy, P.E. (1987) Biochemistry **26**, 7883–7892
42. McPherson, M.J., Jones, K.M. and Gurr, S.J. (1991) in PCR, A Practical Approach (McPherson, M.J., Quirke, P. and Taylor, G.R., eds.), pp. 171–185, Oxford University Press, Oxford
43. Titarenko, E.T., Hargreaves, H., Keon, J. and Gurr, S.J., (1992) in Mechanisms of Plant Defence Responses (Fritig, B. and Legrand, M., eds.), pp. 308–311, Kluwer, The Netherlands
44. Gregerson, P.L., Brandt, J., Thordal-Christensen, H. and Collinge, D.B. (1992) in Mechanisms of Plant Defence Responses (Fritig, B. and Legrand, M., eds.), pp. 304–307, Kluwer, The Netherlands
45. Clark, T.A., Smith, A.G., Bushnell, W.R. and Zeyen, R.J., (1991) Phytopathology **81**, 1139–1140
46. Schweizer, P., Hunizer, W. and Mösinger, E. (1989) Plant Mol. Biol. **12**, 643–654
47. Thordal-Christensen, H., Brandt, J., Cho, B.H., Rasmussen, S.K., Gregersen, P.L., Smedegaard-Petersen, V. and Collinge, D.B. (1994) Physiol. Plant Pathol., in the press
48. Bohlman, H., Clausen, S., Behnke, S., Giese, H., Hiller, C., Reimann-Philipp, Schrader, G., Barkholt, V. and Apel, K. (1988) EMBO J. **7**, 559–1565
49. Smith, L.M., Handley, J., Yi Li., Martin, H., Donovan, L. and Bowles, D.J. (1992) Plant Mol. Biol. **20**, 255–266
50. Smallwood, M.F., Gurr, S.J., McPherson, M.J., Roberts, K and Bowles, D.J. (1992) Biochem. J. **281**, 501–505
51. Brandt, J., Thordal-Christensen, H., Vad, K., Gregerson, P.L. and Collinge, D.B. (1992) Plant J, **2**, 815–820
52. Bowles, D.J. (1990) Annu. Rev. Biochem. **59**, 873–907

Biochem. Soc. Symp. **60**, 89–94
Printed in Great Britain

Oligosaccharins involved in plant growth and host–pathogen interactions

Alan Darvill*, Carl Bergmann, Felice Cervone, Giulia De Lorenzo, Kyung-Sik Ham, Mark D. Spiro, William S. York and Peter Albersheim

The Complex Carbohydrate Research Center, The University of Georgia, 220 Riverbend Road, Athens, GA 30602–4712, U.S.A.

Introduction

Carbohydrates, the building blocks of many structural polymers that give form to living cells and organisms, also play important roles in the interactions of cells with one another and with their environment. Plants and animals have evolved molecular signalling mechanisms to regulate the expression of genes essential for their growth, development and defence against pests. Some of these signals or regulatory molecules are oligosaccharides. Oligosaccharides with regulatory activities are called oligosaccharins. The extensive stereochemistry, multiple hydroxyls and oxygen atoms, and accessible hydrophobic regions characteristic of glycosyl residues make oligosaccharides ideal ligands for specific interactions with recognition sites on proteins. These carbohydrate-binding proteins distinguish among the large number of primary structures and three-dimensional shapes which oligosaccharides can adopt. Thus, protein receptors can distinguish among a range of information-carrying oligosaccharides and transmit this information to the cells to which the receptors are attached. This paper describes recent progress we have made studying oligosaccharins active in plants.

Oligosaccharin elicitors of phytoalexins

The biosynthesis and accumulation of antimicrobial phytoalexins is one of the best-studied plant defence mechanisms. Plants synthesize and accumulate phytoalexins (antibiotics) in response to microbial infection or elicitor treatment. Three oligosaccharin elicitors of phytoalexins have been characterized: a structurally defined hepta-β-glucoside [1]; α-1,4-oligogalacturonides with degrees of polymerization (DPs) from

*To whom correspondence should be addressed.

10–15 [2,3]; and chitosan or chitin oligosaccharides of undefined length [4]. The hepta-β-glucoside (from fungal cell wall β-glucan) and the oligogalacturonides (from the homogalacturonan of plant cell walls) act synergistically in stimulating soybeans to accumulate phytoalexins [5]. Efforts to isolate the physiological receptors of the hepta-β-glucoside and oligogalacturonide elicitors are in progress [6,7].

Pathogenesis-related endo-β-1,3-glucanases can release β-glucan elicitors from fungal cell walls

During experiments to investigate the solubilization of elicitor-active β-glucans from fungal cell walls, we have shown that fungal cell wall oligosaccharide fragments (oligosaccharins) can be solubilized by a pathogenesis-related (PR) endo-β-1,3-glucanase (EC 3.2.1.39) that elicits phytoalexin accumulation in soybean [8]. In these experiments, soybean leaves were treated with salicylic acid, polyacrylic acid or mercuric chloride, or they were infected with *Phytophthora megasperma* H20 (a fungal pathogen of Douglas fir) to which soybean has non-host resistance. Only mercuric chloride and the fungus induced the leaves to synthesize PR proteins. Both endo-β-1,3-glucanase and chitinase (EC 3.2.1.14) activities were induced by treatment with mercuric chloride and by infection with the fungus. During purification to homogeneity of the elicitor-releasing activity, the fractions containing elicitor-releasing activity also contained endo-β-1,3-glucanase activity, providing evidence that endo-β-1,3-glucanase is a principal elicitor-releasing enzyme of the soybean leaf extracts. Antiserum raised against a tobacco PR endo-β-1,3-glucanase cross-reacted with the purified soybean endo-β-1,3-glucanase that accounted for major elicitor-releasing activity in the basic fraction of the soybean leaf extracts. Soybean endo-β-1,3-glucanase, induced by mercuric chloride treatment or pathogenic infection with *P. megasperma* f. sp. *glycinea* (Pmg) race 1 (incompatible with the soybean cultivar used), could not be detected by immunoblots of extracts of control plants, indicating that the endo-β-1,3-glucanase is a PR protein. These results suggest that an endo-β-1,3-glucanase, induced in soybean seedlings by pathogenic infection or by chemical stress, functions in defence by releasing a phytoalexin elicitor from the mycelial walls of pathogenic fungus.

We have also been investigating the role of enzyme inhibitors in the release of oligosaccharin elicitors of phytoalexins from fungal cell walls. We have recently observed that a fungal pathogen secretes a protein that specifically inhibits an endo-β-1,3-glucanase PR protein of its host. As well as the endoglucanase described above (endo A), we have purified a second endo-β-1,3-glucanase (endo B) from soybean seedlings, which is an inducible PR protein and which releases a phytoalexin elicitor from the walls of the fungal pathogen Pmg. We have shown that Pmg secretes a protein that inhibits endo B but not endo A. The inhibitor protein, which has been purified to homogeneity, is heat labile and has a molecular mass of 33.5 ± 0.6 kDa. We have also partially purified a β-1,3-glucanase from Pmg. The Pmg protein that inhibits endo B but not endo A also does not inhibit the Pmg β-1,3-glucanase or a tobacco PR endo-β-1,3-glucanase (PR-O). We also have evidence that Pmg secretes other proteins that inhibit other β-1,3-glucanase(s) of soybean. Thus, pathogens have evolved the ability to inhibit the fungal wall-degrading enzymes produced by their host plants, just as plants have evolved proteins to inhibit plant cell wall-degrading enzymes secreted by their pathogens (see below). It seems likely that pathogens may secrete inhibitors of other PR proteins (e.g. chitinases) and that the interplay of hydrolases and their inhibitors may determine the outcome of plant–pathogen interactions.

Oligogalacturonides, polygalacturonase-inhibiting protein and plant defence

Oligogalacturonide elicitors of phytoalexins are released from host cell walls by microbial enzymes, and this process apparently involves host inhibitor proteins [9]. Fungal endopolygalacturonases hydrolyse polygalacturonic acid to mono-, di-, tri- and tetragalacturonic acid. All dicotyledons examined contain a cell wall-associated polygalacturonase-inhibiting protein (PGIP) which specifically inhibits fungal endopolygalacturonases. The PGIP inhibition of fungal endopolygalacturonases results in an increased stability of α-1,4-oligogalacturonides (DP = 8–20), which are elicitors of phytoalexins and other plant defence responses [9]. Thus it has been proposed that PGIP plays an important role in plant resistance to fungal pathogens by optimizing the formation of elicitor-active oligogalacturonides. Recently, the gene encoding *Phaseolus vulgaris* PGIP has been cloned and characterized [9]. Using the cloned PGIP gene as a probe, we have demonstrated that the transcription of the gene is induced in suspension-cultured bean cells following the addition of elicitor-active oligogalacturonides to the medium. Rabbit polyclonal antibodies specific for PGIP have been generated against a synthetic *N*-terminal peptide coupled to the carrier protein keyhole limpet haemocyanin (KLH). Using the antibodies and the cloned PGIP gene, we have shown that the synthesis of PGIP and its mRNA is induced in *P. vulgaris* hypocotyls in response to wounding or treatment with salicylic acid. We have also demonstrated, using gold-labelled goat anti-(rabbit immunoglobin) secondary antibodies in electron microscopic studies, that in bean hypocotyls infected with *Colletotrichum lindemuthianum* the level of PGIP increases in the cells surrounding the infection site. Our data suggest that synthesis of PGIP represents an active defence mechanism of plants regulated by signal molecules (elicitors) known to induce defence genes in plants.

Oligogalacturonides and organogenesis

As described above, oligogalacturonides are involved in plant defence; however, at a 10-fold lower concentration, they can regulate organogenesis. Most undifferentiated plant cells are totipotent, i.e. each plant cell has the potential to develop into a mature plant. It has been known for some time that hormones can control the ability of plant cells to differentiate and develop into specialized organs. In this regard, oligosaccharins can act like hormones, since oligogalacturonides, with DPs of 10–15, inhibit tobacco explants from forming roots and induce the explants to form flowers when they are grown in media which, without the oligogalacturonides, cause roots or no organs to form [10]. Evidence has been obtained suggesting that plant cell wall-derived oligogalacturonides regulate other developmental processes, including cell elongation and fruit ripening. Furthermore, oligogalacturonides induce a variety of rapid changes in the functions of the plasma membrane of plant cells [11,12], but these rapid effects have yet to be directly associated with any of the demonstrated biological activities of oligogalacturonides. Characterization of oligogalacturonide receptors may facilitate elucidation of the mechanisms by which oligogalacturonides can have so many biological effects.

Oligogalacturonides and their receptors

The characterization of the molecular basis by which α-1,4-oligogalacturonides (DP = 10–16) elicit defence responses and morphogenesis in plants is a major goal of our laboratory. Towards this goal, we have developed procedures for generating and purifying the bioactive oligogalacturonides, including the tridecagalacturonide and its tyramine-coupled derivative. The tyramine derivative will be labelled with ^{125}I and used to locate and isolate oligogalacturonide-specific receptors of plants. Oligogalacturonides (DP > 9) were obtained by ethanol fractionation of the products released by treatment of polygalacturonic acid with an endo-α-1,4-polygalacturonase. Bioactive oligogalacturonides (DP = 10–16) were then purified by Q-Sepharose anion-exchange chromatography, followed by semi-preparative, high performance anion-exchange chromatography with pulsed amperometric detection (h.p.a.e.-p.a.d.). The tridecagalacturonide prepared by this method was shown to be homogeneous by glycosyl-residue composition analysis, h.p.a.e.-p.a.d., f.a.b.-m.s., and ^{1}H-n.m.r. spectroscopy. The reducing end of the tridecagalacturonic acid was coupled to tyramine via reductive amination in the presence of sodium cyanoborohydride and its structure confirmed by the methods described above. The biological activity of the tridecagalacturonide tyramine-coupled derivative is presently being determined.

Xyloglucan and plant growth

In addition to pectic polysaccharides, the cell walls of higher plants also contain a family of highly branched polysaccharides called hemicelluloses. Hemicelluloses are functionally defined as those polysaccharides that form strong, non-covalent associations with cellulose microfibrils. The predominant hemicelluloses in the primary cell walls of higher plants are arabinoxylan and xyloglucan. Xyloglucan is thought to be a load-bearing structural polymer in the primary cell wall because of its role in cross-linking cellulose microfibrils. The dynamic nature of this cross-linking is proposed as the major factor that controls the rate of cell wall expansion, thereby regulating plant cell growth.

Oligosaccharide fragments of xyloglucan are generated by treatment of the polysaccharide with a purified endo-β-1,4-glucanase. The cell wall activity of an enzyme with this substrate specificity was reported to be increased by spraying pea seedlings with the phytohormone auxin [13]. A nonasaccharide product of the action of endo-β-1,4-glucanase on xyloglucan has been shown to inhibit the growth of pea stems [14,15], an observation consistent with a feedback control loop hypothesis in which elevated amounts of auxin promote the formation of xyloglucan-derived oligosaccharides that inhibit the growth-promoting effect of auxin.

Discussion

The oligosaccharins described above, together with others described in the literature [13–18], are able to regulate plant growth, organogenesis, or defence against pathogens. The results of oligosaccharin research provide evidence that plants utilize the structural complexities of oligosaccharides to regulate important physiological processes. Cell wall polysaccharides of plants and microbes are a rich source of oligosaccharins, and the walls also contain glycanases, glycosidases and inhibitors involved in the formation of oligosaccharins.

Considerable evidence supports the hypothesis that oligosaccharins are important regulatory molecules in plants, although much work still remains to evaluate the role of these molecules *in vivo*. Most of the data on the biological activities of oligosaccharins has been obtained in bioassays. Studies with intact plants are needed, perhaps using plants transformed with genes encoding enzymes, receptors, enzyme inhibitors, or other proteins that alter the *in situ* activity of oligosaccharins. Indeed, studies on the enzymes that release and process oligosaccharins, on receptors of oligosaccharins, and on the effects oligosaccharins have on membranes and membrane-associated proteins should elucidate the events that initiate oligosaccharin activities and lead to a better understanding of the signal pathways that transduce the effect of the regulatory molecules.

Progress in this new area of biology has been possible because of the development of sophisticated analytical techniques for purifying and determining the structures of complex carbohydrates and the collaborative research of physiologists, biochemists, molecular biologists and organic chemists. The results of this interdisciplinary research are prompting plant scientists to re-evaluate their concepts of development, defence mechanisms and functions of cell walls. These studies may also lead to biotechnology-based, environmentally friendly approaches to improving resistance to microbial and insect pests and to controlling the growth and development of plants.

This research was supported in part by U.S. Department of Energy (DOE) grants, numbers DE-FG05-93ER20114 and DE-FG05-93ER20115, and by the DOE-funded (grant number DE-FG09-93ER20097) Center for Plant and Microbial Complex Carbohydrates.

References

1. Sharp, J.K., McNeil, M. and Albersheim, P. (1984) J. Biol. Chem. **259**, 11321–11336
2. Davis, K.R., Darvill, A.G., Albersheim, P. and Dell, A. (1986) Z. Naturforsch. **41c**, 39–48
3. Jin, D.F. and West, C.A. (1984) Plant Physiol. **74**, 989–992
4. Hadwiger, L.A. and Beckman, J.M. (1980) Plant Physiol. **66**, 205–211
5. Davis, K.R., Darvill, A.G. and Albersheim, P. (1986) Plant Mol. Biol. **6**, 23–32
6. Cheong, J.-J. and Hahn, M.G. (1991) Plant Cell **3**, 137–147
7. Cosio, E.G., Frey, T. and Ebel, J. (1990) FEBS Lett. **264**, 235–238
8. Ham, K.-S., Kauffmann, S., Albersheim, P. and Darvill, A.G. (1991) Mol. Plant–Microbe Interact. **4**, 545–552
9. Toubart, P., Daroda, L., Desiderio, A., Salvi, G., Cervone, F., De Lorenzo, G., Bergmann, C., Darvill, A.G. and Albersheim, P. (1992) Plant J. **2**, 367–373

10. Marfà, V., Gollin, D.J., Eberhard, S., Mohnen, D., Darvill, A. and Albersheim, P. (1991) Plant J. 1, 217–225
11. Farmer, E.E., Moloshok, T.D., Saxton, M.J. and Ryan, C.A. (1991) J. Biol. Chem. **266**, 3140–3145
12. Mathieu, Y., Kurkdjian, A., Xia, H., Guern, J., Koller, A., Spiro, M.D., O'Neill, M.A., Albersheim, P. and Darvill, A. (1991) Plant J. **1**, 333–343
13. Byrne, H., Christou, N.V., Verma, D.P.S. and Maclachlan, G.A. (1975) J. Biol. Chem. **250**, 1012–1018
14. McDougall, G.J. and Fry, S.C. (1988) Planta **175**, 412–416
15. York, W.S., Darvill, A.G. and Albersheim, P. (1984) Plant Physiol. **75**, 295–297
16. Lerouge, P., Roche, P., Faucher, C., Maillet, F., Truchet, G., Promé, J.C. and Dénarié, J. (1990) Nature (London) **344**, 781–784
17. Schults, M., Quiclet-Sire, B., Kondorosi, E., Virelizier, H., Glushka, J.N., Endre, G., Géro, S.D. and Kondorosi, A. (1992) Proc. Natl. Acad. Sci. U.S.A. **89**, 192–196
18. Spaink, H.P., Sheeley, D.M., van Brussel, A.A.N., Glushka, J., York, W.S., Tak, T., Geiger, O., Kennedy, E.P., Reinhold, V.N. and Lugtenberg, B.J.J. (1991) Nature (London) **354**, 125–130

Biochem. Soc. Symp. **60**, 95–100
Printed in Great Britain

Systemic signals condition plant cells for increased elicitation of diverse defence responses

Heinrich Kauss

FB Biologie der Universität, Postfach 3049, D-67653 Kaiserslautern, Germany

During evolution, higher plants appear to have developed two different strategies for the defence against pathogenic micro-organisms, one of the most severe of their environmental challenges.

In one type of defence strategy, when individual cells are locally attacked by, for example, fungi, they respond by producing a variety of secondary metabolites, such as soluble fungitoxic compounds (phytoalexins), and/or a reinforcement of their cell wall by monomeric and polymeric phenolics, and by the 1,3-β-glucan callose (Fig. 1, dotted arrows). These local responses are generally studied by application of so-called 'elicitors', products of micro-organisms that probably also function in nature as inducers of the local defence responses in plants [1–3]. Transduction of the local signal 'elicitor' includes binding to membrane constituents, followed by a variety of rapid events, such as Ca^{2+}-influx, K^{+}-efflux, internal acidification, membrane depolarization, protein phosphorylation/dephosphorylation and jasmonic acid production [1–4]. These early changes are presumed to be components of the signal transduction mechanism, although the causal relationships and determinants of specificity for the activation of certain genes remain unclear. Nevertheless, as a consequence of elicitor action, mRNA and enyzmes specific for the production of the various phenylpropanoid derivatives are synthesized. Callose deposition can also be elicited, e.g. by chitosan; however, this form of elicitation appears to occur by a direct activation of the 1,3-β-glucan synthase, generally considered to be a constitutive plasma membrane enzyme (Fig. 1). An increase in cytoplasmic [Ca^{2+}] is probably one prerequisite for callose formation, but a requirement for additional unknown signals is likely [3,5]. Parallelling the above responses, with respect to the induction mechanism and biochemistry of defence factors, is the 'hypersensitive reaction', which also occurs locally at single attacked cells and coincides with an extracellular production of reactive oxygen species [3].

The second defence strategy in plants is the 'systemic acquired resistance', which originates from an initial infection by a pathogenic micro-organism and develops within a few days into a resistance in adjacent tissues or plant organs [2,3,6,7]. In this case, signals originating at the site of infection spread systemically over the plant and induce, in a developmental process, the synthesis of many 'pathogenesis-related' proteins

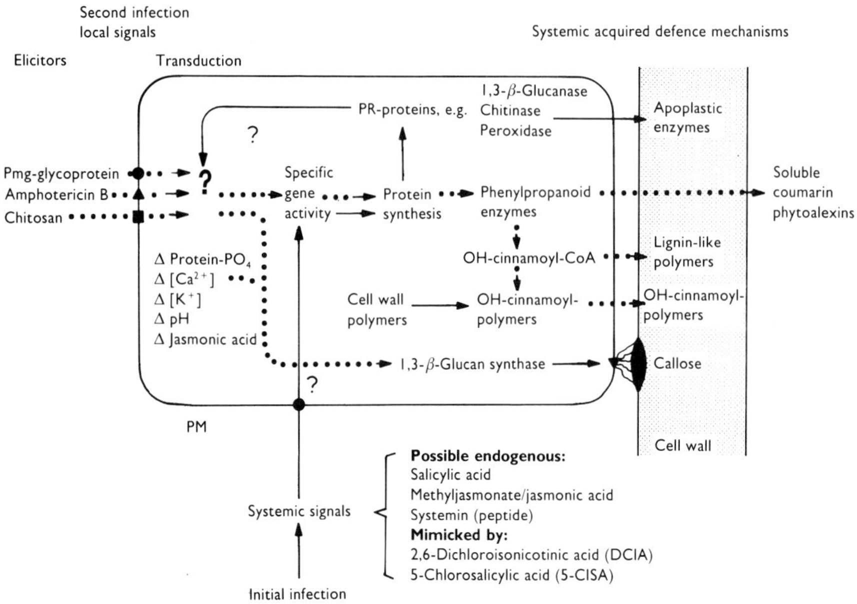

Fig. 1. Cross-links between the systemic acquired defence mechanisms and the elicitor-induced local defence responses. Initial infection of cells causes hypersensitive reactions that generate systemic signals spreading to target cells in other tissues, symbolized by their plasma membrane (PM). These systemic signals induce the synthesis of pathogenesis-related (PR) proteins, some of which are of direct use as intra- or extracellular defence enzymes (solid arrows). In cucumber plants exhibiting systemic acquired resistance the activity of the 1,3-β-glucan synthase [9] is also improved. This enzyme is involved in local callose deposition and can be triggered by, e.g. chitosan. Systemic signals can also condition the target cells for an enhanced sensitivity to elicitors able to trigger the phenylpropanoid defence responses occurring after a local second infection (dotted arrows). The corresponding elicitors and end-products shown are those known for parsley cells [1,3,15] and include H_2O_2 [19]. The conditioning effect suggests that among the PR-proteins there may be unknown components of the elicitor signal transduction system. For more details, see text.

(Fig. 1, solid arrows). The catalytic functions of part of these proteins have been identified, and they include intracellular and extracellular chitinases, 1,3-β-glucanases and peroxidase [2,7]. These enzymes provide an arsenal of weapons for a direct defence against the invading pathogen. Cytological studies [8] suggest, however, that in addition to these obvious defence enzymes, other reactions leading to a reinforcement of the cell wall have to be considered to explain the phenomenon of systemic acquired resistance. If, for example, the increase in apoplastic peroxidase would lead to an increased polymerization of cell wall phenolics, the respective intracellular synthesis of precursors and their export must also be accelerated. Similarly, the specific activity of the 1,3-β-glucan synthase is increased several fold in plasma membranes of cucumber plants exhibiting systemic acquired resistance [9]. This enzyme is involved in callose formation and represents part of the defence reactions induced locally by elicitors [3]. It was desirable, therefore, to have a model system which facilitates investigation of elicitor-triggered responses corresponding to the local defence strategy, and which also combines characteristics of the systemic acquired resistance.

Several lines of evidence suggest that salicylic acid is involved in establishing systemic acquired resistance [10,11]; although it might not be the primary systemic signal [6]. Jasmonic acid [12] and the peptide systemin [13] are related to systemic induction of proteinase inhibitors upon wounding. The universality of these molecules as primary signals in systemic acquired resistance remains to be shown. Nevertheless, 2,6-dichloroisonicotinic acid was shown to mimic the induction of many aspects of systemic resistance in various plants [7]. This non-toxic plant protection compound and salicylic acid have both been employed, therefore, in a model system using suspension-cultured parsley cells to demonstrate cross-links between elicitor-induced local defence reactions and the systemic acquired defence strategies (Fig. 1) [14]. To allow for developmental changes, we have pre-incubated the parsley cells with 20 μM 2,6-dichloroisonicotinic acid (Fig. 2) or salicylic acid [14]. These conditioned cells responded to lower elicitor concentrations, and also exhibited an enhanced overall efficiency towards elicited secretion of coumarin derivatives [14]. The effect was also found for the incorporation of 'lignin-like' polymers into cell walls [15] and for monomeric phenolic derivatives liberated on alkaline hydrolysis from cell walls (Fig. 2). In addition to some as yet unknown compounds, the latter monomers include ferulic acid, *p*-coumaric acid, *p*-hydroxybenzoic acid, *p*-hydroxybenzaldehyde and vanillin, as well as tyrosol and methoxytyrosol [15]. The hydroxycinnamic acids liberated under alkaline conditions might be esterified to cell wall polymers, which appear to be synthesized from the respective CoA-thioesters in endomembranes [16] and exported presumably as preformed macromolecules. The type of binding of the two aldehydes to cell wall polymers remains unclear. As they are fungitoxic, they could possibly serve as 'cell-wall-derived phytoalexins'.

The enhancement effects of pre-incubation with 2,6-dichloroisonicotinic acid and salicylic acid on elicited synthesis of metabolic end-products was also observed for two enzymes involved in the synthesis of phenylpropanoids, phenylalanine ammonia-lyase (PAL) and *S*-adenosyl-L-methionine:xanthotoxol *o*-methyltransferase, as well as for the synthesis of mRNA specific for PAL and for 4-coumarate–CoA ligase [14]. These results suggest that unknown components of the elicitor signal perception/transduction pathway are limiting in unconditioned cells, but are not in cells conditioned by the two substances. We speculate that these limiting compounds of the signal chain are proteins belonging to the, as yet unidentified, portion of the pathogenesis-related (PR) proteins (Fig. 1). We are at present studying which of the many early events of elicitor action occurring before gene activation (Fig. 1) is also affected in the conditioned cells. Current results (D. Lude and H. Kauss, unpublished work) indicate that the conditioning effects on phenylpropanoid responses are also evident when the fungal *Phytophthora megasperma* f. sp. *glycinea* (Pmg) elicitor is replaced by chitosan or amphotericin B. These substances can also elicit coumarin secretion in the parsley system (Fig. 1) [3], but presumably interact with general plasma membrane constituents rather than with a specific receptor assumed necessary for the recognition of the glycoprotein present in

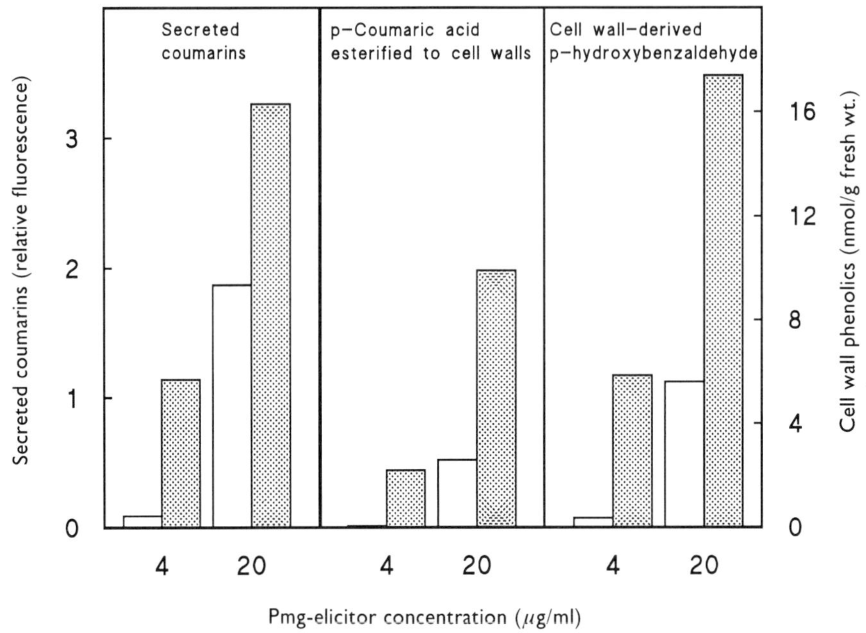

Fig. 2. Parsley suspension cells as a model demonstrating cross-links between the systemic acquired resistance mechanism and elicitor-induced phenylpropanoid defence reactions. Half of a 3-day-old cell suspension remained as a control (open columns) and the other half (dotted columns) was conditioned by incubation for 24 h with 20 μM 2,6-dichloroisonicotinic acid (DCIA), known to mimic the induction of systemic acquired resistance [7]. Both cell samples were than treated with Pmg-elicitor. The secreted coumarin derivatives and the phenolic monomers liberated from cell walls on alkaline hydrolysis were determined 24 h later as described [14,15]. Values from the respective cell samples without elicitor were subtracted.

the crude Pmg elicitor preparation and active on parsley cells [17]. It is also of interest in this context that methyljasmonate-conditioned cells respond to chitosan with increased callose deposition, again indicating that several features of the plasma membrane become altered by the conditioning process.

The enhancement of elicitor-induced coumarin secretion described above has also been used to further investigate the specificity of compounds able to condition the parsley cells [15]. Chlorinated acids (2,6-dichloroisonicotinic acid, 4- or 5-chloro-salicylic acid and 3,5-dichlorosalicylic acid) are most potent, whereas the non-chlorinated analogues showed less activity, possibly because of partial inactivation by the cells. Acetylsalicylic acid and 2,6-dihydroxybenzoic acid are of similar potency to salicylic acid, whereas some isomers (e.g. 2,3-, 2,5-, 3,4-, 3,5-dihydroxybenzoic acid and 3- and 4-hydroxybenzoic acid) are almost inactive.

As mentioned above, the primary systemic signal for the induction of systemic acquired resistance is still unknown. We have used the parsley system, therefore, to determine whether methyljasmonate, which is discussed as an endogenous signalling substance in various plant responses [4,12], can also be employed for conditioning cells. Indeed pre-incubation of parsley cultures with 1–20 μM methyljasmonate greatly enhances the elicitor-induced coumarin secretion and the incorporation of esterified hydroxycinnamic acids and lignin-like polymers into the cell walls [18]. Again, this effect is most pronounced at low Pmg-elicitor concentrations, indicating that part of the effect is due to an increase in the signal perception/transduction system, even though methyljasmonate also leads to a prolongation of the time during which the elicitor stimulus persists (D. Uhl and H. Kauss, unpublished work). The 2,6-dichloro-isonicotinic acid and chlorosalicylic acid given alone were never found to induce any coumarin secretion [14,15]. In contrast, 5 μM methyljasmonate induces a low level of coumarin secretion, in addition to the enhancement response of the Pmg-elicitor [18]. Parsley cells pretreated with methyljasmonate also exhibit a greatly enhanced elicitation of activated oxygen [19]. It is also of interest that in our dark-grown parsley cultures, methyljasmonate does not induce intracellular apiin or the series of other flavonoid glycosides (H. Kauss and W. Jeblick, unpublished work). These compounds are induced instead by u.v.-light in this culture [1]. In contrast, apiin and other flavonoids are found in light-grown parsley cells, and further enhanced by 12-oxophytodienoic acid, a precursor of methyljasmonate [20]. It appears, therefore, that the jasmonic acid derivatives strikingly enhance both the respective elicitor or light responses, but have little effect when given singly.

The above results with the parsley suspension cells prompted us to develop a new assay system using dark-grown cucumber hypocotyl segments. This cucumber system allows external application of various signal substances followed by determination of resistance against fungal infection and cytological investigation of the infection process. When split, the conditioned segments also allow elicitor induction of activated oxygen and phenylpropanoid defence responses [21]. The results obtained up to now indicate that 2,6-dichloroisonicotinic acid, 5-chlorosalicylic acid and also salicylic acid are able to condition the hypocotyl for enhanced phenylpropanoid and activated oxygen defence reactions in a way that is very similar to that described for parsley suspension cells, and can also induce resistance against *Colletotrichum lagenarium*. In contrast, conditioning with methyljasmonate slightly increases elicited phenylpropanoid responses, but appears not to induce resistance against fungal infection. The reason might be that this latter signalling substance may be mainly related to mechanical wounding [22]. In contrast, induction of acquired resistance appears to be a complex process which might require more than one systemic signal.

The experiments reported were financially supported by the Deutsche Forschungsgemeinschaft and the Fonds der Chemischen Industrie.

References

1. Hahlbrock, K. and Scheel, D. (1989) Annu. Rev. Plant Physiol. Plant Mol. Biol. 40, 347–369
2. Bowles, D.J. (1990) Annu. Rev. Biochem. 49, 873–907
3. Kauss, H. (1990) in The Plant Plasma Membrane: Structure, Function and Molecular Biology (Larsson, C. and Müller, I.M., eds.), pp. 320–350, Springer, Berlin
4. Gundlach, H., Müller, M.J., Kutchan, T.M. and Zenk, M.H. (1992) Proc. Natl. Acad. Sci. U.S.A. 89, 2389–2393

5. Kauss, H., Waldmann, T. and Quader, H. (1990) NATO Adv. Sci. Inst. Ser **H47**, 117–131
6. Rasmussen, J.B., Hammerschmidt, R. and Zook, M.N. (1991) Plant Physiol. **97**, 1342–1347
7. Ward, E.R., Uknes, S.J., Williams, S.C., Dincher, S.S., Wiederhold, D.L., Alexander, D.C., Ahl-Goy, P., Métraux, J.-P. and Ryals, J.A. (1991) Plant Cell **3**, 1085–1094
8. Hammerschmidt, R. and Kuc, J. (1982) Physiol. Plant Pathol. **20**, 61–71
9. Schmele, I. and Kauss, H. (1990) Physiol. Mol. Plant Pathol. **37**, 221–228
10. Malamy, J. and Klessig, D.F. (1992) Plant J. **2**, 643–654
11. Raskin, I. (1992) Plant Physiol. **99**, 799–803
12. Staswick, P.E. (1992) Plant Physiol. **99**, 804–807
13. Pearce, G., Strydom, D., Johnson, S. and Ryan, C.A. (1991) Science **253**, 895–898
14. Kauss, H., Theisinger-Hinkel, E., Mindermann, R. and Conrath, U. (1992) Plant J. **2**, 655–600
15. Kauss, H., Franke, R., Krause, K., Conrath, U., Jeblick, W., Grimmig, B. and Matern, U. (1993) Plant Physiol. **102**, 459–466
16. Meyer, K., Kohler, A. and Kauss, H. (1991) FEBS Lett. **290**, 209–212
17. Parker, J.E., Schulte, W., Hahlbrock, K. and Scheel, D. (1991) Mol. Plant–Microbe Interact. **4**, 19–27
18. Kauss, H., Krause, K. and Jeblick, W. (1992) Biochem. Biophys. Res. Commun. **189**, 304–308
19. Kauss, H., Jeblick, W., Ziegler, J. and Krabler, W. (1994) Plant Physiol. **105**, 89–94
20. Dittrich, H., Kutchan, T.M. and Zenk, M.H. (1992) FEBS Lett. **309**, 33–36
21. Siegrist, J., Jeblick, W. and Kauss, H. (1994) Plant Physiol. **105**, in the press
22. Creelman, R.A., Tierney, M.L. and Mullet, J.E. (1992) Proc. Natl. Acad. Sci. U.S.A. **89**, 4938–4941

Biochem. Soc. Symp. **60**, 101–112
Printed in Great Britain

Characterization of hepta-β-glucoside elicitor-binding protein(s) in soybean

Michael G. Hahn*§†, Jong-Joo Cheong§‡, Rob Alba§† and François Côté§

§Complex Carbohydrates Research Centre and †Department of Botany, ‡Department of Biochemistry, The University of Georgia, 220 Riverbend Road, Athens, GA 30602-4712, U.S.A.

Synopsis

We are studying the cellular signalling pathway leading to pterocarpan phytoalexin biosynthesis in soybean that is induced by a branched hepta-β-glucoside originally isolated from the mycelial walls of the phytopathogenic oomycete, *Phytophthora megasperma* f. sp. *glycinea*. Our research has focused on the first step in this signal pathway, namely the specific recognition of the hepta-β-glucoside elicitor by plasma-membrane-localized binding protein(s) in soybean cells. Binding of a radio-iodinated derivative of the elicitor-active hepta-β-glucoside by membrane elicitor-binding proteins is specific, reversible, saturable and of high affinity (K_d = 0.75 nM). After solubilization using the non-ionic detergent n-dodecylsucrose, the elicitor-binding proteins retain the binding affinity (K_d = 1.8 nM) for the radiolabelled elicitor and the binding specificity for elicitor-active oligoglucosides. A direct correlation is observed between the ability of elicitor-active and structurally related inactive oligoglucosides to displace labelled elicitor from the elicitor-binding proteins and the elicitor activity of the oligosaccharides. Thus, the elicitor-binding proteins recognize the same structural elements of the hepta-β-glucoside elicitor that are essential for its phytoalexin-inducing activity, suggesting that the elicitor-binding proteins are physiological receptors for the elicitor. Current research is directed toward the purification and cloning of the hepta-β-glucoside elicitor-binding proteins. Purification and characterization of the hepta-β-glucoside-binding protein(s) or their corresponding cDNAs is a first step toward elucidating how the hepta-β-glucoside elicitor triggers the signal transduction pathway that ultimately leads to the synthesis of phytoalexins in soybean.

*To whom correspondence should be addressed at the Complex Carbohydrate Research Centre.

Introduction

Living organisms utilize a large number of signal molecules to regulate their growth and development. Furthermore, the cells that make up an organism have evolved complex and diverse mechanisms for perceiving and responding to signal molecules originating not only from within the organism, but also from the external environment. We are utilizing plant responses to one external stimulus, namely the activation of plant defence mechanisms upon infection with micro-organisms, in an effort to gain a better understanding of the molecular basis for signal perception and transduction in plants.

Plants utilize a multi-faceted array of defence responses when confronted by invasive micro-organisms [1,2]. These defence responses include the synthesis and accumulation of anti-microbial compounds (phytoalexins) [3,4], the production of glycosylhydrolases capable of attacking surface polymers of pathogens [5], the synthesis of proteins that inhibit degradative enzymes produced by pathogens [2,6] and the modification of plant cell walls by deposition of callose [7,8], hydroxyproline-rich glycoproteins [9] and/or lignin [10]. Detailed investigation of these defence responses has led to the discovery of new classes of signal molecules and provided useful model systems for molecular studies on signal perception, signal transduction and gene regulation in plants. (For recent reviews, see [11–13].)

Biochemical analysis of the induction of plant defence responses has been facilitated by the recognition that cell-free extracts of microbial and plant origin are capable of inducing defence responses when applied to plant tissues. The active components in the extracts are commonly referred to as 'elicitors.' The term elicitor was originally used to refer to molecules and other stimuli that induce the synthesis and accumulation of phytoalexins in plant cells [14], but is now commonly used for molecules that stimulate any plant defence mechanism [4,11,15,16]. A number of different biotic elicitors, including oligosaccharides isolated from fungal and plant cell walls [17], proteins [18] and lipids [19], as well as abiotic elicitors (such as heavy metal salts or u.v. light) are known. Abiotic elicitors are thought to result in the release of biotic elicitors from the cell walls of the plant [20,21]. Several recent reviews [2,17,22] provide a broad overview of the structures of diverse elicitors and their activities.

We are attempting to elucidate the cellular signalling pathway that is triggered by oligoglucoside elicitors and results in the biosynthesis and accumulation of pterocarpan phytoalexins in soybean (Fig. 1). The extent of biochemical information about this cellular signalling pathway (reviewed in [4,11,12,16,23]) makes this experimental system particularly attractive for studies on plant signalling mechanisms. To date, our research has focused on one of the first steps in this signalling pathway, namely the recognition of the oligoglucoside elicitors by receptors in soybean root cells. This article will first provide a brief history of the identification and purification of oligoglucoside elicitors. This will be followed by a summary of our studies to identify structural features of elicitor-active oligoglucosides that are essential for their activity. Finally, we will review the results of investigations to identify, purify and characterize binding proteins whose properties suggest that they are physiological receptors for oligoglucoside elicitors.

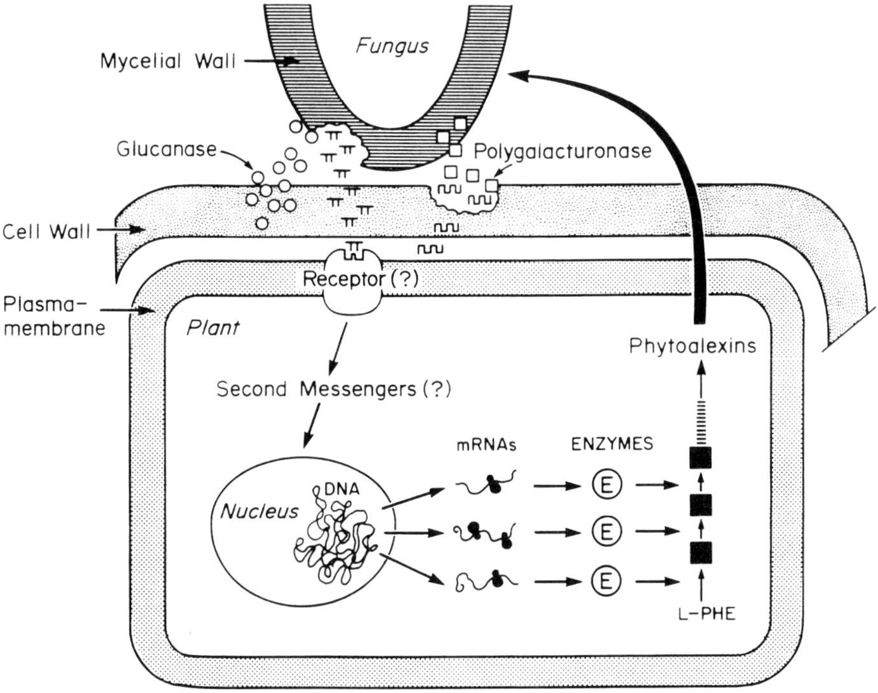

Fig. 1. Hypothetical model for the cellular signalling pathway leading to the synthesis of phytoalexins in soybean.

Purification of glucan elicitors

Elicitor-active glucans were first detected in the culture filtrates of *Phytophthora megasperma* f. sp. *glycinea*, a phytopathogen of soybean [24], and were later also purified from commercial yeast extract [25]. These elicitors were shown to be composed of 3-, 6- and 3,6-linked glucosyl residues [24,25], a composition very similar to β-glucans, which are important constituents of various mycelial walls [26]. Thus, subsequent research focused on elicitors released from mycelial walls.

Partial hydrolysis of the mycelial walls of *P. megasperma* using either hot water [27] or 2-*N*-trifluoroacetic acid [28] releases a very complex mixture of oligo- and polysaccharides. This mixture contains β-glucan fragments ranging in size from short oligomers up to polysaccharides [relative molecular mass (M_r) > 100000] [28,29]. Oligoglucans having two different biological activities, inducing phytoalexin accumulation [28] and inducing resistance to viruses [30], are known to be present in this mixture. There may be other, as yet unidentified, activities in the oligoglucan mixture as well. In addition, the oligoglucan mixture contains a large number of biologically inactive molecules.

Fractionation of the mixture of oligoglucosides generated by partial acid hydrolysis of mycelial walls of *P. megasperma* on a gel permeation column revealed that elicitor-active oligosaccharides are present in all fractions containing oligomers of degree of polymerization (DP) ≥ 6 [28]. The heptamer-enriched fraction was further fractionated on a series of normal- and reversed-phase h.p.l.c. [28]. A branched hepta-β-glucoside (compound 1, Fig. 2) was the only heptaglucoside that had the ability to elicit the accumulation of phytoalexins in soybean cotyledon tissue [28]. It was estimated that the original heptaglucoside mixture contained approx. 300 structurally distinct hepta-β-glucosides, based on the number of peaks observed in the various chromatographic steps. Homogeneous preparations of the aldehyde-reduced forms (i.e. the hexa-β-glucosyl glucitols) of the elicitor-active hepta-β-glucoside and of seven other elicitor-inactive hepta-β-glucosides were obtained in amounts sufficient to determine their structures [28,31]. The structure of the elicitor-active hepta-β-glucoside [32] was subsequently confirmed by its chemical synthesis [33–35].

The ability of the chemically synthesized, unreduced hepta-β-glucoside elicitor to induce phytoalexin accumulation in soybean cotyledons is identical to that of the corresponding hexa-β-glucosyl glucitol purified from fungal wall hydrolysates [32]. They are active at concentrations of approx. 10 nM, making them the most active elicitors of phytoalexin accumulation yet observed. The seven other hexa-β-glucosyl glucitols that were purified from the partial hydrolysates of fungal cell walls had no elicitor activity over the limited concentration range (≥ 400 μM) that could be tested

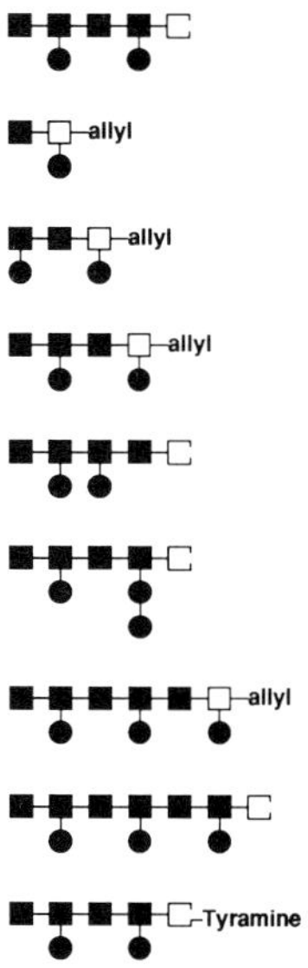

Compound	Structure	Relative Elicitor Activity	Relative Binding Activity
1		1000	1000
2	allyl	0.16	0.12
3	allyl	0.31	1.2
4	allyl	730	960
5		1.2	1.3
6		270	93
7	allyl	570	900
8		420	1380
15	Tyramine	730	3600

Fig. 2. Structure, relative elicitor activity, and relative binding activity of oligo-β-glucosides. The relative elicitor activity is calculated with respect to the phytoalexin-inducing activity of hepta-β-glucoside 1 on soybean cotyledons (= 1000). The relative binding activity is calculated with respect to the ability of the hepta-β-glucoside elicitor (compound 1) to displace radiolabelled ligand (iodinated compound 15) from soybean root membranes (= 1000). Key: ■-, Glc *p*-β-1,6-; □-, Glc *p*-α-1,1-; ●-, Glc *p*-β-1,3-; ⊏ , reducing Glc. Redrawn from [52] with permission.

[28]. These results provided the first evidence that specific structural features are required for an oligo-β-glucoside to be an effective elicitor of phytoalexin accumulation. Structure–activity studies carried out in our laboratory have confirmed and expanded upon these initial findings.

Structure–activity studies

Thirteen oligo-β-glucosides (compounds 2–14, see Fig. 2 and Fig. 3), structurally related to the elicitor-active hepta-β-glucoside (compound 1, Fig. 2) were chemically synthesized [33,34,36–38] (N. Hong and T. Ogawa, unpublished work). This group of oligoglucosides was instrumental in allowing the identification of structural features essential for effective elicitation of phytoalexin accumulation in soybean cotyledon tissue [39,40]. The four most active oligo-β-glucosides (compounds 1, 4, 7, and 8, Fig. 2) have the same branching pattern as that of elicitor-active hepta-β-glucoside 1 [31], and require a concentration of approx. 10 nM for half-maximal induction of phytoalexin accumulation (EC_{50}). Hexa-β-glucoside 4 (Fig. 2) is the minimum fully elicitor-active structure [39]. Increasing the length of this hexaglucoside, by addition of glucosyl residues at the reducing end of the molecule, has no significant effect on its elicitor

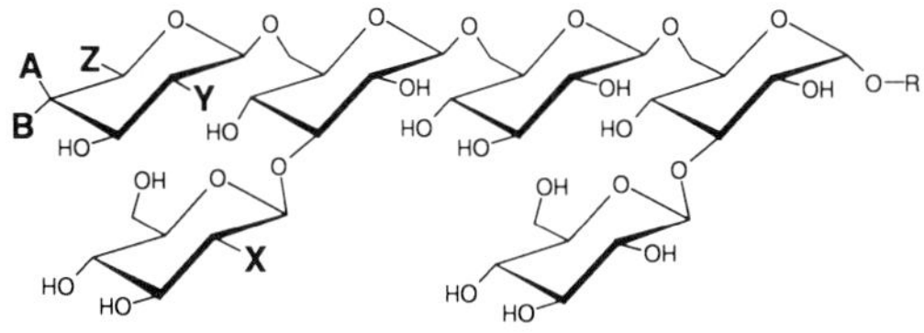

Compound	A	B	X	Y	Z	Relative Elicitor Activity	Relative Binding Activity
4	-H	-OH	-OH	-OH	$-CH_2OH$	730	960
9	-H	-OH	-OH	-OH	-H	130	530
10	-OH	-H	-OH	-OH	$-CH_2OH$	11	26
11	-H	-OH	$-NH_2$	-OH	$-CH_2OH$	130	180
12	-H	-OH	-NHAc	-OH	$-CH_2OH$	1.2	3.8
13	-H	-OH	-OH	$-NH_2$	$-CH_2OH$	4.4	21
14	-H	-OH	-OH	-NHAc	$-CH_2OH$	<0.08	0.22

Fig. 3. Structure, relative elicitor activity and relative binding activity of structural variants of hexa-β-glucoside 4. The relative elicitor activity is calculated with respect to the phytoalexin-inducing activity of the hepta-β-glucoside elicitor (compound 1, see Fig. 2) on soybean cotyledons (= 1000). The relative binding activity is calculated with respect to the ability of the hepta-β-glucoside elicitor (compound 1) to displace radiolabelled ligand from soybean root membranes (= 1000). R = allyl for compounds 4, 9, 10, 12 and 14; R = propyl for compounds 11 and 13. The nature of R at the reducing end of these hexasaccharides has little effect on elicitor or binding activities.

activity. In contrast, removing glucosyl residues from the hexaglucoside (compounds 2 and 3) or rearranging its side-chains (compound 5) results in molecules with significantly lower elicitor activity (Fig. 2). For example, removal of the non-reducing, terminal, backbone glucosyl residue (compound 3) resulted in a 4000-fold reduction in elicitor activity, suggesting that this glucosyl residue has a particularly important function [40].

Several recently synthesized oligoglucosides (R. Verduyn and J. van Boom, unpublished work) have permitted the direct assessment of the importance of the side-chain glucosyl residue closest to the reducing end of the elicitor-active oligoglucosides. These oligoglucosides range in size from tetramer to hexamer. All three of these oligoglucosides are significantly less active (1000- to 10 000-fold) than the elicitor-active hepta-β-glucoside (J-J. Cheong and M.G. Hahn, unpublished work). Additional confirmation of the importance of the side-chain glucosyl residues for the elicitor activity of oligoglucosides was provided by the demonstration that a linear, 6-linked hepta-β-glucoside is inactive [39].

An additional set of hexasaccharides was synthesized in which one or the other of the terminal glucosyl residues at the non-reducing end of hexa-β-glucoside 4 was modified (Fig. 3). Thus, replacement of the side-chain glucosyl residue of the terminal trisaccharide with a β-glucosaminyl (compound 11) or *N*-acetyl-β-glucosaminyl residue (compound 12) reduces the elicitor activity approx. 10- and 1000-fold, respectively. The corresponding modifications of the non-reducing terminal backbone glucosyl residue (compounds 13 and 14, respectively) results in even greater decreases in elicitor activity (approx. 100- and 10 000-fold, respectively). Substitution of the same glucosyl residue with a xylosyl residue (compound 9) or a galactosyl residue (compound 10) reduces the activity about 10- and 100-fold, respectively [40]. These data, together with those described earlier, prove that all three non-reducing terminal, glucosyl residues present in the hepta-β-glucoside elicitor are essential for the ability of the molecule to effectively elicit phytoalexin accumulation in soybean.

In contrast, the reducing terminal glucosyl residue of the hepta-β-glucoside elicitor does not appear to be essential for biological activity. Phytoalexin elicitor assays of reducing-end derivatives of hepta-β-glucoside-1 demonstrated that attachment of an alkyl or aromatic group to the oligosaccharide (e.g. compound 15) does not have a significant effect on its EC_{50} [39]. A tyramine-coupled derivative of hepta-β-glucoside-5 is slightly more active (approx. 2.5-fold) than underivatized hepta-β-glucoside-5. Coupling tyramine or benzylhydroxylamine to maltoheptaose, a structurally unrelated hepta-α-glucoside, yielded derivatives with no detectable elicitor activity. These results establish that coupling of aromatic groups to biologically inactive oligoglucosides does not endow those oligosaccharides with phytoalexin elicitor activity. Thus, it was possible to prepare a fully active, radio-iodinated form of compound 15 for use as a labelled ligand to search for the presence of elicitor-specific, high-affinity binding sites in soybean membranes.

Ligand binding studies

The first step in the signal transduction pathway induced by the hepta-β-glucoside elicitor is likely to be its recognition by a specific receptor. Indeed, the specificity of the response of soybean tissue to oligoglucoside elicitors of phytoalexin accumulation [28,39] described in the previous section suggests that a specific receptor for the hepta-β-glucoside elicitor exists in soybean cells. Several earlier studies utilizing heterogeneous mixtures of fungal glucan fragments indicated that binding sites for such fragments exist in plant membranes [41–44]. The existence of membrane-localized binding sites was demonstrated for the hepta-β-glucoside elicitor coupled to radio-iodinated aminophenethylamine [45] or tyramine [46]. Binding sites are found in membranes isolated from every major organ of young soybean plants [46]. The hepta-β-glucoside elicitor-binding sites co-migrates with an enzyme marker (vanadate-sensitive ATPase) for plasma membranes in isopycnic sucrose density gradients [40], confirming earlier results obtained with partially purified labelled glucan fragments [43,44]. Binding of the radiolabelled hepta-β-glucoside elicitor to the root membranes is saturable over a concentration range of 0.1–5 nM, which is somewhat lower than the range of concentrations (6–200 nM) required to saturate the bioassay for phytoalexin accumulation [28,32,39]. The root membranes possess only a single class of high-affinity, hepta-β-glucoside-binding sites (apparent K_d approx. 1 nM), which are inactivated by heat or Pronase treatment [46], suggesting that the molecule(s) responsible for the binding are proteinaceous. Binding of the active hepta-β-glucoside to the membrane preparation is reversible, indicating that the elicitor does not become covalently attached to the binding protein(s).

The membrane-localized, elicitor-binding proteins exhibit a high degree of specificity with respect to the oligoglucosides that they bind. More importantly, the ability of an oligoglucoside to bind to soybean root membranes correlates with its ability to induce phytoalexin accumulation (Fig. 4) [40,46]. Oligo-β-glucosides with high elicitor activity are efficient competitors of the radiolabelled elicitor, while biologically less active oligo-β-glucosides are less efficient. Thus, four oligo-β-glucosides, ranging in size from hexamer to decamer (compounds 1, 4, 7 and 8), indistinguishable in their abilities to induce phytoalexin accumulation [39], are equally effective competitive inhibitors of binding of radiolabelled hepta-β-glucoside-15 to soybean root membranes (Fig. 2). The abilities of structurally modified oligo-β-glucosides to compete with radiolabelled 15 are reduced in proportion to the reduction in the biological activities of these oligo-β-glucosides (Fig. 2 and Fig. 3).

The results of the structure–activity [39] and ligand-binding [40,46] studies demonstrate that those structural elements of the hepta-β-glucoside required to elicit phytoalexin synthesis are also essential for efficient binding of the elicitor to its putative receptor. These essential structural features include the non-reducing, terminal, backbone glucosyl residue and the two side-chain, terminal glucosyl residues of the hepta-β-glucoside elicitor. The distribution of side-chain glucosyl residues along the backbone of the molecule is also important for recognition of the elicitor by the binding proteins. The combined results of the biological assays [39] and the binding studies [40,46] provide strong evidence that the binding proteins are physiological receptors for the hepta-β-glucoside elicitor.

Purification studies

Characterization of the hepta-β-glucoside elicitor-binding proteins will require their solubilization from the membranes and subsequent purification. The low abundance of the elicitor-binding proteins in the soybean root membranes (approx.

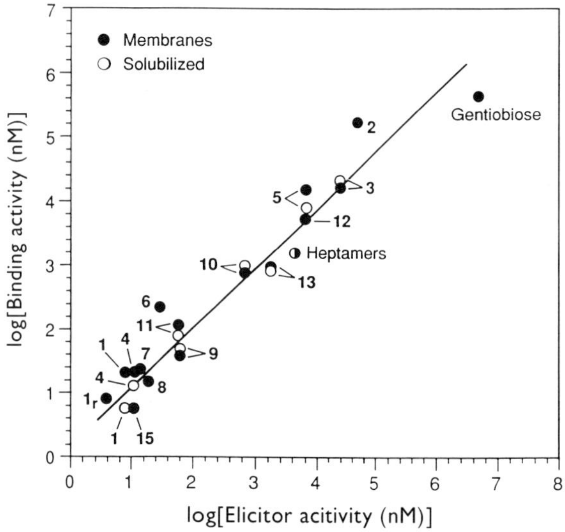

Fig. 4. Correlation of the elicitor activities of oligoglucosides with their affinities for the membrane-localized (●) or detergent-solubilized (○) hepta-β-glucoside-binding protein(s). The relative elicitor activity is defined as the concentration of an oligosaccharide required to give half-maximum induction of phytoalexin accumulation ($A/A_{std} = 0.5$) in the soybean cotyledon bioassay corrected to the standard curve for hepta-β-glucoside I [39]. The binding activity is defined as the concentration of oligosaccharide required to give 50% inhibition of the binding of radiolabelled hepta-β-glucoside 15 to its binding protein(s). Data points are identified with numbers identifying the oligoglucosides (see Fig. 2 and Fig. 3); I_r = reduced hepta-β-glucoside I; Heptamers = mixture of heptaglucosides prepared from mycelial wall hydrolysates of *P. megasperma* [28].

1 pmol/mg of protein) [45,46] suggests that ligand affinity chromatography will be required to achieve purification to homogeneity. The following paragraphs summarize the results from solubilization studies and describe recent progress toward purification of elicitor-binding proteins.

Solubilization of fully functional elicitor-binding proteins from soybean root microsomal membranes has been achieved with the aid of several detergents [40,47]. The non-ionic detergents n-dodecylmaltoside, n-dodecylsucrose and Triton X-114 each solubilize between 40% and 60% of the elicitor-binding activity from the membranes in a single extraction. The n-dodecylsucrose-solubilized binding protein preparation is reasonably stable upon storage after solubilization, losing about half of its binding activity after 3 weeks at 4°C. The zwitterionic detergent *N*-dodecyl-*N*,*N*-dimethyl-3-ammonio-1-propane-sulphonate (ZW 3–12) is almost as effective at low detergent concentrations (approx. 0.3%) as the non-ionic detergents, but the zwitterionic detergent inactivates a significant proportion of the elicitor-binding protein(s).

The solubilized elicitor-binding proteins retain their high affinity for the hepta-β-glucoside elicitor, having an apparent K_d of 1.8 nM for the n-dodecylsucrose-solubilized binding proteins. The apparent dissociation constant determined for ZW 3–12-solubilized binding proteins is very similar [48,49]. More importantly, the solubilized hepta-β-glucoside binding proteins retain the specificity for elicitor-active oligoglucosides characteristic of the membrane-localized proteins (Fig. 4), regardless of which detergent is used for solubilization [40]. The successful solubilization of the elicitor-binding proteins, apparently with retention of their elicitor-binding properties, is a crucial first step toward purification of these proteins, since it will permit the application of affinity techniques to this problem.

Recently, progress has been achieved toward identification and purification of hepta-β-glucoside elicitor-binding proteins. Cosio *et al.* [47] reported initial purification (approx. 4-fold) of ZW 3–12-solubilized elicitor-binding proteins using a combination of ion-exchange chromatography, poly(ethylene glycol) precipitation and gel permeation chromatography, although with significant losses of binding activity (approx. 70%). We have also obtained significant purification (approx. 8-fold) of n-dodecylsucrose-solubilized elicitor-binding proteins in a single gel permeation chromatography with lower losses of elicitor-binding activity (30–50%) [40]. Gel permeation chromatography of detergent-solubilized hepta-β-glucoside elicitor-binding proteins indicated that elicitor-binding activity is associated primarily with large detergent–protein micelles ($M_r > 200\,000$) [40,47]. Indeed, we have found that the highest specific elicitor-binding activity was present in detergent–protein micelles having a $M_r > 660\,000$. Little or no elicitor-binding activity was associated with the smallest detergent–protein micelles [40]. These data suggest that the elicitor-binding activity requires either a multimeric protein complex or a protein conformation preferentially stabilized in large detergent micelles. Alternatively, the elicitor-binding protein could consist of a single large polypeptide ($M_r > 200\,000$). However, photo-affinity labelling studies [49] resulted in the identification of several protein bands, most prominently one of about 70 kDa, whose labelling properties suggest that they are elicitor-binding proteins. Further purification and characterization of the elicitor-binding protein(s) is required to resolve these issues.

Most recently, ligand-affinity chromatography has been applied in an effort to achieve a more dramatic enrichment of elicitor-binding proteins. Frey *et al.* [50] have prepared an affinity matrix consisting of partially purified, elicitor-active oligoglucosides (DP = 15–25) coupled to porous glass beads. We have coupled synthetic hepta-β-glucoside elicitor to a cross-linked agarose matrix via a 10-atom spacer arm (F. Côté, J.-J. Cheong and M.G. Hahn, unpublished work). Both of these affinity matrices appeared to retain elicitor-binding proteins specifically, while control matrices that contain no immobilized elicitors do not bind elicitor-binding proteins. However, the low yield of elicitor-binding protein activity after affinity chromatography [50] suggests that additional improvements in chromatography conditions will be necessary to obtain sufficient amounts of elicitor-binding protein for detailed characterization.

Conclusions

Progress in elucidating the mechanisms by which oligoglucoside elicitors exert their effects on soybean cells has been hampered by the great molecular heterogeneity of the elicitor preparations. Detailed biochemical investigations of the cellular signalling pathways triggered by oligosaccharide elicitors require the preparation of homogeneous preparations of these elicitors to unambiguously assign observed effects to single elicitor-stimulated pathways. Recent improvements in the techniques for the purification of oligosaccharides, as well as contributions from synthetic organic chemists, have made available homogeneous preparations of oligoglucoside elicitors, which have greatly facilitated these biochemical studies. Indeed, the availability of chemically synthesized hepta-β-glucoside elicitor, and structurally related, less-active agonists, has led to the tentative identification of binding proteins (putative receptors) for this elicitor. The binding characteristics of these proteins (high affinity, saturability, ligand specificity) strongly suggest that the elicitor-binding proteins are physiological receptors for the hepta-β-glucoside elicitor.

Purification and characterization of elicitor-binding proteins and/or isolation of the gene(s) encoding these proteins will be required to prove that the binding proteins function as physiological receptors. Evidence supporting a role for the elicitor-binding protein(s) in the cellular signalling pathway leading to phytoalexin accumulation could be obtained by reconstitution of an elicitor-responsive system using purified binding protein(s) or by functional expression of genes encoding the binding protein(s). The recent development of photo-activatable elicitor ligands and affinity chromatography matrices will undoubtedly greatly facilitate efforts to obtain sufficient amounts of the hepta-β-glucoside elicitor-binding proteins for detailed characterization and reconstitution studies. Characterization of the glucan elicitor-binding protein(s) and/or their genes may also give important insights into the mechanism by which the elicitor signal is transmitted into the cell interior.

The oligoglucoside elicitors discussed in this article belong to a recently discovered class of signal compounds, the oligosaccharins (oligosaccharides with regulatory properties [51]). This class of compound includes oligosaccharides isolated from plants, fungi and, most recently, bacteria (reviewed in [52–54]). Oligosaccharins have been shown to play important roles in the regulation of plant defence responses, plant development and plant–symbiont interactions. Additional biochemical analysis of the cellular signalling pathways induced by these signal molecules is likely to yield valuable insights into how plants perceive and respond to stimuli from their environment.

Our research on glucan elicitor-binding proteins is supported by a grant from the National Science Foundation (MCB-9206882), and in part by the Department of Energy-funded Center for Plant and Microbial Complex Carbohydrates (DE-FG09-93ER20097). We are grateful to P. Garegg (University of Stockholm, Sweden), T. Ogawa (RIKEN, Japan), R. Verduyn (Leiden University, The Netherlands) and their colleagues for their generous gifts of synthetic oligoglucosides. The technical assistance of Anna-Maria Sult and the drawing skills of Carol L. Gubbins Hahn are also gratefully acknowledged.

References

1. Bell, A.A. (1981) Annu. Rev. Plant Physiol. **32**, 21–81
2. Hahn, M.G., Bucheli, P., Cervone, F., Doares, S.H., O'Neill, R.A., Darvill, A. and Albersheim, P. (1989) Plant–Microbe Interact. **3**, 131–181
3. Dixon, R.A., Dey, P.M. and Lamb, C.J. (1983) Adv. Enzymol. **55**, 1–136

4. Ebel, J. (1986) Annu. Rev. Phytopathol. **24**, 235–264
5. Boller, T. (1987) Plant–Microbe Interact. **2**, 385–413
6. Ryan, C.A. (1990) Annu. Rev. Phytopathol. **28**, 425–449
7. Aist, J.R. (1976) Annu. Rev. Phytopathol. **14**, 145–163
8. Kauss, H. (1987) Naturwissenschaften **74**, 275–281
9. Showalter, A.M. and Varner, J.E. (1989) **15**, 485–520
10. Ride, J.P. (1983) in Biochemical Plant Pathology (Callow, J.A., ed.), pp. 215–236, Wiley, New York
11. Dixon, R.A. (1986) Biol. Rev. **61**, 239–291
12. Lamb, C.J., Lawton, M.A., Dron, M. and Dixon, R.A. (1989) Cell **56**, 215–224
13. Scheel, D. and Parker, J.E. (1990) Z. Naturforsch. **45c**, 569–575
14. Keen, N.T. (1975) Science **187**, 74–75
15. Hahlbrock, K. and Scheel, D. (1987) in Innovative Approaches to Plant Disease Control (Chet, I., ed.), pp. 229–254, Wiley, New York
16. Dixon, R.A. and Lamb, C.J. (1990) Annu. Rev. Plant Physiol. Plant Mol. Biol. **41**, 339–367
17. Darvill, A.G. and Albersheim, P. (1984) Annu. Rev. Plant Physiol. **35**, 243–275
18. Anderson, A.J. (1989) Plant–Microbe Interact. **3**, 87–130
19. Bostock, R.M., Kuc, J.A. and Laine, R.A. (1981) Science **212**, 67–69
20. Hahn, M.G., Darvill, A.G. and Albersheim, P. (1981) Plant Physiol. **68**, 1161–1169
21. Bailey, J.A. (1980) Ann. Phytopathol. **12**, 395–402
22. Ryan, C.A. (1988) Biochemistry **27**, 8879–8883
23. Hahlbrock, K. and Scheel, D. (1989) Annu. Rev. Plant Physiol. Plant Mol. Biol. **40**, 347–369
24. Ayers, A.R., Ebel, J., Finelli, F., Berger, N. and Albersheim, P. (1976) Plant Physiol. **57**, 751–759
25. Hahn, M.G. and Albersheim, P. (1978) Plant Physiol. **62**, 107–111
26. Bartnicki-Garcia, S. (1968) Annu. Rev. Microbiol. **22**, 87–108
27. Ayers, A.R., Ebel, J., Valent, B. and Albersheim, P. (1976) Plant Physiol. **57**, 760–765
28. Sharp, J.K., Valent, B. and Albersheim, P. (1984) J. Biol. Chem. **259**, 11312–11320
29. Ayers, A.R., Valent, B., Ebel, J. and Albersheim, P. (1976) Plant Physiol. **57**, 766–774
30. Kopp, M., Rouster, J., Fritig, B., Darvill, A. and Albersheim, P. (1989) Plant Physiol. **90**, 208–216
31. Sharp, J.K., McNeil, M. and Albersheim, P. (1984) J. Biol. Chem. **259**, 11321–11336
32. Sharp, J.K., Albersheim, P., Ossowski, P., Pilotti, Å., Garegg, P.J. and Lindberg, B. (1984) J. Biol. Chem. **259**, 11341–11345
33. Ossowski, P., Pilotti, Å., Garegg, P.J. and Lindberg, B. (1984) J. Biol. Chem. **259**, 11337–11340
34. Fügedi, P., Birberg, W., Garegg, P.J. and Pilotti, Å. (1987) Carbohydr. Res. **164**, 297–312
35. Lorentzen, J.P., Helpap, B. and Lockhoff, O. (1991) Angew. Chem. **103**, 1731–1732
36. Fügedi, P., Garegg, P.J., Kvarnström, I. and Svansson, L. (1988) J. Carbohydr. Chem. **7**, 389–397
37. Birberg, W., Fügedi, P., Garegg, P.J. and Pilotti, Å. (1989) J. Carbohydr. Chem. **8**, 47–57
38. Hong, N. and Ogawa, T. (1990) Tetrahedron Lett. **31**, 3179–3182
39. Cheong, J.-J., Birberg, W., Fügedi, P., Pilotti, Å., Garegg, P.J., Hong, N., Ogawa, T. and Hahn, M.G. (1991) Plant Cell **3**, 127–136
40. Cheong, J.-J., Alba, R., Côté, F., Enkerli, J. and Hahn, M.G. (1993) Plant Physiol. **103**, 1173–1182
41. Peters, B.M., Cribbs, D.H. and Stelzig, D.A. (1978) Science **201**, 364–365
42. Yoshikawa, M., Keen, N.T. and Wang, M.-C. (1983) Plant Physiol. **73**, 497–506

43. Schmidt, W.E. and Ebel, J. (1987) Proc. Natl. Acad. Sci. U.S.A. **84**, 4117–4121
44. Cosio, E.G., Pöpperl, H., Schmidt, W.E. and Ebel, J. (1988) Eur. J. Biochem. **175**, 309–315
45. Cosio, E.G., Frey, T., Verduyn, R., Van Boom, J. and Ebel, J. (1990) FEBS Lett. **271**, 223–226
46. Cheong, J.-J. and Hahn, M.G. (1991) Plant Cell **3**, 137–147
47. Cosio, E.G., Frey, T. and Ebel, J. (1990) FEBS Lett. **264**, 235–238
48. Cheong, J.-J. and Hahn, M.G. (1991) Plant Physiol. **96**, 70
49. Cosio, E.G., Frey, T. and Ebel, J. (1992) Eur. J. Biochem. **204**, 1115–1123
50. Frey, T., Cosio, E.G. and Ebel, J. (1993) Phytochemistry **32**, 543–550
51. Albersheim, P., Darvill, A.G., McNeil, M., Valent, B.S., Sharp, J.K., Nothnagel, E.A., Davis, K.R., Yamazaki, N., Gollin, D.J., York, W.S., *et al.* (1983) in Structure and Function of Plant Genomes (Ciferri, O. and Dure, L., III, eds.), pp. 293–312, Plenum, New York
52. Darvill, A., Augur, C., Bergmann, C., Carlson, R.W., Cheong, J.-J., Eberhard, S., Hahn, M.G., Lo, V.-M., *et al.* (1992) Glycobiology **2**, 181–198
53. Dénarié, J. and Roche, P. (1992) in Molecular signals in Plant–Microbe Communications (Verma, D.P.S., ed.), pp. 295–324, CRC Press, Boca Raton.
54. Mohnen, D. and Hahn, M.G. (1993) Semin. Cell Biol. **4**, 93–102

Biochem. Soc. Symp. 60, 113–130
Printed in Great Britain

Cytosolic protons as secondary messengers in elicitor-induced defence responses

Y. Mathieu*, J.-P. Jouanneau†, S. Thomine, D. Lapous and J. Guern

C.N.R.S. Institut des Sciences Végétales, 22 Avenue de la Terrasse, F-91198 Gif-sur-Yvette Cedex, France

Synopsis

A variety of early elicitor-induced membrane responses have been described, and their possible role in the generation of second messengers involved in the cascades of events leading to the activation of defence genes is actively investigated. Treatment of tobacco cells with a crude elicitor preparation from *Phytophthora megasperma*, purified oligouronides and a commercial pectate lyase, induce a common set of membrane reactions similar to those described in a variety of plant material, i.e. efflux of K^+, extracellular alkalinization, net Ca^{2+} uptake and membrane depolarization. In the same conditions the three elicitors stimulate the activity of phenylalanine ammonia-lyase (PAL) and O-diphenol methyltransferase (OMT), two enzymes of the phenylpropanoid pathway. A good correlation between the intensity of the membrane response and the extent of enzyme stimulation has been observed.

Cytosolic acidifications have also been measured as a rapid response to the different elicitor preparations used. These results show that plant cells (which usually succeed in counteracting pH-perturbing processes associated with their metabolism, with the transport of solutes or with the effect of various factors from the environment) display significant variation in the concentration of cytosolic protons in specific physiological circumstances, such as the perception of signals inducing defence reactions. Direct evidence that these cytosolic pH changes could be interpreted by plant cells as messages involved in triggering defence responses is provided by experiments showing that artificial acidifications of the cytoplasm lead to a co-ordinated stimulation of PAL and OMT. These results stress the need to explore in more detail the role played by cytoplasmic mechanisms underlying those pH changes.

*To whom correspondence should be addressed.
†Deceased 1993.

Introduction

In simplified biological systems, such as plant organs or cell cultures, a variety of effectors elicit reactions usually displayed by plants in response to pathogen injury. The large spectrum of active biotic or abiotic elicitors, as well as the wide array of induced defence reactions, are now well-documented (see [1, 2] and references therein).

Chemically unrelated elicitors, most probably recognized by distinct specific receptors, induce a common set of defence reactions. This suggests that the signalling pathways initially generated by these effectors probably converge at one (or several) common step(s) of transduction, leading to the activation of a common set of defence genes.

In fact, little is known about the transduction cascades initiated by elicitors [2,3], and even less with regard to the existence of possible cross-points between different elicitor-specific transduction pathways.

In several instances, early elicitor-induced modifications of membrane properties, such as changes in membrane polarization [4–6], K^+ efflux [6–8], changes in extra- and/or intracellular pH [6–9] and net Ca^{2+} uptake [6,8,10,11] have been reported and shown to precede the expression of defence reactions. These reactions, or at least some of them, can be induced by chemically unrelated molecules, such as glycoproteins, oligosaccharides, glycosidases, Hg^{2+} salts, vanadate, and so on.

The general objective of this work has been to determine whether a relationship exists between the rapid elicitor-induced syndrome of membrane responses and the regulation of defence reactions. We present here results of a study investigating the question of whether modifications of plasma membrane properties by elicitors might be a way to generate second messengers involved in the cascade of events leading to the activation of key enzyme activities of the phenylpropanoid pathway. Three elicitor preparations have been used: a crude preparation from *P. megasperma* culture medium (where the active components have been shown to be proteins), purified oligouronides obtained from Professor P. Albersheim, and a commercial pectate lyase. A special emphasis is put on the origin and possible roles of extra- and intracellular pH changes induced by elicitors.

Materials and methods

Tobacco cell suspension cultures

Tobacco cells (*Nicotiana tabacum* cv. *Xanthi*) were cultured as described previously [6], either in B5 Gamborg medium with 1 μM 2,4-dichlorophenoxyacetic acid (2,4-D) and 60 nM kinetin, or in a modified 'low K^+' medium containing the same hormone concentrations, but where 7 mM KNO_3, 2 mM NH_4NO_3 and 2 mM glutamine were substituted for the initial 25 mM KNO_3. Cells were used after 4–6 days of culture.

Origin of the elicitor signals used

A crude fungal elicitor (Pmg) was prepared from the culture medium of *P. megasperma Drechs* f.sp. *glycinea* race 1 (obtained from P. Ricci, INRA, Antibes, France) by concentration, extensive dialysis, centrifugation and lyophilization of the medium. The active components of this preparation were shown to be proteins or glycoproteins, as revealed by their sensitivity to proteases.

Oligogalacturonides (OGs) (obtained from Professor P. Albersheim, CCRC, Athens, GA, U.S.A) were used as a mixture of oligomers with degrees of polymerization (DP) of 6–15. Their activity as a function of DP has been described in [6].

The pectolyase preparation used as elicitor is the commercial pectolyase Y23 (Seishin, Tokyo, Japan), dissolved in water and adjusted to pH 5.6 before use.

Measurement of external and intracellular pH Cell suspensions (4 ml) dispensed in 20 ml glass vials, and weakly buffered by the addition of Mes-Tris buffer, pH 5.7, 1 mM final concentration, were equilibrated for 2 h in open vials with continuous agitation (100 r.p.m.) at 26°C. Their extracellular pH was measured using a glass combined pH electrode (Heito) immersed directly into the agitated cell suspensions.

The cytosolic pH was evaluated by ^{31}P-n.m.r. spectroscopy essentially as described previously [6,12]. Briefly, about 6 g (fresh wt.) of packed cells were continuously perfused by aerated culture medium at a flow rate of 8 ml min^{-1}. ^{31}P-n.m.r. spectra were obtained from a Brucker spectrometer operating in a pulsed Fourier transform mode at 161 MHz. Total acquisition time was 10 min, with a repetition time of 1.2 s. $^{2}H_2O$ (20%) was used to lock the field frequency, and methylene diphosphonate (100 mM), contained in a capillary tube, served as a reference. Cytosolic pH was determined by measuring the chemical shift of the glucose 6-phosphate ^{31}P peak, using a standard calibration curve.

Measurement of ion fluxes

Elicitor-induced net K^+ fluxes have been measured using K^+-specific electrodes (F2312K Radiometer, Copenhagen) as described in [6]. For that purpose, cells were either grown in the standard medium, washed and resuspended in a medium without K^+ [6], or cultivated in a medium with an initial concentration of K^+ limited to 7 mM where, after a 5 day culture period, the K^+ concentration was lowered to about 1.5 mM. Cell suspension samples were then equilibrated for 2 h with agitation, before treatment with elicitors. After elicitation, aliquots (100–200 μl) were withdrawn at intervals and rapidly filtered. K^+ concentrations were measured after dilution in a K^+-free buffer.

Net Ca^{2+} exchanges between the cells and their extracellular medium were measured essentially as described in [6]. Briefly, the external Ca^{2+} concentration was monitored on aliquot fractions of the incubation medium using the fluorescence of Fluo-3 as a probe. Fluorescence units were converted to Ca^{2+} concentration with a standard calibration curve. Appropriate dilutions were done to perform the Ca^{2+} determinations in the range 0.2–2 μM, where Fluo-3 fluorescence is a linear function of Ca^{2+} concentration.

Measurement of PAL and OMT enzyme activities

Cells were collected 16–18 h after elicitation, filtered, frozen and ground in liquid nitrogen and then extracted with a 0.1 mM phosphate buffer, pH 7.5, 15 mM β-mercaptoethanol (3 ml g^{-1} initial fresh wt.). Enzyme extracts were recovered by centrifugation (10 000 *g*, 15 min, 4°C) and used immediately to determine PAL and OMT activities, respectively, in 350 μl of a reaction mixture composed of 130 μM L-[U-^{14}C]-phenylalanine (117 KBq μmol^{-1}) (Amersham), 15 mM β-mercaptoethanol in 0.1 M borate buffer, pH 8.8, or 1 ml of 50 μM *S*-adenosyl-L-[methyl-^{3}H]-methionine (88 kBq μmol^{-1}) (Amersham), 3 mM caffeic acid and 15 mM β-mercaptoethanol in 0.1 M phosphate buffer, pH 7.5. Enzymic reactions, initiated by the addition of 75 μl of enzyme extract, were continued for 1 h at 37°C and stopped by the addition of

75 μl of 18M H_2SO_4. Reaction mixtures were then completed to 2 ml final volume with water and partitioned with 2 ml of 1:1 (v/v) cyclohexane–diethyl ether.

The radioactivity of the organic phase containing [^{14}C]cinnamic acid and [^{3}H]ferulic acid was measured by scintillation counting (Packard Tri-Carb 2200 CA). Enzyme activities have been expressed in nkat.kg^{-1} of cells (fresh wt.).

Results and discussion

Elicitor-induced changes in plasma membrane properties of tobacco cells

As shown in Fig. 1, tobacco cells respond to the three elicitor preparations by an alkalinization of their incubation medium, the intensity and the kinetics of these extracellular pH modifications differing according to the elicitor used. The extracellular alkalinization induced by the OGs is characterized by its transient character. A rapid elevation of medium pH is first observed, resulting in a maximum alkalinization after about 50 min. Thereafter, a progressive reacidification is observed until the extracellular pH returns to its initial value after about 150–200 min. Contrary to what was observed with OGs, the Pmg preparation induced a long-lasting extracellular alkalinization. The

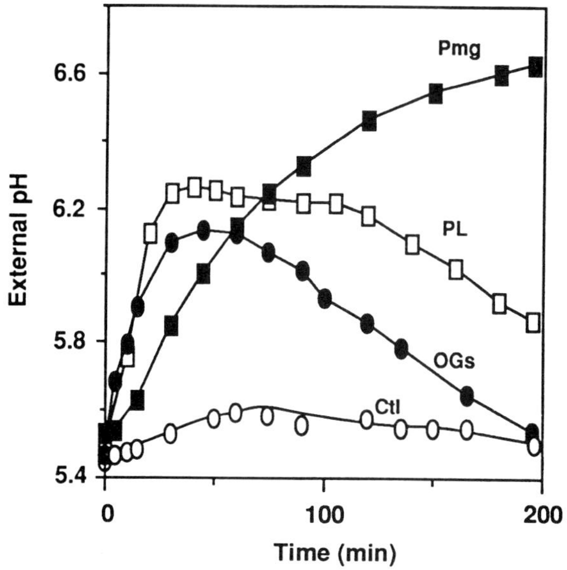

Fig. 1. Extracellular alkalinization induced by a *P. megasperma* elicitor preparation (Pmg) (■), by oligogalacturonides (OGs) (●) and by a commercial pectate lyase preparation (PL) (□). Cells (about 160 mg·ml^{-1}) were treated at zero time with Pmg, a semi-purified M7 *P. megasperma* preparation (1.5 μg·ml^{-1}); with PL, a commercial pectolyase from *Aspergillus japonicus* (1.5 μg·ml^{-1}); or OGs, a mixture of active oligogalacturonides of DP = 6–15 (15 μg·ml^{-1}). (○) represent the pH values for unelicited cell suspensions.

pectolyase effect was intermediate between the other two, exhibiting as for the OGs a transient behaviour but with a more sustained alkalinization that was only reversed after 300 min at a concentration of 1 $\mu g \cdot ml^{-1}$.

The effects of the effectors studied on changes in K^+ concentration of the incubation medium are illustrated in Fig. 2. Each elicitor induced a net efflux of K^+ with specific characteristics, in terms of intensity and kinetics of effects, quite similar to those exhibited by the extracellular alkalinization.

The elicitor activity of the three preparations studied was monitored by their ability to increase the activity of PAL and OMT. Fig. 3 presents the results of several independent experiments in which the efficiency of Pmg elicitors, pectate lyase and OGs in stimulating PAL and OMT activities was compared. The Pmg preparation was a strong inducer of both enzyme activities: stimulations from 2–3-fold at 1 $\mu g \cdot ml^{-1}$ to 3–4-fold at 3 $\mu g \cdot ml^{-1}$ being measured. Conversely, OGs only weakly stimulated both defence reactions, even at concentrations which saturate the induction of membrane responses (i.e. 15 $\mu g \cdot ml^{-1}$). In good agreement with previously reported results [7,13], the pectate lyase preparation was a good inducer of defence reactions, as revealed by the marked stimulation of the two enzymes. Interestingly, in each case, PAL and OMT

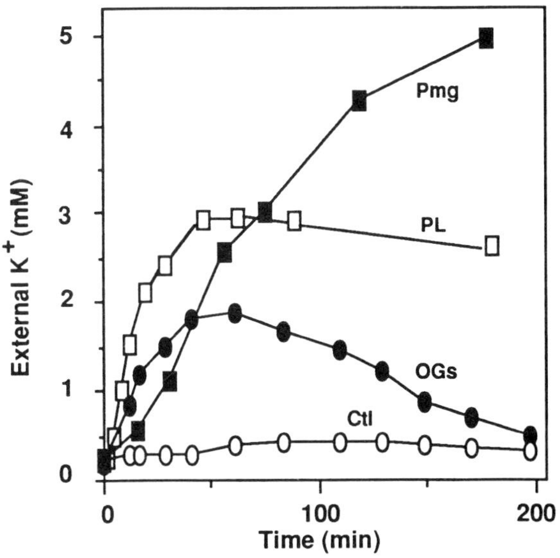

Fig. 2. Induction of K^+ efflux in tobacco cells by Pmg elicitors (■) (1.5 $\mu g \cdot ml^{-1}$ M7 fraction), pectate lyase (□) (1.5 $\mu g \cdot ml^{-1}$) and OGs (●)(DP = 6–15, 15 $\mu g \cdot ml^{-1}$). Cells (about 160 $mg \cdot ml^{-1}$) were first washed in a medium deprived of K^+ as described in [6] and re-equilibrated in the same medium for 2 h before adding the elicitors. The extracellular K^+ concentration was determined at intervals from aliquots of the incubation medium by using a K^+-specific electrode, as described in the Materials and methods section. (○) represent the extracellular K^+ concentration in unelicited cell suspensions.

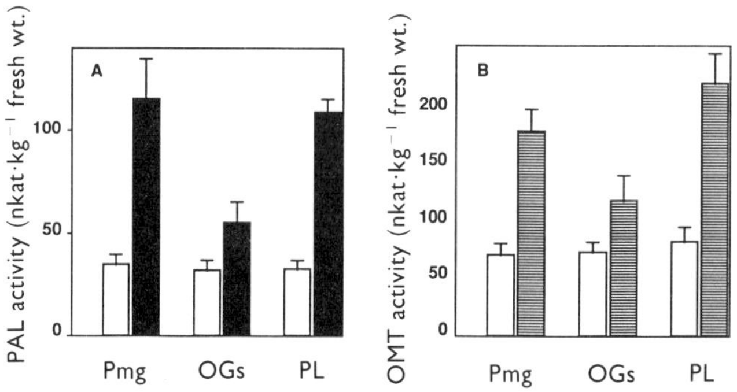

Fig. 3. Stimulation of PAL (A) and OMT (B) activities in tobacco cells treated by a crude *P. megasperma* elicitor preparation (Pmg, 1–3 μg·ml^{-1}), a commercial *Aspergillus japonicus* pectate lyase (PL, 1 μg·ml^{-1}) or purified oligouronides (OGs, DP = 6–15, 10 μg·ml^{-1}). Cells were extracted 18 h after elicitation and the PAL and OMT activities measured as described in the Materials and methods section. Results are expressed in nkat.kg^{-1} fresh wt. and S.E.M. indicated for n = 10, 11 and 11 respectively for Pmg, OG and PL elicitations. White bars correspond to unelicited cell suspensions.

activities increased to the same extent, indicating co-ordinated stimulations of different enzymes of the same pathway.

The fact that OGs which induced weak and transient membrane responses were poorly effective in stimulating PAL and OMT activities, whereas the two other elicitors which induced stronger and more sustained membrane responses were better stimulators of the phenylpropanoid metabolism, strongly suggested that the low efficiency of OG elicitors on PAL and OMT activities could be related to the transient character of their action on the membrane responses. In agreement with this idea, five successive OG additions (at the limiting concentration of 2 μg·ml^{-1}, which induced successive waves of extracellular alkalinization — maintaining the external pH increase for about 5 h — led to an 85 ± 27% (n = 4) increase in PAL activity, whereas a single addition of OG at this concentration stimulated PAL activity only by 15 ± 7% (n = 4).

Furthermore, when tobacco cells were treated with the Pmg preparation at different concentrations in the range of 0.1–2 μg·ml^{-1} a positive correlation was observed between the intensity of the moderate elicitor-induced K^+ efflux and the degree of stimulation of PAL activity. At higher concentrations, non-specific, large K^+ effluxes occur in association with cell death.

The results described above show that the extent of stimulation of PAL and OMT appears to be correlated with both the amplitude and the duration of the activation of early membrane responses. They support the hypothesis that the elicitor-induced changes in plasma membrane properties could be a way to generate intermediate cellular messages involved in the control of more integrated, long-term defence responses [6,8].

Putative second messengers generated at the plasma membrane of elicitor-treated cells

In fact, the elicitor-induced membrane responses are more diverse than described above. Aside from the K^+ efflux and external pH changes, tobacco cells also responded to the addition of OGs by transient membrane depolarization, cytosolic acidification and net influx of Ca^{2+} ions (data not shown, detailed in [6]) and to Pmg by cytosolic acidification and net influx of Ca^{2+}. Both OGs and Pmg elicitors thus induced common membrane responses which were similar to those observed in parsley cells treated with different elicitors [8]. As described by Scheel and Parker [3], cells from species as diverse as soybean, parsley or potato recognize different components, either proteins, carbohydrates or lipids, from crude elicitor preparations from *P. megasperma*. The authors thus favour the idea of different target sites for elicitor recognition on the cell surface. However, the elicitor-recognition step appears immediately, followed by a rather common set of plasma membrane responses which differ, according to the elicitor used, in the intensity and kinetics of the ionic and electrical changes induced. These modified plasma membrane properties, as well as the induction of oxidative reactions (for example, see [14]), represent a source of information for the cell to trigger long-term reactions.

With respect to the possible role of the ionic and electrical changes in the activation of defence reactions, it has been suggested that changes in cytosolic Ca^{2+} [6,8,15–17], extracellular pH [18,19], intracellular pH [6,20], membrane potential [4–6,21] or intracellular phosphate concentration [22] could act as secondary signals transmitted to the plant genome.

The idea that Ca^{2+} could be involved in the transduction of elicitor signals has been favoured by several groups and evidence has been obtained to support this idea. The absence of extracellular Ca^{2+} significantly reduces the response to elicitors of carrot [15], parsley (see [8] and references therein), and soybean cells [16,17]. On the contrary, the Ca^{2+} ionophore A23187 has been shown to stimulate phytoalexin accumulation in soybean [16] and carrot cells [16], but not in parsley [3]. Callose synthesis in *Glycine max* (soybean) and *Catharanthus roseus* cells is slightly (only 10% of the stimulation induced by chitosan) promoted by A23187 and ionomycin [10]. Furthermore, several elicitors have been shown to induce a stimulation of external Ca^{2+} uptake by the treated cells. This is the case for *G. max* and *C. roseus* cells induced by digitonin [10], for tobacco cells elicited by *Pseudomonas syringae* pv. *syringae* [23], for parsley cells treated with Pmg proteins [8] and for tobacco cells treated with oligouronides [6]. As shown in Fig. 4, Pmg and OGs induced decreases in the external Ca^{2+} concentration with relative intensities and kinetics in good agreement with those of the other membrane responses observed.

In fact, the intensity of the net Ca^{2+} uptake is quite high, far above what could represent simple cytosolic increases (a rise in the cytosolic Ca^{2+} concentration from 0.1 μM to 1 μM would only induce a 0.01 μM change in external concentration in the experimental conditions used). In fact, the net Ca^{2+} uptake induced in 1 h by OGs and Pmg corresponds to potential rises in intracellular Ca^{2+} concentration of 50 μM and 170 μM, respectively (Fig. 4). This suggests that Ca^{2+} ions taken up at the plasma membrane are removed from the cytosol by being transported into the vacuole and/or trapped by cytosolic calcium buffers. Thus these results, as well as those obtained with equivalent techniques [8,23], only provide indirect evidence in favour of elicitor-

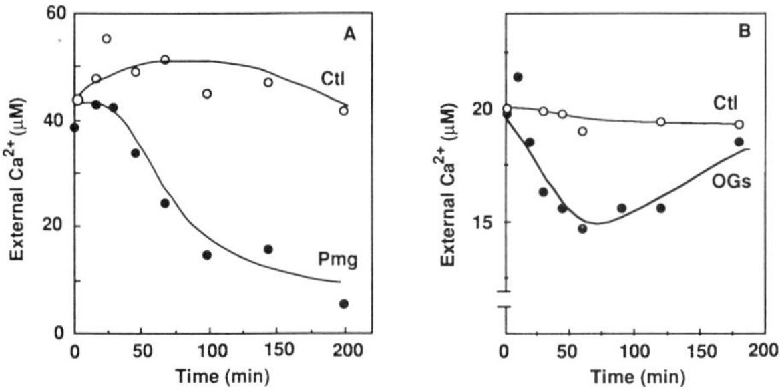

Fig. 4. Net Ca^{2+} uptake by tobacco cells treated with a *P. megasperma* elicitor preparation (A) or with purified OGs (B). Cells (about 200 μg·ml^{-1} fresh wt.) were washed and resuspended (as described in [6]) in a modified incubation medium where the Ca^{2+}concentration was lowered to 10–40 μM and the $MgCl_2$ concentration set to 1 mM. At zero time they were treated with either 20 μg·ml^{-1} OGs (DP = 6–15) or with the Pmg preparation (15 μg·ml^{-1}). Aliquots of the incubation medium of treated and control suspensions were collected at intervals and their Ca^{2+} content measured as described in the Materials and methods section.

induced cytosolic Ca^{2+} increases, and call for a direct monitoring of the cytosolic Ca^{2+}. The first evidence along this line has been provided recently for tobacco plants treated with a yeast elicitor preparation [24].

Membrane depolarization which results from the activity of different elicitors [4–6] has also been proposed as a message used by plant cells to elicit defence reactions. Low and Heinstein [21] suggested as early as 1986 that changes in membrane potential could be involved in the transduction of elicitor signals produced by *Verticillium dahliae* in a variety of cells. Oligouronides rapidly depolarize tomato leaf cells [5] or suspension-cultured tobacco cells [6]. The recent demonstration that action potentials are transmitted over a distance in tomato plants and interpreted as messages to trigger the induction of defence genes is quite interesting [25].

The depolarizing effect of vanadate, through the inhibition of the plasma membrane ATPase coupled to its ability to induce defence reactions in several materials, as revealed by PAL stimulation [18,19,26], has been taken as an argument suggesting that changes in membrane potential could be intermediates in converting an extracellular signal into an intracellular messenger [26]. We think, in fact, that the situation is more complex. We have, for example, checked that in our material vanadate is also able to induce PAL and OMT stimulations (Fig. 5D) in a dose-dependent manner. Increases in PAL activity of 100 ± 17% (n = 4) over the control have been measured at 30 μM, higher concentrations being clearly supra-optimal. This eliciting activity cannot be unambiguously attributed to changes in membrane potential, as vanadate induces the whole set of membrane responses described for the other elicitors, i.e. K^+ efflux and extracellular alkalinization (Fig. 5A), membrane depolarization (Fig. 5B), net Ca^{2+} influx (Fig. 5C) and cytosolic acidification (data not shown). As discussed above, any of these modifications could be at the origin of a second message interpreted by the cells to

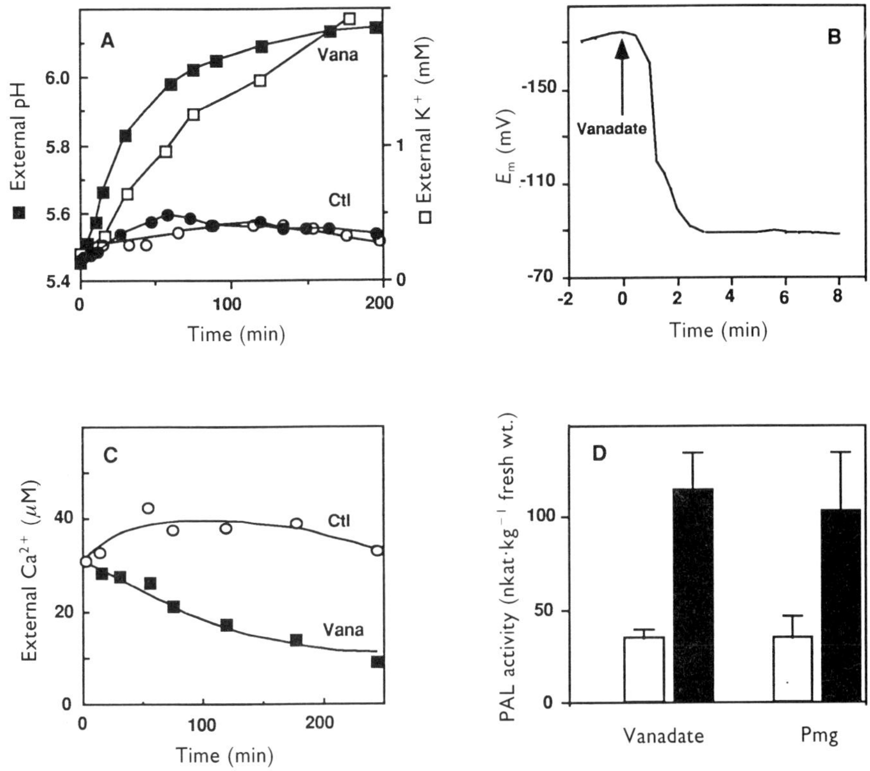

Fig. 5. Vanadate-induced extracellular alkalinization (A), K^+ efflux (A), membrane depolarization (B), net Ca^{2+} uptake (C) and PAL stimulation (D). (A) Tobacco cells (about 160 mg·ml^{-1} fresh wt.) have been treated with 100 μM vanadate and the pH (■) and K^+ content (□) of the extracellular medium was monitored as described in the Materials and methods section in comparison with control suspensions (● and ○). (B) Changes in membrane potential of tobacco cells treated with 200 μM vanadate (↑). The electrophysiological measurements have been performed essentially as described in [6] in a perfusion medium composed of 0.5 mM Mes, 0.5 mM $CaSO_4$, 10 mM saccharose, pH 5.8, with about 0.2 mM K^+. The trace has been chosen from nine experiments. (C) Changes in extracellular concentration of Ca^{2+} in cells treated at zero time with 200 μM vanadate (■) compared with control cells (○). The Ca^{2+} content of aliquot fractions from the incubation medium was monitored at intervals as described in the legend of Fig. 4. (D) Vanadate-induced stimulation of PAL activity. Tobacco cells were treated at zero time with either 10 μM vanadate or 1.5 μg.ml^{-1} Pmg, and their PAL activity determined after 18 h. Essentially similar results were observed for OMT activity (stimulations over controls of 135% for vanadate and 212% for Pmg). Results are means ± S.E.M. (n = 4). Abbreviations used: Ctl, control; Vana, vanadate.

trigger the activation of the phenylpropanoid pathway. Furthermore, the supra-optimal effects observed most probably come from the multi-targeted action of vanadate which, besides its effect on the proton pump ATPase, has been reported to inhibit — at rather high concentrations — acid phosphatases and protein kinases [27,28].

Variations in cytosolic pH (pHc) probably represent second messengers in the transduction of elicitors. This is supported by the observation that various elicitors have been shown to induce a rapid cytosolic acidification [6,20,29]. Moreover, several groups have demonstrated that the inhibition of the proton pump by vanadate or the nigericin-induced H^+/K^+ exchange, both putative inducers of cytoplasmic acidification, led to the activation of defence reactions [18,19,26,30].

We have shown that the three types of elicitor used in this study induce a marked cytosolic acidification (Fig. 6). Changes in pHc were measured by ^{31}P-n.m.r. applied to analysis of the chemical shift of ^{31}P in cytosolic glucose 6-phosphate. Changes in the extracellular pH were monitored in order to check that the cells kept their reactivity to elicitors in the conditions used to measure the intracellular pH and exhibited a typical elicitor-dependent alkalinization of the perfusion medium whose kinetics and amplitude were similar to those observed in elicitor-treated cell suspensions.

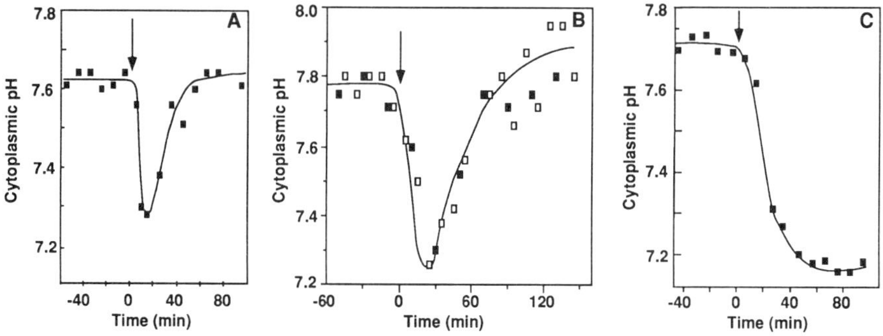

Fig. 6. Cytosolic acidification induced in tobacco cells by OGs (A), pectate lyase from *Aspergillus japonicus* (B) and crude elicitors from *P. megasperma* (C). Tobacco cells, about 6 g fresh wt., were packed in an n.m.r. tube (20 mm diam.) and perfused with a well-aerated incubation medium at a rate of 8 ml min^{-1}. ^{31}P-n.m.r. spectra were recorded by cumulating 512 scans of 1.2 s each. pHc was determined from the chemical shift corresponding to the glucose 6-phosphate peak, using a calibration curve as described previously. When necessary, spectra corresponding to the sum of two successive 512 scan cumulations were used. After equilibration in the n.m.r. tube, cells were treated with (A) 25 μg·ml^{-1} OGs (DP = 6–15); (B) 1.5 μg·ml^{-1} of pectolyase (Pectolyase Y23 from Seishin Pharmaceutical Co., Japan); or (C) with a crude *P. megasperma* elicitor preparation (7.5 μg·ml^{-1}). Changes in pH of the perfusion medium have been recorded in each case to monitor the elicitor-induced extracellular alkalinization to ensure that, in the conditions used for measuring pHc changes, cells were really reacting to the elicitors studied.

Fig. 6A shows that the treatment of a tobacco cell suspension with OGs induced a rapid but transient acidification of the cytoplasm. In the same conditions, a transient efflux of K^+ and an extracellular alkalinization were also observed (data not shown). The *P. megasperma* preparation also induced a strong acidification of the cytosol which, in contrast to that obtained with the OGs, was sustained for hours (Fig. 6C). When cells were treated with the pectate lyase preparation, a marked transient cytoplasmic acidification was also observed (Fig. 6B). Interestingly, vanadate, which (as shown in Fig. 4) was a potent stimulator of the phenylpropanoid pathway in tobacco cells, shared with the other three elicitors studied the ability to induce a significant (0.2–0.25 pH units at 150 μM) acidification of the cytosol (data not shown).

In contrast with these results and those reported for suspension-cultured bean and parsley cells [20,29], no pHc changes were detected in soybean cells treated with either polygalacturonic acids or with an elicitor preparations from *V. dahliae* [31]. These authors proposed that this discrepancy could result either from the material used or from changes in intracellular pH that are too small to be detected by the ^{31}P-n.m.r. technique. Another explanation to be considered is that the environmental conditions imposed on cells by the n.m.r. technique possibly interfered with the responsiveness of cells to elicitors. As a matter of fact, the cells, processed in the conditions described by the authors, behaved in a rather peculiar way — their intracellular pH being quite sensitive to changes in the external pH, contrary to what has been classically described [32] for cells in good energy status. This calls for the necessity to monitor, independently of the pHc changes, an elicitor-dependent parameter (external pH or K^+) to control whether the cells do respond to elicitors under the conditions used for the n.m.r. measurements.

Other proposals have been made concerning the role of elicitor-induced pH changes in the cascades of transduction of elicitor signals. For example, Ojalvo *et al.* [20] have reported changes in pHc, vacuolar pH, ATP and cytoplasmic P_i levels induced by a glucan elicitor in *Phaseolus vulgaris* cells. Only the vacuolar pH change was specific for the glucan elicitor, the other three changes being also triggered by decreases in cell aeration. This was considered as an argument for the vacuolar pH acting as a specific signal for the induction of phytoalexin formation. On the other hand, Haggendoorn *et al.* [30] have shown that *Petunia hybrida* cells react to various ionophores modifying the activity of the plasma membrane ATPase by a stimulation of the phenylpropanoid pathway, as revealed by an increase in PAL activity. Their idea is that alterations in the transmembrane ΔpH represent the parameter sensed by the cells as a key step in signal transduction to increase PAL activity.

The idea that increases in extracellular pH could also be used as messages by plant cells comes from experiments showing that simply increasing extracellular pH stimulates PAL activity in *Vigna angularis* cells [18] and stimulates PAL activity and lignin production in *P. hybrida* cells [19]

This analysis shows that a strong correlation emerges between the induction of extra- and intracellular pH changes and the activation of the cell responses. However, from the data reported, no clear decision can be made as to the localization of the H^+ concentration (external pH, pHc, transmembrane ΔpH) sensed by the cells and used to trigger elicitor-induced defence reactions. More direct approaches are thus needed.

Activation of the phenylpropanoid pathway, independently of elicitor treatment, by articial acidification of the cytoplasm

In animal systems, the role of pHc variations as second messengers in the action of hormones or mitogenic factors has been rather extensively explored. It is generally admitted that these signals raise intracellular pH by stimulating the Na^{+}/H^{+} antiporter (for example, see [33,34]).

The question of intracellular pH changes acting as second messengers mediating the action of hormones or related signals in plants is not really well-clarified [35–37], despite extensive efforts devoted to the hormone auxin and to the toxin fusicoccin.

The possibility for elicitors of defence reactions to use intracellular pH variations as second messengers is being investigated at present, with the aim of bringing more direct evidence for the role of cytosolic acidification in elicitor signalling.

The strategy used has been to study the consequences of a cell acidification in the absence of elicitor treatment. A classic procedure widely used to study the mechanisms involved in intracellular pH regulation in animal cells is to perturbate intracellular pH — by submitting cells or organs to an acid load — and look for the short-term reactions of the cells to the so-induced acidification (for a review, see [38]). This approach has been used more recently in plant cells, where it gave some interesting information (for reviews, see [36,37]). Briefly, the technique consists of using a weak lipophilic acid, essentially permeating the membranes in its protonated form. The acid dissociation in the different cell compartments induces a release of protons and a corresponding acidification.

The reaction of plant cells to such an acid load has been described in several previously published papers [12,36,37]. In short, the injection of propionic acid (PA) in cell suspensions induces an immediate drop in the pHc. The strong initial pHc decrease is followed by a partial recovery, elevating pHc to a new sub-stable pHc value, lower than the initial pH, after about 30 min. This demonstrates the operation of strong mechanisms able to compensate for the proton load brought about by the entry of PA. The intensity of the pHc drop is a direct function of the concentration of the protonated form of the acid used (i.e. a function of the external pH and of the total concentration of acid).

The effects of moderate concentrations of PA, supplied to tobacco cell suspensions at external pH 5.6, on pHc and PAL stimulation are reported in Fig. 7. Fig. 7(A) shows that the PA treatment induced a dose-dependent stimulation of PAL and OMT with an optimum around 1 mM (stimulation $141 \pm 15\%$, $n = 10$), whereas increasing PA concentration to 5 mM led to a supra-optimal effect ($79 \pm 11\%$, $n = 10$). As observed for cells treated with the biotic elicitors used, OMT and PAL activities were stimulated to the same extent by the acid load, indicating a co-ordinated effect on these enzymes. In the same range of concentration, PA induced a pHc decrease, associated with an external pH increase as shown in Fig. 7(B). We also checked that acetic acid was able to induce the same PAL and OMT activations (data not shown).

These results clearly demonstrate that acid-loading tobacco cells stimulates the phenylpropanoid pathway. An ambiguity was left by this first type of experiment, as the treatment modified simultaneously pHc, external pH and transmembrane ΔpH. The external alkalinization being due to the uptake of the protonated form of PA, the same type of experiment was conducted in a highly buffered medium (50 mM Tris-Mes), where the external alkalinization was suppressed. In these conditions, 1 mM PA induced

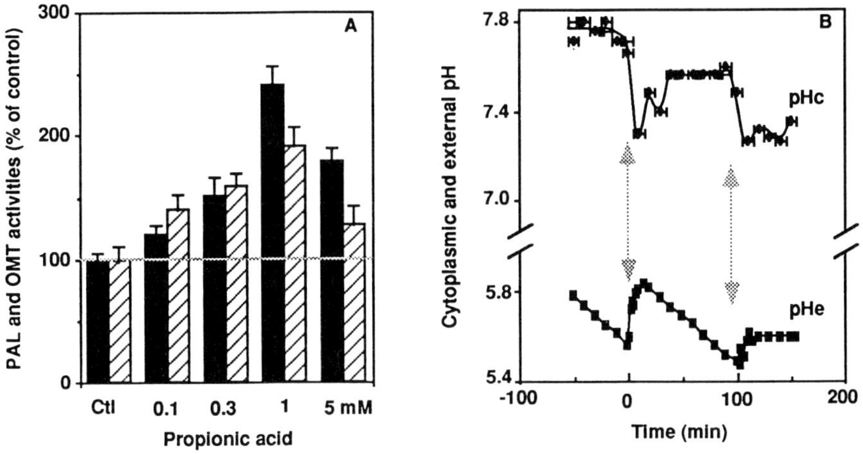

Fig. 7. PAL and OMT activations (A) and cytoplasmic and extracellular pH changes (B) induced by acid-loading tobacco cells with PA. (A) PAL (black bars) and OMT (hatched bars) activities of tobacco cells treated with different concentrations of PA at pH 5.6 and harvested after 18 h. The enzyme activities are expressed as a percentage of controls. Results are means ± S.E.M. with *n* = 3, 3, 16 and 10 for 0.1, 0.3, 1 and 5 mM PA, respectively. Control values for PAL and OMT are, respectively, 112 ± 9 nmol·h^{-1}·g^{-1} fresh wt. (*n* = 16) and 181 ± 21 nmol·h^{-1}·g^{-1} fresh wt. (*n* = 12). (B) Tobacco cells (about 6 g fresh wt.) were processed (as described in the Materials and methods section and in the legend of Fig. 6) to monitor both pHc with the ^{31}P-n.m.r. technique and the extracellular pH, with a pH electrode in the reservoir of the continuously circulating perfusion medium. At times indicated by arrows, PA (as sodium propionate, pH 5.6) was injected to give a final concentration of 3 mM (first arrow) or 6 mM (second arrow).

a 113 ± 27% (*n* = 3) stimulation of PAL over the control. This approach was also extended to the effects of pectate lyase on PAL and OMT activations. In 1 mM Tris-Mes, pH 5.4, pectate lyase (1 μg·ml^{-1}) induced a 0.9 unit increase in external pH in 60 min and a PAL stimulation of 130 ± 6% (*n* = 2), whereas in 50 mM Tris-Mes, a PAL stimulation of 182 ± 24% (*n* = 2) was induced without external pH change. These results strongly suggest that pHc modifications, and not changes in external pH, are the messengers used by the cell to trigger the activation of defence reactions.

The response to the acid load can be markedly amplified when a stronger proton load is given for only 2 h, cells being then washed out of the PA and reincubated on a normal medium. In these conditions, PAL activities as high as 438 ± 21 μmol h^{-1}·g^{-1} fresh wt. (*n* = 6) could be recorded for 5 mM PA corresponding to a 302 ± 14% stimulation over the control (109 ± 4 μmol h^{-1}·g^{-1} fresh wt.; *n* = 6). In the same conditions, continuous treatments with 1 or 5 mM PA gave stimulations of 438 ± 5% and 98 ± 5%, respectively. The high level of PAL activation reached almost the level triggered in the same conditions by a pectate lyase treatment (484 ± 22 μmol h^{-1}·g^{-1}

fresh wt.; *n*@ = @6). This shows that the intensity and the duration of the cytosolic acidification are very important factors in the induction of defence reactions.

These results raise the question of the mechanisms by which changes in the cytosolic concentration of protons can initiate a chain of events leading to a modulation of the expression of defence genes.

Most proteins react to pH changes, but this general effect can be specifically intense for some of them in the range of the usual pHc variations. Among the key enzymes which could react to these pH shifts are those involved in protein phosphorylation–dephosphorylation. As a matter of fact, aside from the well-known regulatory effect of Ca^{2+}, pH-induced changes in protein phosphorylation have been described in animal systems (for example, see [39,40]). In plants, a pH-dependent phosphorylation of a 33 kDa protein in *Acer pseudoplatanus* cells *in vivo* [41] has also been reported as stimulated by agents alkalinizing the cytoplasm and inhibited by decreases in the pHc.

Several recent studies have shown that fungal elicitors can trigger rapid and specific protein phosphorylation [42], and that the early membrane responses, as well as the expression of defence genes induced by elicitors, can be rapidly inhibited by staurosporine and related inhibitors of protein kinases [11,43,44]. Our working hypothesis is that elicitor-induced modulations of protein phosphorylation might result from the interplay between simultaneously occurring Ca^{2+} and H^+ changes in the cytosol. Such associated changes have already been described in animal and plant systems [45–47].

Possible mechanisms of elicitor-triggered extra- and intracellular pH changes

The evidence, described above, strongly suggests that cytosolic protons, in association with other cellular messengers, could be involved in the signalling cascades initiated by the perception of elicitors at the plasma membrane. Understanding the mechanisms by which pHc changes could result from elicitor recognition thus represents an important step in the understanding of their mechanism of action.

In all cases investigated so far, the elicitor-induced K^+ efflux has been found to be associated with an extracellular alkalinization, indicating that opposite fluxes of K^+ and protons, or 'proton equivalents', are somehow coupled. The stoichiometry between K^+ and proton fluxes in tobacco cells reacting to *P. syringae* has been estimated to be approximately 1 H^+:1 K^+ [48]. We have shown that in tobacco cells treated with either oligouronides or the commercial pectate lyase, a 1 H^+:1 K^+ ratio is observed during the first 20 min of elicitor treament, with a slight decrease afterwards, leading to an approximate ratio of 0.8 H^+:1 K^+ at 60 min (data not shown). Much lower and more complex stoichiometries (0.17 H^+:1 K^+) have, however, been described for parsley cells treated with a glycoprotein elicitor [8].

Various hypotheses have been proposed to account for the extracellular alkalinization and its coupling with the K^+ efflux. A survey of several possible mechanisms is illustrated in Fig. 8. A K^+/H^+ antiport could catalyse coupled K^+ and H^+ fluxes with a stoichiometry of 1. Physiological evidence in favour of the presence of such an antiport at the plasma membrane of rape seed has been reported recently [49] but nothing is known of its regulation.

Electrical coupling between K^+ channels (which could open as the result of elicitor-induced membrane depolarization and mediate the efflux of K^+) and proton

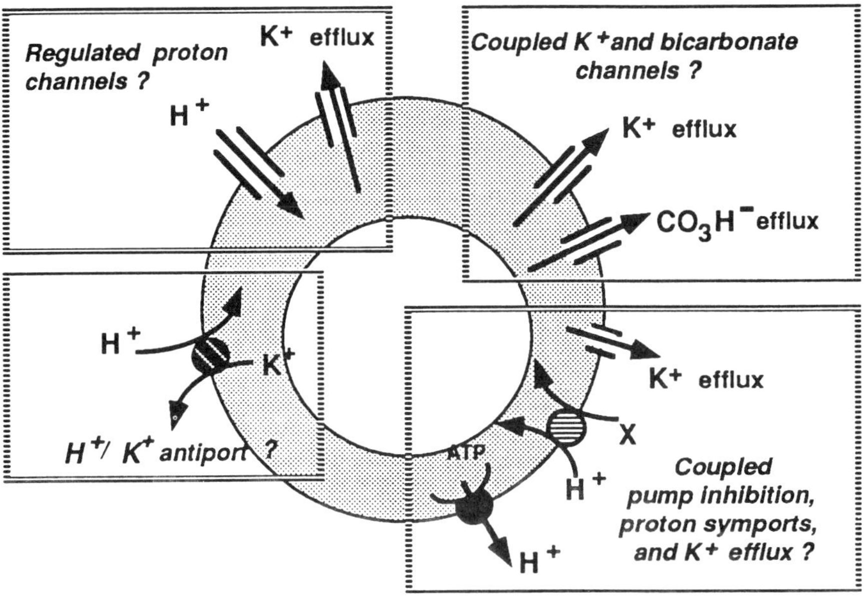

Fig. 8. Possible mechanisms of coupling between elicitor-induced K^+ efflux and proton, or proton equivalents, influx. (Bottom left) H^+/K^+ antiport; (top left) electrically coupled K^+ and H^+ channels; (bottom right) coupled operation of the proton pump ATPase, H^+/solute co-transporters and K^+ channels; (top right) electrically coupled K^+ and bicarbonate channels.

channels responsible for the entry of H^+ would also account for the observed fluxes of equivalent intensities. However, owing to the large difference in K^+ and H^+ concentrations, the putative proton channels should have a conductance for protons at least 10^4-fold higher than that of K^+ channels for K^+. No evidence has been provided for such proton channels at the plasma membrane.

A more complex coupling, between the proton pump ATPase, proton short-circuits corresponding to H^+/solute co-transport and K^+ channels, would also account for associated proton influx and K^+ efflux resulting from the inhibition of the pump. It is, however, difficult in such a case to predict what could be the final stoichiometry and the kinetics of the coupled fluxes.

Finally, one should stress the fact that the extracellular alkalinization and the cytosolic acidification observed do not automatically mean that protons are exchanged across the plasma membrane. An elicitor-induced efflux of bicarbonate ions associated with the K^+ release would leave an excess of protons in the cytosol and would trap protons in the external medium to form carbonic acid, which is then released in the atmosphere as CO_2. Thus an electrical coupling between K^+ and bicarbonate channels, both activated by elicitor-induced membrane depolarization, would account for the pH changes observed and their quantitative association with the K^+ efflux. Murphy and co-workers [50,51] have shown that the u.v. irradiation of rose cells in suspension culture resulted in a K^+ efflux matched by an equivalent increase in the amount of extracellular bicarbonate, supporting their hypothesis that CO_3H^- transport would serve to balance the charge corresponding to the release of K^+. We also measured recently an elicitor-induced increase in the bicarbonate concentration in the culture medium of tobacco cells treated with OGs or pectate lyase. The extracellular CO_3H^- accumulation is of about the same magnitude as the extracellular K^+ increase (data not shown). Further experiments are, however, necessary to decide whether this bicarbonate change simply results from an enhanced trapping of CO_3H^- ions owing to the medium alkalinization, or whether it corresponds to transport of bicarbonate across the plasma membrane.

This review shows that elucidating the mechanisms by which elicitors induce large fluxes of protons, or 'proton equivalents', remains an important goal of the studies on plant cell signalling. The classic idea that they originate from the inhibition of the proton pump ATPase, resulting in an intracellular accumulation of H^+, has to be considered with caution because, as we have stressed before [37], pH regulation cannot be simply discussed in terms of primary H^+-pumping ATPases. The transmembrane proton fluxes driven by the primary proton pumps should be associated with fluxes of strong cations in the opposite direction, or of strong anions in the same direction, to induce significant acidifications and alkalinizations on the respective sides of the membrane. Thus modifications in the activity of protons in any cell compartment result from changes in the concentrations of strong ions (K^+, Cl^-, NO_3^- and so on) or ionized species of weak acids and bases. A special interest should be devoted now to the role played by CO_2–HCO_3^- exchanges which could be involved in the responses to elicitors, either through a specific elicitor-induced bicarbonate efflux or, more generally, through the role of CO_2 and bicarbonate in the regulation of intra- and extracellular pH [36,37].

Conclusions

How extracellular signals are converted to intracellular signals, which are then transmitted to the nucleus, is not really well-understood in plants. This problem is particularly critical for elicitor signals, for which evidence has accumulated that the recognition of the primary inducing signal probably occurs at the cell surface — inducing a subsequent cascade of events transducing the signal to the nucleus with the consequent activation of defence genes.

A variety of early elicitor-induced membrane responses have been described in different plant material. In fact, a rather common set of reactions appears to be induced by totally different elicitor molecules. These common reactions probably constitute cross-points in the action of different effectors, from where common second messengers involved in the cascades of events leading to the activation of defence genes are generated. We have shown here that the increase in the concentration of cytosolic protons is most likely to be one of the second messengers rapidly generated in elicitor-treated cells, plant cells extracting information from these signal-induced pH transients. Interestingly, the efficiency of such messengers, in eliciting the co-ordinated stimulation of PAL and OMT, is a function of the intensity and duration of the H^+ messengers, as revealed by the effects of repeated additions of OGs and of transient exposures to PA loads.

Cytosolic protons probably represent only one component of the network of information sensed by cells responding to elicitors, this component interacting with other types of second messenger. Cytoplasmic free protons exhibit many functional similarities with cytoplasmic free Ca^{2+} ions in terms of mean concentration, range of variation, ability to affect structure and activity of proteins. Ca^{2+} and pH signals often appear to be associated with responses of animal cells to hormones, and a few reports in plants demonstrate associated changes in H^+ and Ca^{2+} concentrations. These considerations stress the interest to explore in more detail the role played by cytoplasmic free protons in plant cell signalling and how they interact with the cytosolic Ca^{2+} messengers. It will be particularly important now to determine whether or not other defence reactions, independent of the phenylpropanoid pathway, are also controlled by changes in pHc.

References

1. Dixon, R.A. and Lamb, C.J. (1990) Annu. Rev. Plant Physiol. Plant Mol. Biol. **41**, 339–367
2. Lamb, C.J., Lawton, M.A., Dron, M. and Dixon, R.A. (1989) Cell **56**, 215–224
3. Scheel, D. and Parker, J.E. (1990) Z. Naturforsch **45c**, 569–575
4. Pélissier, B., Thibaud, J.B., Grignon, C. and Esquerré-Tugayé, M.T. (1986) Plant Sci. **46**, 103–109
5. Thain, J.F., Doherty, H.M., Bowles, D.J. and Wildon, D.C. (1990) Plant Cell Environ. **13**, 569–574
6. Mathieu, Y., Kurkdjian, A., Hua Xia, Guern, J., Koller, A., Spiro, M.D., O'Neill, M., Albersheim, P. and Darvill, A. (1991) Plant J. **1**, 333–343
7. Atkinson, M.M., Baker, C.J. and Collmer, A. (1986) Plant Physiol. **82**, 142–146
8. Scheel, D., Colling, C., Hedrich, R., Kawallek, P., Parker, J.E., Sacks, W.R., Somssich, I.E. and Hahlbrock, K. (1991) Adv. Mol. Genet. Plant–Microbe Interact. **1**, 373–380
9. Blein, J.P., Milat, M.L. and Ricci, P. (1991) Plant Physiol. **95**, 486–491
10. Waldmann T., Jeblick, W. and Kauss,H. (1988) Planta **173**, 88–95
11. Conrath, U., Jeblick, W. and Kauss, H. (1991) FEBS Lett. **279**, 141–144
12. Guern, J., Mathieu, Y., Péan, M., Pasquier, C., Beloeil, J.-C. and Lallemand, J.-Y. (1986) Plant Physiol. **82**, 840–845
13. Walker-Simmons, M., Jin, D., West, C.A., Hadwiger, L. and Ryan, C.A. (1984) Plant Physiol. **76**, 833–836
14. Apostol, I., Heinstein, P.F. and Low, P. (1989) Plant Physiol. **90**, 109–116
15. Kurosaki, F., Tsurusawa, Y. and Nishi, A.L (1987) Phytochemistry **26**, 1919–1923
16. Stab, M.R. and Ebel, J. (1987) Arch. Biochem. Biophys. **257**, 416–423
17. Köhle, H., Jeblick, W., Poten, F., Blascheck, W. and Kauss, H. (1985) Plant Physiol. **77**, 544–551
18. Hattori, T. and Ohta, Y. (1985) Plant Cell Physiol. **26**, 1101–1110
19. Hagendoorn, M.J.M., Traas, T.P., Boon, J.J. and Van Der Plas, L.H.V. (1990) J. Plant Physiol. **137**, 72–80
20. Ojalvo, I., Rokem, J.S., Navon, G. and Goldberg, I. (1987) Plant Physiol. **85**, 716–719
21. Low, P.S. and Heinstein, P.F. (1986) Arch. Biochem. Biophys. **249**, 472–479
22. Strasser, H., Tietjen, K.G., Himmelspach, K. and Matern U. (1983) Plant Cell Rep. **2**, 140–143
23. Atkinson, M.M., Keppler, L.D., Orlandi, E.W., Baker, C.J. and Mischke, C.F. (1990) Plant Physiol. **92**, 215–221
24. Knight, M.R., Campbell, A.K., Smith, S.M. and Trewavas, A.J. (1991) Nature (London) **352**, 524–526
25. Wildon, D.C., Thain, J.F., Minchin, P.E.H., Gubb, I.R., Reilly, A.J., Skipper, Y.D., Doherty, H.M., O'Donnell, P.J. and Bowles, D.J. (1992) Nature (London) **360**, 62–65
26. Steffens, M., Ettl, F., Kranz, D. and Kindl, H. (1989) Planta **177**, 160–168

27. Crans, D.C., Simone, C.M., Saha, A.K. and Glew, R.H. (1989) Biochem. Biophys. Res. Commun. **165**, 246–250
28. Villar-Palasi, C., Guinovart, J.J., Gomez-Foix, A.M., Rodriguez-Gil, J.E. and Bosch, F. (1989) Biochem. J. **262**, 563–567
29. Kneusel, R.E., Matern, U. and Nicolay, K. (1989) Arch. Biochem. Biophys., **269**, 455–462
30. Hagendoorn, M.J.M., Poortinga, A.M., Wong Fong Sang, H.W., Van Der Plas, L.H.W. and Van Walraven, H.S. (1991) Plant Physiol. **96**, 1261–1267
31. Horn, M.A., Meadows, R.P., Apostol, I., Jones, C.R., Gorenstein, D.G., Heinstein, P.F. and Low, P.S. (1992) Plant Physiol. **98**, 680–686
32. Goot, E., Bligny, R. and Douce, R. (1992) J. Biol. Chem. **262**, 13903–13909
33. Pouyssegur, J., Franchi, A., Kohno, M., L'Allemain, G. and Paris, S. (1985) Curr. Top. Membr. Trans. **26**, 201–220
34. Pouyssegur, J., Franchi, A. and L'Allemain, G. (1985) FEBS Lett. **190**, 115–119
35. Felle, H. (1989) in Second Messengers in Plant Growth and Development. (Boss, W.F. and Morré, D.J., eds.), pp. 145–166, Liss, New York
36. Kurkdjian, A. and Guern, J. (1989) Annu. Rev. Plant Physiol. Plant Mol. Biol. **40**, 271–303
37. Guern, J., Felle, H., Mathieu, Y. and Kurkdjian, A. (1991). Int. Rev. Cytol. **127**, 111–173.
38. Roos, A. and Boron, W.F. (1981) Physiol. Rev. **61**, 296–434
39. Carr, D.W. and Acott, T.S. (1989) Biol. Reprod. **41**, 907–920
40. Hyrc, K. and Rose, B. (1990) J. Cell Biol. **110**, 1217–1226
41. Tognoli, L. and Basso, B. (1987) Plant Cell Environ. **10**, 233–239
42. Dietrich, A., Mayer, J.E. and Hahlbrock, K. (1990) J. Biol. Chem. **265**, 6360–6368
43. Schwacke, R. and Hager, A. (1992) Planta **187**, 136–141
44. Felix, G., Grosskopf, D.G., Regenass, M. and Boller, T. (1991) Proc. Natl. Acad. Sci. U.S.A., **88**, 8831–8834
45. Grinstein, S. and Goetz, J.D. (1985) Biochim. Biophys. Acta **819**, 267–270
46. Felle, H. (1988) Planta **174**, 495–499
47. Felle, H. (1988) Planta **176**, 248–255
48. Atkinson, M.M., Huang, J.S. and Knopp, J.A. (1985) Plant Physiol. **79**, 843–847
49. Cooper, S., Lerner, H.R. and Reinhold, L. (1991) Plant Physiol. **97**, 1212–1220
50. Murphy, T.M. and Wilson, C. (1982) Plant Physiol. **70**, 709–713
51. Huerta, A.J. and Murphy, T.M. (1989) Plant Physiol. **90**, 749–753

Biochem. Soc. Symp. **60**, 131–141
Printed in Great Britain

Signal transduction pathways in plant pathogenesis response

Robert Fluhr*, Yoram Eyal, Yael Meller, Vered Raz, Xiao-Qing Yang and Guido Sessa

Department of Plant Genetics, Weizmann Institute of Science, Rehovot 76100, Israel

Introduction

Plant cells have evolved the ability to detect pathogen ingress and subsequently activate defence-related functions as part of the plant pathogenesis response. A most vivid example of elicited plant reaction is the hypersensitive response (HR). It is a result of specific genetic-based recognition between the pathogen and the host. Concomitant with the appearance of HR, plants initiate broad-range defence mechanisms, including phytoalexin production, lignin accumulation and the synthesis of host-encoded polypeptides called pathogenesis-related (PR) proteins. These proteins are encoded by gene families that are co-ordinately expressed as part of the HR and are examples of the non-specific host reaction to pathogen invasion [1]. Many chemical and environmental cues elicit the same defences, or subsets therein. Elicitors could either interact with putative receptors or directly modify signal transduction pathways. A prevailing accepted assumption is that chemical elicitors in some way mimic plant–pathogen interactions and are valid model systems to study general responses of plants to pathogens.

PR proteins as paradigms of the plant pathogenesis response

PR proteins were initially defined as a group of acidic-type proteins that accumulate in the extracellular space of tobacaco mosaic virus (TMV)-infected leaves of hypersensitively reacting tobacco plants [1,2]. Functions were assigned to some of these proteins, such as endoglucanases (PR-2a, b and c) [3] and endochitinases (PR-3a and b) [4]. Other PR proteins have suspected functions based on sequence similarity. One of them, PR-5, is homologous to a bifunctional α-amylase/trypsin inhibitor from maize [5] and to an osmotic stress-inducible tobacco protein termed osmotin [6]. In contrast,

*To whom correspondence should be addressed.

the PR-1 protein family has no assigned biological function and no significant homology to any plant gene.

Serological and molecular comparisons revealed that most acidic-type PR proteins have basic-type homologues as well [1–7]. These proteins differ in some cases from the acidic-type counterparts in their induction pattern [7,8] and, in the case of basic glucanases, in their intracellular distribution [9]. They appear, in a fashion similar to their acidic-type counterparts, to be encoded by a gene family of up to eight members [10]. The isolation and sequence of a basic-type *PR-1* genomic clone have been reported [11]. Studies involving the expression of the basic-type *PR-1* cDNA revealed that the transcript accumulates in virus-infected or ethephon-treated leaves, in roots and in leaves of axenically cultured plants [12].

A novel basic-type *PR-1* gene serves as a paradigm for the pathogenesis response

Low-stringency screening, using an acidic-type *PR-1* probe, was used to isolate a basic-type *PR-1* clone (*PRB-1b*) [7,12a]. The open reading frame of the gene codes for a 179 amino acid protein which has 87% sequence similarity with the deduced amino acid sequence of PRB-1a [11] and 64% sequence identity with the acidic-type protein PR-1b [12,13]. The *PRB-1b* upstream promoter region is approximately 70% similar to the *PRB-1a* upstream region (allowing for large gaps in the comparison) up to position −654 where similarity abruptly stops. Sequence identity with the upstream region of acidic-type *PR-1* genes is very low and limited to several boxes in the first 150 bases of the promoter [14]. Several sequences resembling putative nuclear protein-binding sites were found in the *PRB-1b* promoter region. The most outstanding of these includes a G-box core motif [15], two AT-1-like sequences and a GCC motif [12a, 16]. The gene is rapidly induced by the plant hormone ethylene as shown in Fig. 1.

Differential expression pattern of basic- and acidic-type *PR-1*

Plants were exposed to a constant stream of hormone-like physiological concentrations of ethylene (18 μl/l) in the light and examined for *PRB-1b* transcripts.

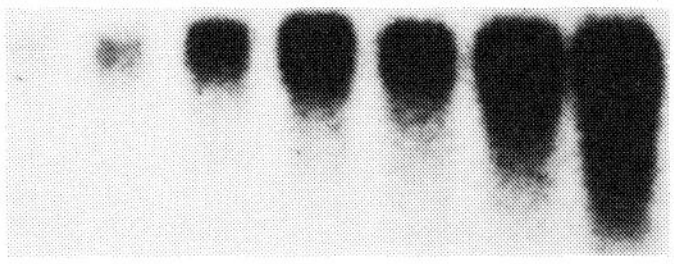

Fig. 1. Northern analysis of RNA from leaves treated by ethylene in the light. Each lane contains 20 μg of total RNA extracted from leaves treated with 18 μl/l ethylene from 0 h to 6 h and hybridized with *PRB-1b* probe.

Based on densitometric measurements, *PRB-1b* transcripts increased 3-fold within 1 h and accumulated 100-fold after 6 h of exposure to ethylene (Fig. 1). The hybridization results represent the total accumulation of basic-type *PRB-1* gene family transcripts and do not differentiate between genes within this family. Under these conditions, little or no induction of acidic-type *PR-1* transcript was detected (Table 1).

Basic-type and acidic-type *PR-1* genes were found to be induced following infection by TMV. In both cases accumulation of transcript began between 24 h and 48 h after inoculation (Table 1). Salicylic acid and ethephon were relatively rapid inducers of *PRB-1b*. However, their inductive effect was differential. Ethylene induced the expression of the basic-type *PRB-1b* exclusively, while salicylic acid induced both basic and acidic types of gene. Application of α-aminobutyric acid was shown to cause ethylene evolution and induce microlesions on tobacco leaves [17]. It was an efficient inducer of both *PRB-1b* and acidic *PR-1* transcripts.

S1-nuclease analysis of *PRB-1b* transgene promoter deletions in leaves using tagged transcripts

We identified promoter regions necessary for ethylene induction and transcriptional activation of the *PRB-1b* gene by S1 analysis of transgenes containing promoter deletions. A tagged transcript was created by inserting a fragment of 76 bp of foreign sequence at the 3′ transcribed non-translated region of *PRB-1b*. The tag served as a marker to differentiate between RNA molecules originating from endogenous *PRB-1b* transcripts and transgene transcripts (Fig. 2a). This system caused minimal changes to the original *PRB-1b* gene and enabled direct correlation to endogenous gene activity. Detection of transcript was achieved by use of a 50-nucleotide-long probe that included 30- and 40-base-long sequences similar to the endogenous and introduced transgenes, respectively (Fig. 2b). After end-labelling of the oligonucleotide and hybridization to RNA extracts from ethylene-treated control plants or transgenic

Table 1. Transcript accumulation of basic- and acidic-type genes in the presence of elicitors and light.

Elicitor treatment	Basic-type *PR-1*	Acidic-type *PR-1*
Tobacco mosaic virus	+	+
Salicylic acid	−, +[1]	+
Ethylene	+	−
α-Aminobutyric acid	+	+
Xylanase	+	+
Dark, 24 h	−, +	−
Xylanase: dark, 24 h	+	+
Ethylene: dark, 24 h	−, +	−

[1]Intermediate level.

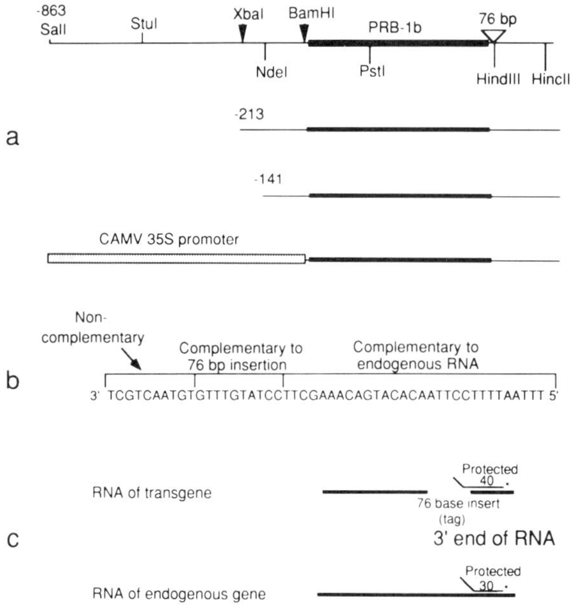

Fig. 2. *PRB-1b*-tagged gene constructs for S1-nuclease analysis of *PRB-1b* transcripts. (a) A *Hind*III 76 bp reporter fragment was inserted downstream of the *PRB-1b* open reading frame and attached to *PRB-1b* promoter deletions or 35S promoter. (b) The 50-base oligonucleotide which served as a probe for S1 analysis. The first 30 bases from the 5′ end are complementary to *PRB-1b* sequences, followed by 10 bases complementary to part of the 76 bp insertion sequence. The final 10 bases were chosen at random to be non-complementary. (c) Schematic representation of S1-nuclease analysis of tagged genes. The oligonucleotide was end-labelled and annealed to total RNA of transgenic and control plants. After digestion with S1 nuclease, protected fragments were detected by gel electrophoresis. A protected region of 40 bases corresponds to RNA of an introduced gene, while a protected region of 30 bases corresponds to RNA of the endogenous gene.

plants, the S1-nuclease digestion should yield protected fragments of 30 or both 30 and 40 bases, respectively (Fig. 2c).

S1 analysis of leaf RNA from non-treated control (Fig. 3a) and transgenic plants yielded no specific signal. Control plants treated with ethylene resulted in the expected protected region centred around 30 bases, corresponding to RNA of the endogenous gene (Fig. 3a). Analysis of leaf RNA, from ethylene-treated transgenic plants harbouring the 863 bp *PRB-1b* promoter fused to the tagged gene (−863 plants), resulted in protected regions centred around 30 and 40 bases — indicating transcription activity of endogenous and introduced genes, respectively. Transcript, albeit at lower levels, was detected in transgenic plants harbouring −213 bp, but not −141 bp, of promoter region. These results clearly indicate the presence of sequences necessary for

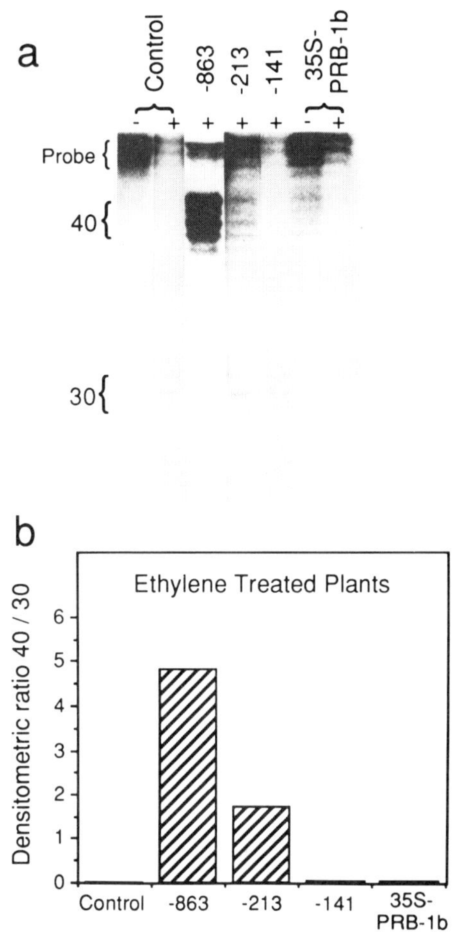

Fig. 3. Promoter analysis of tagged *PRB-1b* transgene. (a) S1 analysis of RNA isolated from non-treated (−) and 24 h ethylene-treated (+) control and transgenic plants. Mobility corresponding to protected regions of 40 and 30 bases is indicated. Exposure of the autoradiogram was adjusted so that the intensity of the endogenous RNA-protected region (30 base), which serves as reference, would appear in all lanes. (b) Densitometric scan of S1 analysis of ethylene-treated plants. Protected regions of 40 and 30 bases were scanned by a densitometer and the results are presented as a ratio of their respective intensities.

ethylene induction between positions −213 and −141. Basal level of transgene expression of non-treated plants was not detected in the S1 analysis, making fold-induction comparisons impossible. Owing to the influence of oligonucleotide length on DNA–RNA duplex stability, the absolute level of transcript cannot be ascertained. However, transcriptional activity of the endogenous gene can serve as a reference for transgene activity, and the densitometric ratio (40/30) enables normalized comparisons between deletion constructs (Fig. 3b). The average densitometric ratio (40/30) of activity for ethylene-treated plants was 4.8 for −863 and 1.7 for −213 plants. Thus the overall transgene activity of −213 plants was reduced by more than half compared with −863 plants. Since −213 plants still retain inducibility, we conclude that sequences between positions −863 and −213 contribute to the overall level of expression, but they do not mediate responses to external signals.

To examine the influence of downstream sequences on *PRB-1b* transcription regulation, the CAMV-35S promoter was used to replace the *PRB-1b* promoter (Fig. 2a). S1 analysis of RNA of transgenic plants harbouring the 35S promoter fused to the tagged gene (35S-*PRB-1b*) showed a constitutive protected region of 40 bases (Fig. 3a). After ethylene treatment, these plants showed an actual decrease in transcript level. The results suggest that downstream sequences in the context of the CAMV-35S promoter are not significantly involved in ethylene-induced transcriptional activation.

Dark-induced accumulation of basic-type *PRB-1b* transcripts

Several hours of 'natural' darkness were found to be sufficient for the accumulation of detectable amounts of transcript [7]. When greenhouse plants were transferred to complete darkness for up to 72 h *PRB-1* transcript accumulated and reached maximum levels of 50-fold induction [7]. The steady-state, dark-induced accumulation was 30% of the maximum achieved by induction in light with α-aminobutyric acid when Northern-blot analysis was carried out coincidentally. Decrease in the relatively high transcript levels occurred after the return of plants to the light, so that after 12 h the transcript level was barely detectable. Dark-induced accumulation of transcript was not detected in the acidic-type *PR-1* family (Table 1). The dark-induced accumulation is physiologically significant as it appears during normal diurnal growth cycles. Indeed, the polypeptide gene product was detected by immunoblot analysis of dark-grown plant as well (X.-G. Yang, unpublished work). The activity of the *PRB-1b* gene in the dark suggests a requirement by the plant for the *PRB-1b* gene product during dark growth, which is advantageously enhanced during pathogenesis. It may indicate non-stress-related functions for *PRB-1b*, as have been proposed for other PR proteins [17,18]. Recently, a senescence-associated gene that showed limited homology to the *PR-1* class was reported to be dark-induced [19]. Whether or not the senescence-related gene is induced by pathogens is not known. The dark accumulation of transcript may involve enhanced RNA stability or direct transcriptional activation. Recent results, using a chimaeric transgene of the 35S promoter fused to the *PRB-1b* coding sequence, show marked dark induction indicating that the former mode of regulation is the case (G. Sessa and R. Fluhr, unpublished work).

Light requirement of the ethylene-dependent signal transduction pathway

Light has been shown to be necessary for induction of acidic-type PR proteins via the ethylene-dependent signal transduction pathway [17]. A second induction pathway, which does not require ethylene, was shown to be light independent. The elicitor ethylene did not induce the basic-type *PRB-1* transcript in the dark above the basal dark level. However, xylanase, in contrast to ethylene, was clearly capable of inducing transcript accumulation of acidic-type *PR-1* in the dark (Table 1).

The co-ordinated accumulation of the various PR protein classes in response to pathogens might give the impression, albeit misleading, of common induction pathways. However, analysis of induction by different elicitors and the influence of light reveals a complex expression pattern. *PRB-1b* can be directly induced by ethylene. Indeed, the rapid accumulation of *PRB-1b* transcript in the presence of hormone-like physiological quantities of ethylene is the fastest response shown by PR genes with any elicitor. The time sequence of events is similar to the rapid appearance of senescence-related gene products in ethylene-treated carnation petals [20], and of transcripts that accumulate in ethylene-treated green tomato fruit [21]. In contrast, the acidic-type *PR-1* transcript level did not respond directly to ethylene, although it was induced with α-aminobutyric acid which depends on ethylene evolution for elicitation [17]. Hence ethylene is necessary, but not sufficient, for induction of acidic-type *PR-1* expression. The data are consistent with a dual signal requirement for induction of acidic *PR-1* via ethylene-dependent elicitation. One component required is ethylene, as inhibitors of ethylene biosynthesis and action obstruct this induction pathway. The nature of the second component is unknown. Thus, even closely related genes in the *PR-1* family show differential induction, and alternative mediators exist in the interaction between pathogens and plants.

Not only do the genes display differential activation, but the elicitors themselves have differential light requirements. Light-dependent induction of acidic-type PR proteins has been previously noted [17,22]. The light requirement is not an inherent characteristic of the genes, but apparently influences the signal transduction pathway. Thus the ethylene-requiring pathway and ethylene do not induce acidic- or basic-type *PR-1* transcript accumulation in the dark. In contrast, xylanase, an ethylene-independent elicitor, can induce acidic-type *PR-1* and basic-type *PRB-1b* transcript accumulation in the dark. A light requirement for *PR* transcript accumulation during ethylene-mediated induction is unexpected, since exogenously applied ethylene was shown to be active in the dark during abscission, fruit ripening and tissue senescence. Note that our experiments are conducted in relatively young leaves and that the light dependence for PR protein induction via ethylene may be limited to this tissue type.

The ethylene-dependent and -independent pathways for pathogenesis response show differential responses to calcium

Elicitors of pathogenesis response show differential sensitivity to ethylene inhibitors and a differential requirement for light. Based on these observations, we have defined two pathways. One pathway is ethylene dependent and is exemplified by ethylene itself [24,25]. It also includes elicitors, such as α-aminobutyric acid, that promote ethylene production and require ethylene in their mode of action [14–17]. The other pathway is ethylene independent and is exemplified by the fungal endoxylanase [26]. The enzyme apparently degrades β-1,4-xylan linkages in the plant cell wall and is an exceptionally potent inducer of chitinase accumulation. It also promotes ethylene evolution, but ethylene is not required in its mode of action [17].

We asked whether additional requirements differentiate the two pathways. Cellular transduction of external signals provided by hormones is typically a chain of consecutive events. In some cases, extracellular hormonal or environmental signals were shown to alter cytosolic levels of calcium, which is considered to be a primary event in triggering cellular responses. In plant cells, the concentration of free calcium in the cytosol is very low (100 nM), whereas in extra- and intracellular pools it is 10^4–10^5 times higher. Release of free calcium from these sources to the cytosol can play the role of a second messenger in signal transduction. In animal cells, elevated levels of cytosolic free calcium were shown to regulate target proteins, directly or via Ca^{2+}-binding proteins. In plant cells, there is a growing list of stimuli–response coupling in which cytosolic calcium appears to play an essential role. Examples are stomata opening, gravitropism, and certain light-stimulated chloroplast movements [27–30].

The physiological role of calcium in promoting the pathogenesis response was examined by treating plant leaves with the calcium chelator EGTA [31]. As shown in Table 2, elicitors of the ethylene-requiring pathway were inhibited in the presence of EGTA. The effect of EGTA was not pleiotropic, since the induction by xylanase was not affected. Plants can be grown in a calcium-free medium; they grow slowly but, nevertheless, appear normal. Calcium-poor plants contain 20-fold less extractable calcium in their leaves. Their ability to respond to ethylene-dependent inhibitors was stymied; however, their ability to respond to xylanase was not affected (Table 2) [31].

The ethylene-dependent pathway for pathogenesis response shows a requirement for phosphorylation

Cellular transduction of external signal initiated by ethylene is presumed to be an ordered process of consecutive events. One component in the pathway was shown to require calcium, as illustrated above [31]. Other possible components, such as protein phosphorylation, are thought to play a key role in diverse biological signal transduction systems (for review, see [32,33]). In plants, protein phosphorylation was shown to play a direct regulatory role in light response (for review, see [34,35])and circadian rhythms [36]. Correlations between protein phosphorylation and physiological response were demonstrated in pollen embryogenesis [37] and elicitor treatment of cultured cells [38].

Table 2. Chitinase accumulation in water- or elicitor-treated leaves.

	EGTA		Calcium poor plants	
Treatment	None	10 mM	No addition	Plus calcium[1]
Water	–	–	–, +[2]	
Xylanase	+	+	+	+
α-Aminobutyric acid	+	–	+	
Ethylene	+	–	–	+

[1]Leaves imbibed in 10 mM calcium in addition to elicitor. [2]Intermediate accumulation of PR proteins detected.

Ethylene application induced very rapid and transient protein phosphorylation in tobacco leaves [38a]. In the presence of the kinase inhibitors H-7 and K252a, the transient rise in phosphorylation and the induced expression of *PR* genes were abolished. Similarly, these inhibitors blocked the response induced by an ethylene-dependent elicitor, α-aminobutyric acid. Reciprocally, application of okadaic acid, a specific inhibitor of phosphatases type-1 and type-2A, enhanced total protein phosphorylation and, by itself, elicited the accumulation of PR proteins. In the presence of H-7 and K252a, the accumulation of PR proteins induced by okadaic acid was blocked. In contrast to the action of ethylene and α-aminobutyric acid, xylanase elicits the accumulation of PR proteins by an ethylene-independent pathway. Xylanase-induced accumulation of PR proteins was not affected by H-7 and K252a.

Okadaic acid, a specific inhibitor of animal and plant phosphatases, type 1 and 2a, increased the general level of protein phosphorylation in intact leaves, similar to its effect in animal cell cultures [39,40]. In addition, it elicited the rapid accumulation of a *PR*-specific transcript and PR proteins in intact leaves. In animal systems, okadaic acid was shown to be capable of inducing specific genes at the transcriptional level [41,42]. In leaves, the induction of PR genes by okadaic acid was abrogated by the presence of kinase inhibitors, suggesting that the transduction events requiring kinase and phosphatase activity are probably acting through the same pathway. Hence, at least one stage of the transduction pathway requires kinase activity, which is modified by a

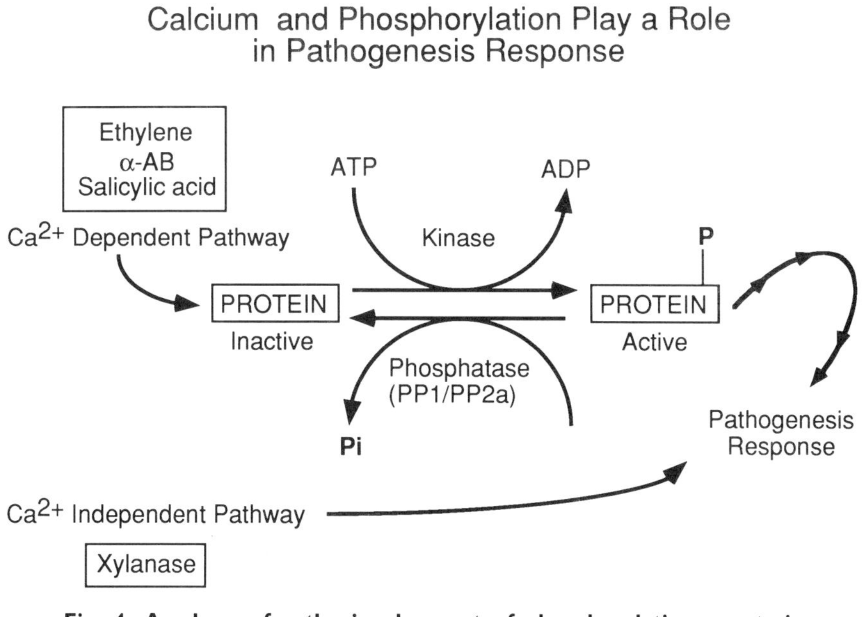

Fig. 4. A scheme for the involvement of phosphorylation events in the induction of ethylene-dependent pathogenesis response. Ethylene, α-aminobutyric acid (α-AB) and salicylic acid utilize the ethylene- and calcium-dependent pathway, while xylanase utilizes a different pathway [31]. The transducing activity of a putative control protein is determined by its phosphorylation level. The level of phosphorylation is determined by the kinetics of kinase and phosphatase activity.

mechanism that maintains phosphorylation equilibrium, as shown schematically in Fig. 4 [38a]. Alternatively, or in addition, the induction of PR proteins by ethylene could be transduced via negative regulation of PP1/PP2A phosphatase activity.

Elicitors of pathogenesis utilize at least two pathways for the induction of PR proteins, which differ in their requirement for ethylene [43] and calcium [31]. The ethylene-independent pathway, which is induced by xylanase, was not affected by the kinase inhibitors we have used. Thus, these pathways differ in their initial cellular signal-transducing machinery as well. Apparently, downstream from the transduction events depicted here, the pathways leading to gene activation converge to give, in each case, co-ordinate regulation of PR protein accumulation (Fig. 4).

We gratefully acknowledge support of the Israeli National Council for Research and Development, the Gesellschaft fuer Biotechnologische Forschung MBH, Braunschweig and the Forchheimer Center for Molecular Biology. R.F. is a recipient of the Jack and Florence Goodman Career Development Chair. V.R. is a recipient of the Levi Eshkol Scholastic Fund.

References

1. Bol, J.F., Linthorst, H.J.M. and Cornelissen, B.J.C. (1990) Annu. Rev. Phytopathol. **28**, 113–138
2. Van Loon, L.C. and Van Kammen, A. (1970) Virology **40**, 199–211
3. Kauffmann , S., Legrand, M., Geoffroy, P. and Fritig, B. (1987) EMBO J. **6**, 3209–3212
4. Legrand, M., Kauffmann, S., Geoffroy, P. and Fritig, B. (1987) Proc. Natl. Acad. Sci. U.S.A. **84**, 6750–6754
5. Richardson, M., Valdes-Rodriguez, S. and Blanco-Labra, A. (1987) Nature (London) **327**, 432–434
6. Singh, N.K., Bracker, C.A., Hasegawa, P.M., Handa, A.K., Buckel, S., Hermodson, M.A., Pfankoch, E., Regnier, F.E. and Bressan, R.A. (1987) Plant Physiol. **85**, 529–536
7. Eyal, Y., Sagee, O. and Fluhr, R. (1992) Plant Mol. Biol. **19**, 589–599
8. Ward, E.R., Payne, G.B., Moyer, M.B., Williams, S.C., Dincher, S.S., Sharkey, K.C., Beck, J.J., Taylor, H.T., Ahl-Goy, P., Meins, F., Jr, and Ryals, J.A. (1991) Plant Physiol. **96**, 390–397
9. Van den Bulcke, M., Bauw, G., Castresana, C., Van Montagu, M. and Vandekerckhove, J. (1989) Proc. Natl. Acad. Sci. U.S.A.. **86**, 2673–2677
10. Cornelissen, B.J.C., Horowitz, J., Van Kan, J.A.L., Goldberg, R.B. and Bol, J.F. (1987) Nucleic Acids Res. **15**, 6799–6811
11. Van Kan, G., Middlesteadt, W., Desai, N., Williams, S., Dincher, S. and Carnes, M. (1989) Plant Mol. Biol. **12**, 595–596
12. Cornelissen, B.J.C., Hooft van Huijsduijnen, R.A.M., Van Loon, L.C. and Bol, J.F (1986). EMBO J. **5**, 37–40

12a. Eyal, Y., Meller, Y., Lev-Yadun, S. and Fluhr, R. (1993) Plant J. **4**, 225–235

13. Fitzner, U.M. and Goodman, H.M. (1987) Nucleic Acids Res. **15**, 4449–4465
14. Eyal, Y. and Fluhr, R. (1991) in Oxford Surveys of Plant Molecular and Cellular Biology (B.J. Miflin, ed.), pp. 223–254, Oxford University Press, Oxford
15. Giuliano, G., Pichersky, E., Malik, V.S., Timko, M.P., Scolnik, P.A. and Cashmore, A.R. (1988) Proc. Natl. Acad. Sci. U.S.A. **85**, 7089–7093
16. Datta, N. and Cashmore, A.R. (1989) Plant Cell **1**, 1069–1077
17. Lotan, T. and Fluhr, R. (1990b) Plant Physiol. **93**, 811–817
18. Ori, N., Sessa, G., Lotan, T., Himmelhoch, S. and Fluhr, R. (1990) EMBO J. **9**, 3429–3436

19. Azumi, Y. and Watanabe, A. (1991) Plant Physiol. **95**, 577–583
20. Lawton, K.A., Raghothama, K.G., Goldsbrough, P.B. and Woodson, W.R. (1990) Plant Physiol. **93**, 1370–1375
21. Lincoln, J.E., Cordes, S., Read, E. and Fischer, R.L. (1987) Proc. Natl. Acad. Sci. U.S.A. **84**, 2793–2797
22. Asselin, A., Grenier, J. and Cote, F. (1985) Can. J. Bot. **63**, 1276–1283
23. Reference deleted
24. Boller, T., Gehri, A., Mauch, F. and Vögeli, U. (1983) Planta **157**, 22–31
25. Ecker, J.R. and Davis, R.W. (1987) Proc. Natl. Acad. Sci. U.S.A. **84**, 5202–5206
26. Mauch, F., Hadwiger, L.A., and Boller, T. (1984) Plant Physiol. **76**, 607–611
27. Hepler, P.K. and Wayne, R.O. (1985) Annu. Rev. Plant Physiol. **36**, 397–439
28. Gilroy, S., Blowers, D.P. and Trewavas, A.J. (1987) Development **100**, 181–184
29. Pooviah, B.W. and Reddy, A.S.N. (1987) CRC Crit. Rev. Plant Sci. **6**, 47–103
30. Trewavas, A. and Gilory, S. (1991). Trends Genet. **7**, 356–361
31. Raz, V. and Fluhr, R. (1992) Plant Cell **4**, 1123–1130
32. Cohen, P. (1982) Nature (London) **296**, 613–620
33. Hardie, D.G. (1990) Symp. Soc. Exp. Biol. **44**, 241–255
34. Bennett, J. (1991) Annu. Rev. Plant Physiol. Plant Mol. Biol. **42**, 281–311
35. Allen, J.F. (1992) Trends Biol. Sci. **17**, 12–17
36. Carter, P.J., Nimmo, H.G., Fewson, C.A. and Wilkins, M.B. (1991) EMBO J. **10**, 2063–2068
37. Kyo, M. and Harada, H. (1990) Planta **182**, 58–63
38. Dietrich, A., Mayer, J.E. and Hahlbrock, K. (1990) J. Biol. Chem. **265**, 6360–6368
38a. Raz. V. and Fluhr, R. (1993) Plant Cell **5**, 523–530
39. Cohen, P. (1989) Annu. Rev. Biochem. **58**, 453–508
40. Heystead, T.A.J., Sim, A.T.R., Carling, D., Honnor, R.C., Tsukitani, Y., Cohen, P. and Hardie, D.G. (1989) Nature (London) **337**, 78–81
41. Nagamine, Y. and Ziegler, A. (1991) EMBO J. **10**, 117–122
42. Guy, G.R., Cao, X., Chua, S.P. and Tan, Y.H. (1992) J. Biol. Chem. **25**, 1846–1852
43. Lotan, T. and Fluhr, R. (1990) Symbiosis **8**, 33–46

Biochem. Soc. Symp. **60**, 143–148
Printed in Great Britain

Signals involved in the wound-induced expression of the proteinase inhibitor II gene of potato

Hugo Peña-Cortés, José Sánchez-Serrano, Salomé Prat and Lothar Willmitzer*

Institut für Genbiologische Forschung Berlin GmbH, Ihnestrasse 63, 14195 Berlin, Germany

Introduction

Plants respond to mechanical injury by inducing a defence response characterized by the expression of a set of proteins mainly aimed at wound healing and prevention of pathogen invasion. These responses include reinforcement of the cell wall by deposition of callose, lignin and hydroxyproline-rich glycoprotein; synthesis of the antimicrobial compounds phytoalexins; and production of proteinase inhibitors and lytic enzymes, such as chitinases and glucanases [1–3]. Induction of most defence proteins involves transcriptional activation of the corresponding genes. Their activation is often confined to the immediate vicinity of the wound site, but some of them show a systemic response — being also transcribed in distal tissues, which themselves are not damaged [3]. The mechanisms underlying this different, spatial activation are likely to be connected to the distribution of a 'wound signal'.

Potato and tomato plants accumulate two non-homologous proteinase inhibitors, proteinase inhibitor I (pin1) and proteinase inhibitor II (pin2), as a direct consequence of wounding by chewing insects or severe mechanical damage [4–6]. Mechanical wounding of the leaves results in the accumulation of these inhibitors in the directly injured tissue, as well as in other aerial organs of the plant, far away from the primary wound site [7]. In potato plants, products of both gene families accumulate to very high levels in tubers, where they can comprise more than 15% of the soluble protein [8]. In addition, pin2 has been found to be expressed during early stages of floral development; young floral buds accumulate pin2 mRNA which, thereafter, is absent in the fully developed flower [9]. A single *pin2* promoter was shown to be active both in leaves, upon wounding, and constitutively in tubers [10]. This gene might thus be responding

*To whom correspondence should be addressed.

to a signal common to the wounding and tuberization processes or, alternatively, this complex pattern of expression could be modulated by at least two different *cis*-regulatory elements: one reacting to the signals associated with tuberization and during flower formation, the other being wound responsive.

The mechanism by which the plant regulates *pin2* expression is, at present, not well understood. An inducing factor, or 'wound hormone', has been postulated which, when released at the wound site, would be transported through the vascular system to the rest of the plant, thus leading to the systemic activation of these genes. Several different stimuli have been shown to induce *pin2* expression and have thus been postulated as putative wound signals. Oligogalacturonides isolated from plant cell walls were shown to induce the expression of *pin* genes when supplied through cut petioles of excised leaves [11,12]. The uronides, however, may not be the systemic signals, because their mobility through the phloem is limited [13]. Rather, these compounds are thought to be released from the wounded tissues as early signals in the pathway that ultimately leads to both localized and systemic wound-induced expression of the *pin* genes. Several other compounds, such as the peptide systemin, were shown to induce pin2 mRNA accumulation in both tomato and potato leaves [14]. Recently, it has also been shown that a propagated electrical signal may be the messenger for the systemic induction of pin2 [15].

Here we describe our evidence suggesting that both abscisic acid (ABA) and jasmonic acid (JA) play a role as signals for wound-induced expression of *pin2* and other genes. Furthermore, evidence is given demonstrating that the developmental-specific expression does not require signals involved in its wound-induced expression.

Results and discussion

Expression of the *pin2* gene is under both developmental and environmental influence

Pin2 is constitutively expressed in potato and tomato flowers and potato tubers. In potato, pin2 mRNA accumulates in young flower buds but it is below the limits of detection in fully developed flowers. Analysis of plants transgenic for a *pin2* promoter–β-glucuronidase gene fusion indicated that *pin2* promoter activity was primarily restricted to developing ovules. In contrast to this stage-specific activity, pin2 mRNA is present throughout development of the tomato flower [9]. This discrepancy in the expression pattern may reflect the differential activity of members of the tomato *pin2* gene family or, alternatively, differences in persistence of the putative signals responsible for *pin2* gene activation in flowers, which might be related to differences in the metabolism of potato and tomato. *Pin2* belongs to a small multigene family and, therefore, its different modes of expression (constitutive in tubers and flowers, and inducible in leaves upon wounding) might reflect differential activity of members of the gene family. Alternatively, one or several members of the gene family may be active in part or all of these situations. This issue was addressed by introducing into potato a gene fusion consisting of a *pin2* promoter hooked to a β-glucuronidase (GUS) reporter. This construction endowed the transgenic plants with constitutive GUS activity in tubers and floral buds, and wound-induced activity in leaves [10]. A single promoter element, therefore, mirrors the expression pattern of the whole gene family, indicating the presence within this promoter of all *cis*-acting sequences responsible for the complex regulation of *pin2* by environmental and developmental factors.

To define *cis*-regulatory elements involved in *pin2* promoter activity, deletion analysis of a potato *pin2* promoter has been performed in stably and transiently transformed potato and tobacco plants. Two different elements, a quantitative enhancer and a regulatory element, are required for promoter activity. While functional promoter elements required for *pin2* activity in tubers and wounded leaves could not be separated and were located between positions −624 and −405, the expression of *pin2* in flowers is mediated by different *cis*-acting sequences which are located upstream of position −928. Induction of *pin2* expression in leaves by treatment with the plant growth regulators ABA and JA, and the general metabolite sucrose, depends on the presence of the regulatory element involved in expression in tubers and wounded leaves. Thus, *pin2* expression in tubers and wounded leaves apparently results from the action of signals on closely linked promoter elements, while a different signal pathway leads to its constitutive expression in flowers.

ABA and JA play a pivotal role in wound-induced expression of the *pin2* genes

The plant hormone ABA has often been associated with plant responses to different stress situations [16]. Several lines of evidence indicate that ABA also plays a role in *pin2* induction upon wounding. First, the expression of the *pin2* gene family is not induced in wounded leaves of tomato and potato mutants deficient in ABA synthesis, and this lack of expression is lifted by supplying ABA through the petioles of detached leaves. Secondly, ABA treatment can, by itself, induce pin2 accumulation in the foliage of ABA-sprayed potato plants, in the absence of any damage [17]. Determination of endogenous ABA concentrations upon wounding revealed that this ABA-induced pin2 activation is physiologically meaningful. There is a six-fold rise in endogenous ABA concentration upon wounding in both tomato and potato leaves. These levels peak before maximal pin2 mRNA accumulation is reached.

While ABA is involved at some stage during wound-induced gene activation, constitutive *pin2* expression in potato flowers and tubers is, apparently, not affected by the block in the ABA biosynthetic pathway. Thus the ABA-deficient, droopy mutants have wild-type pin2 mRNA levels in tubers and flower buds, suggesting the presence of different, or additional, factors for *pin2* gene activation in tubers and flowers [9]. *Pin2* expression is not induced in water-deficit situations, in spite of an 8–10-fold increase in the endogenous ABA concentration. This encourages the argument for the existence of some kind of modulation of the effect of ABA on *pin2* gene activation. This could either be different subcellular compartmentation of ABA produced in response to wounding, water stress or modification of the ABA, respectively.

The association of JA and its methyl ester with wound responses has recently been established. JA was reported to stimulate the accumulation of two wound-inducible soybean genes when applied to fully expanded leaves or cell cultures [18,19]. Moreover, this plant hormone has been shown to induce the accumulation of proteinase inhibitor proteins to even higher levels than those obtained by wounding [20]. To establish whether this compound can also induce the expression of the *pin2* gene, detached potato leaves were incubated through the petiole in solutions of ABA or methyljasmonic acid (MeJA) in the dark. At the same time, leaves were incubated in water to determine the levels of expression originated by wounding during leaf excision. RNA samples were isolated from leaves of control and water-stressed plants, and locally and systemically wound-induced leaves were included for comparison. The transcript corresponding to *pin2* accumulates dramatically upon wounding, ABA or MeJA treatments.

Aspirin prevents wound-induced gene expression in tomato leaves by blocking JA biosynthesis

Aspirin or related benzoic compounds were shown to repress the wound-induced *pin2* gene expression in tomato plants [21]. In animal systems, almost all aspirin-like drugs (generally termed non-steroidal anti-inflammatory drugs, or NSAIDs) inhibit the cyclo-oxygenase activity (COX) of the prostaglandin-endoperoxide synthase enzyme [22]. Prostaglandin synthase and lipoxygenase (LOX) represent key enzymes in the oxidation of polyunsaturated fatty acids in animal cells. COX catalyses the first enzymic step in the conversion of polyunsaturated fatty acids to prostaglandins. LOX, on the other hand, is involved in the synthesis of leukotrienes which, like prostaglandins, are released during inflammation and injury reactions [23]. Eicosanoids (prostaglandins and leukotrienes) and the phytohormone JA display some similarity in both chemical structure and biosynthetic pathways. Eicosanoids and JA are derived from lipids via LOX- or COX-dependent pathways [24,25]. LOX catalyses the oxidation of linolenic acid, the first step in the synthesis of JA, whereas hydroperoxide dehydratase mediates cyclization of 13-hydroperoxylinolenic acid (13-HPLA) to 12-oxo-phytodienoic acid (12-oxo-PDA) [26]. Based on this resemblance among eicosanoids and JA biosynthetic pathways, we decided to test whether or not mammalian COX-inhibitors, such as aspirin, affect wound-induced accumulation of pin2 mRNA in tomato plants using the previously described system of detached leaves [7]. More specifically, we asked the following questions. (i) Does aspirin affect the wound-induced or the systemically induced expression of *pin2*? (ii) Does aspirin affect the synthesis of JA in tomato plants; and, if so, which step is blocked by aspirin?

Aspirin blocks both the local and the systemically wound-induced *pin2* gene expression in tomato plants

To test the effect of aspirin on the systemically induced expression of *pin2*, whole tomato plants were excised at the base and allowed to take up either water alone or aspirin solutions through the cut stems. The effect of aspirin on the expression of other wound-induced plant genes, those encoding the potato cathepsin D inhibitor (*Cdi*) and the threonine deaminase (*Td*) was also analysed. Both *Cdi* and *Td* gene expression are induced by either wounding, ABA or MeJA in tomato leaves. However, whereas Cdi mRNA, like pin2, accumulates both in the directly wounded and systemic tissues, Td mRNA only accumulates in the wounded leaves [27]. Treatment of plants with aspirin led to a significant reduction of pin2, Cdi and Td mRNA accumulation upon wounding. Besides local wound-response inhibition, aspirin repressed the systemically induced *pin2* and *Cdi* gene activation. Hence, these results suggest a common feature in both local and systemic activation of these wound-induced genes.

Aspirin represses the wound-induced increase of endogenous JA levels

It has recently been reported that wounding increases endogenous levels of JA in soybean [28]. We therefore decided to examine the changes of endogenous levels of JA in tomato plants as a result of wounding. Mechanical wounding leads to an increase in JA levels after 6 h. Thereafter, levels of JA decline, being slightly higher than the control level after 24 h. Plants pre-treated with aspirin and subsequently wounded showed the same levels of JA as the control, non-wounded plants. These results strongly suggest that aspirin blocks some step of the JA biosynthetic pathway, thus preventing the synthesis of this compound upon wounding.

Effect of aspirin on ABA- and MeJA-induced gene activity and on the JA-synthesis pathway

To obtain more information about the effect of aspirin on gene expression, we investigated the inhibitory effect of aspirin on the JA-synthesis pathway using the intermediate compounds α-linolenic acid (LA), 13-HPLA and 12-oxo-PDA. To this end, detached tomato leaves were pre-treated with aspirin and after 3 h either ABA, LA, 13-HPLA, 12-oxo-PDA or JA was added. As controls, tomato leaves were supplied with the same substances without aspirin. Neither LA nor 13-HPLA was able to complement the inhibition mediated by aspirin. Contrary to this, 12-oxo-PDA and JA — which were the strongest inducers of *pin2* gene expression in tomato plants — were also able to suppress the inhibitory effect of aspirin.

Conclusion

Wounding results in the expression of a large set of new proteins. In many cases, this is owing to the transcriptional activation of the corresponding genes. A detailed study of the wound-induced expression of *pin2* genes in tomato and potato plants has provided conclusive evidence that both ABA and JA represent important signals for their activation as a result of wounding.

References

1. Collinge, D.B. and Slusarenko, A.J. (1987) Plant Mol. Biol. **9**, 389–410
2. Hahlbrock, K. and Scheel, D. (1987) in Innovative Approaches to Plant Disease Control (Chet, I., ed.), pp. 229–254, Wiley-Interscience, New York
3. Bowles, D.J. (1990) Annu. Rev. Biochem. **59**, 873–907
4. Graham, J., Pearce, G., Merryweather, J., Titani, K., Ericsson, L. and Ryan, C.A. (1985) J. Biol. Chem. **260**, 6555–6560
5. Graham, J., Pearce, G., Merryweather, J., Titani, K., Ericsson, L. and Ryan, C.A. (1985) J. Biol. Chem. **260**, 6561–6564
6. Sánchez-Serrano, J., Schmidt, R., Schell, J. and Willmitzer, L. (1986) Mol. Gen. Genet. **203**, 15–20
7. Peña-Cortés, H., Sánchez-Serrano, J., Rocha-Sosa, M. and Willmitzer, L. (1988) Planta **174**, 84–89
8. Ryan, C.A. (1984) in Plant Gene Research: Genes Involved in Microbe–Plant Interactions (Verma, D.P.S. and Hihn, T.H., eds.), pp. 375–386, Springer-Verlag, Vienna
9. Peña-Cortés, H., Willmitzer, L. and Sánchez-Serrano, J. (1991) Plant Cell **3**, 963–972
10. Keil, M., Sánchez-Serrano, J. and Willmitzer, L. (1989) EMBO J. **8**, 1323–1330
11. Bishop, P.D., Makus, D.J., Pearce, G. and Ryan, C.A. (1981) Proc. Natl. Acad. Sci. U.S.A. **78**, 3536–3540

12. Walker-Simmon, M., Jin, D., West, C.A., Hadwiger, L. and Ryan, C.A. (1984) Plant Physiol. **76**, 833–836
13. Baydoun, E.A. and Fry, S.C. (1985) Planta **165**, 269–276
14. Pearce, G., Strydom, D., Johnson, S. and Ryan, C.A. (1991) Science **253**, 895–898
15. Wildon, D.C., Thain, J.F., Minchin, P.E.H., Gubb, I.R., Reilly, A.J., Skipper, Y.D., Doherty, H.M., ODonnell, P.J. and Bowles, D.J. (1992) Nature (London) **360**, 62–65
16. Skriver, K. and Mundry, J. (1990) Plant Cell **2**, 503–512
17. Peña-Cortés, H., Sánchez-Serrano, J., Mertens, R., Willmitzer, L. and Prat, S. (1989) Proc. Natl. Acad. Sci. U.S.A. **86**, 9851–9855
18. Anderson, J.M., Spilatro, S.R., Klauer, S.F. and Franceschi, V.R. (1989) Plant Sci. **62**, 45–52
19. Staswick, P.E. (1990) Plant Cell **2**, 1–6
20. Farmer, E.E. and Ryan, C.A. (1990) Proc. Natl. Acad. Sci. U.S.A. **87**, 7713–7716
21. Doherty, H.M., Selvendran, R.R. and Bowles, D.J. (1988) Physiol. Mol. Plant Pathol. **33**, 377–384
22. Smith, W.L. and Marnett, L.J. (1991) Biochim. Biophys. Acta **1083**, 1–17
23. Pace-Asciak, C.R. and Smith, W.L. (1983) Enzymes **16**, 543–603
24. Vick, B.A. and Zimmermann, D.C. (1979) Plant Physiol. **64**, 203–205
25. Needleman, P., Turk, J., Jakschik, B.A., Morrison, A.R. and Lefkowith, J.B. (1986) Annu. Rev. Biochem. **55**, 69–102
26. Vick, B.A. and Zimmermann, D.C. (1983) Biochem. Biophys. Res. Commun. **111**, 470–477
27. Hildmann, T., Ebneth, M., Peña-Cortés, H., Sánchez-Serrano, J., Willmitzer, L. and Prat, S. (1992) Plant Cell **4**, 1157–1170
28. Creelman, R.E., Tierney, M.L. and Mullet, J.E. (1992) Proc. Natl. Acad. Sci. U.S.A. **89**, 4938–4941

Biochem. Soc. Symp. **60**, 149–154
Printed in Great Britain

Polypeptide signalling for plant defence genes

Barry McGurl, Gregory Pearce and Clarence A. Ryan*

Institute of Biological Chemistry, Washington State University, Pullman, WA 99164–6340, U.S.A.

Synopsis

The synthesis of proteinase inhibitor proteins in response to wounding is a defensive response against pest and pathogen attacks. Wounding of leaves results in the release of a mobile wound signal which induces proteinase inhibitor synthesis throughout the plant. The signal transduction pathway regulating this response is not fully understood, but several compounds have been identified which are capable of inducing proteinase inhibitor synthesis in tomato and potato leaves. These compounds include cell wall fragments of both plant and pathogen origin [1–3], abscisic acid [4], jasmonic acid [5] and, most recently, an 18-amino-acid polypeptide, called systemin, isolated from tomato leaves [6]. In this chapter, we describe the properties of systemin and its precursor prosystemin, and we summarize the evidence supporting a role for systemin as an initial signal that regulates proteinase inhibitor synthesis in response to wounding.

Proteinase inhibitors

Proteinase inhibitor proteins occur throughout the plant kingdom, as members of about 12 non-homologous families based on primary structure and specificity [7]. Proteinase inhibitors are typically encoded by small gene families that are expressed either developmentally, or in response to environmental cues [8,9]. The induction of proteinase inhibitor synthesis in response to mechanical wounding has been intensively studied using the Inhibitor I and II families of tomato and potato as model systems [10]. Wounding triggers the release of at least one mobile wound signal, which results in the transcriptional activation of the proteinase inhibitor genes [11]. A single wound causes proteinase inhibitor accumulation over a 24 h period. The proteinase inhibitors are synthesized as larger precursors on the endoplasmic reticulum and are targeted to the vacuole where they are stored [12,13].

*To whom correspondence should be addressed.

Proteinase inhibitors are potent anti-nutrients to animals, including insects. The adverse effects of proteinase inhibitors on insect growth have been convincingly demonstrated in transgenic tobacco plants expressing proteinase inhibitor genes [14].

The nature of the mobile wound signal

Irrespective of the age or variety of tomato plant, the wound signal moves relatively slowly, taking up to 90 min to leave the wounded leaf after wounding. Evidence from both indirect physiological measurements and direct analysis of phloem content suggests that the wound signal is transported through the phloem [6,15].

To identify components of the signal transduction pathway, a biological assay was developed in which small quantities of the potential inducer were supplied to young, excised tomato plants through their cut stems for a few minutes and then the plants were transferred to water. After a 24 h incubation period under light the leaves were assayed immunologically to quantify the induction of the proteinase inhibitors [16–18]. Cutting the stems of the young plants does not itself trigger significant proteinase inhibitor synthesis.

Using this assay, the first putative wound signals to be identified were oligouronides derived from the plant cell wall [1]. Wounding the leaf was thought to liberate endogenous, cell-wall-degrading enzymes, such as polygalacturonase, which released mobile pectic fragments from the plant cell wall. This idea is no longer accepted, since the necessary hydrolytic enzymes have not been found in tomato or potato leaves; in addition, experiments with radiolabelled oligosaccharides showed that these molecules do not move throughout the plant [19]. Cell wall fragments are now regarded as localized inducers of proteinase inhibitor synthesis, being released by enzymes produced by pathogenic micro-organisms at the site of infection. Cell wall fragments of both plant and pathogen origin are known to induce a variety of plant defence responses [20].

The well-characterized plant hormone abscisic acid (ABA) has also been proposed as the mobile wound signal. Studies using tomato and potato plants deficient in ABA production have shown that these mutants do not produce proteinase inhibitors in response to wounding, but they do respond to exogenously applied ABA [4]. In contrast, wild-type tomato plants do not respond to exogenously applied ABA, although wild-type potato plants do respond. At this time, it appears unlikely that ABA is a primary signal in the wound response in tomato plants, but it probably acts by indirectly perturbing the system.

Evidence for the wound signal being electrical, rather than chemical, in nature comes from studies in which the rapid release (less than 5 min) of a wound signal from tomato cotyledons was correlated with a transient action potential [21]. Compounds which perturb or dissipate the electrochemical gradient across the plasma membrane also inhibited the wound induction of proteinase inhibitor synthesis [22]. These results are incompatible at this time with the extensive literature describing a mobile, chemical wound signal. In addition, there is no evidence demonstrating a causal link between electrical activity and proteinase inhibitor synthesis.

More recently, an 18-amino-acid polypeptide, which was named systemin, was isolated from tomato leaves and was shown to be the most potent inducer of proteinase

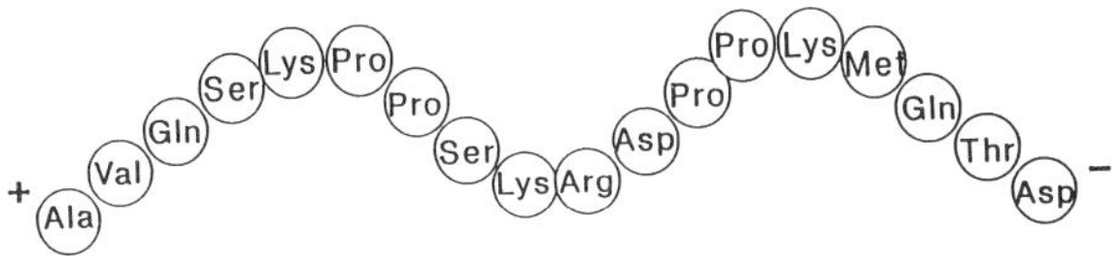

Fig. 1. The amino acid sequence of systemin.

inhibitor synthesis identified to date [6]. The sequence of systemin is shown in Fig. 1. It is the first candidate peptide hormone identified in plants. The properties of systemin, described below, are consistent with its proposed role as the mobile wound signal.

Systemin and its precursor prosystemin

Systemin is present at very low concentration in the leaves of unwounded tomato plants; 60 lb (27.216 kg) of tomato leaves were required to isolate 1 μg of pure systemin [6]. Synthetic systemin is fully bioactive and provides an abundant supply of the molecule for experimentation. When fed to young, excised tomato plants, synthetic systemin maximally induces proteinase inhibitor synthesis in femtomolar quantities. To determine whether systemin can move throughout the tomato plant, a ^{14}C-systemin derivative was synthesized and applied to the wound site of a wounded tomato leaf. The radiolabelled systemin was highly mobile and its movement in the phloem exactly mimicked the behaviour of the systemic wound signal [6].

Deletion–substitution studies have shown that the biological activity of systemin is sensitive to changes in primary structure, especially at the *N*- and *C*-termini. In particular, *C*-terminal deletion of the aspartic acid at position 18, and substitution of the penultimate threonine, resulted in complete inactivation of systemin [23]. It is possible that Thr-17 may be chemically modified, *in vivo*, in a manner which may be essential for biological activity.

Based on the sequences of the cDNA and gene encoding systemin, it was found that systemin is derived from a larger precursor protein, prosystemin, which is composed of 200 amino acids [24]. Systemin is represented only once within the prosystemin sequence, being located close to the *C*-terminus from which it must be processed. Three other polypeptide sequences, unrelated to systemin, are duplicated within prosystemin, each occurring once within the *N*-terminal half of prosystemin and once within the *C*-terminal half (Fig. 2). These conserved sequence elements may be

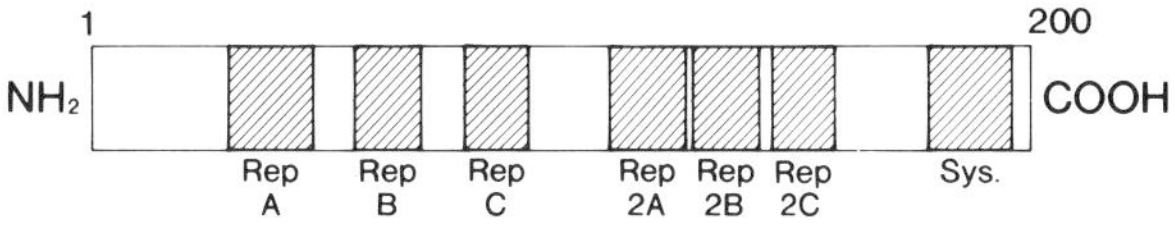

Fig. 2. Repeated sequence elements within prosystemin. Conserved sequence elements Rep A, Rep B and Rep C, and their repeats Rep 2A, Rep 2B and Rep 2C, are represented by hatched bars. Systemin is represented by a hatched bar labelled Sys.

other biologically active oligopeptides which are processed from the precursor and which may regulate other physiological responses of the plant. Biologically active polypeptides, peptide hormones, are common in animals and are typically derived from a larger precursor which is often processed to yield more than one type of biologically active peptide [25].

At present little is known about the synthesis and processing of prosystemin. Northern blot analysis using the prosystemin cDNA revealed that the prosystemin gene is constitutively expressed throughout the plant except in the roots. In addition, the prosystemin gene is systemically wound inducible, providing an amplification mechanism whereby the wounded plant produces more systemin, allowing an even more vigourous response to subsequent attack.

It is not known in which cell types prosystemin is synthesized, nor whether it is immediately processed after synthesis or whether it is processed in response to wounding. Prosystemin is a relatively hydrophilic molecule, apparently lacking a signal sequence, and so it is probably not targeted to the vacuole or to the plasma membrane. There are, however, examples of proteins which are exported by a pathway not involving a classical signal sequence [26], so this question is still open.

We are currently investigating the synthesis and processing of prosystemin using antibodies raised against prosystemin purified from a bacterial expression system. In addition, a β-glucuronidase (GUS) reporter gene, under the regulation of the prosystemin gene promoter, has been stably transformed into tobacco and tomato plants which are now being analysed histochemically.

While wound-inducible proteinase inhibitors are found in a wide variety of plant species, Northern and Southern blot analysis has identified a putative prosystemin gene homologue only in potato, a close relative of tomato. An inducer of proteinase inhibitor synthesis that appears to be a polypeptide has recently been purified from tobacco leaves and may provide additional information in our attempts to understand the distribution and evolution of systemin-like signalling molecules throughout the plant kingdom.

Molecular genetic evidence for the role of systemin in the wound signalling pathway

To demonstrate that systemin has a role in the wound induction of proteinase inhibitor synthesis *in vivo*, an antisense prosystemin gene, driven by the constitutive 35S promoter, was stably transformed into tomato plants [24]. The resulting transgenic plants (T1) exhibited a severely reduced systemic induction of proteinase inhibitor synthesis in response to wounding. The Inhibitor-I and -II levels were approximately 20% and 10%, respectively, of those measured in control plants transformed with the binary vector alone.

Recently, transgenic tomato plants which contain a stably integrated prosystemin cDNA under the regulation of the 35S promoter have been regenerated (McGurl, B., Orozco-Cardenas, M. and Ryan, C.A., unpublished work) These plants continuously synthesize proteinase inhibitor proteins, even in the absence of wounding. The levels of proteinase inhibitors increase with age: in one experiment 2-week-old plants had Inhibitor-I levels of about 80 μg/g of leaf tissue, compared with a level of about 500 μg/g of leaf tissue at 6 weeks. Previously, the highest obtainable Inhibitor-I levels were approximately 250 μg/g of leaf tissue in response to prolonged exposure to methyljasmonic acid [27].

The observation that overexpression of prosystemin results in constitutive expression of proteinase inhibitors suggests that the processing enzymes necessary for cleaving systemin from its precursor are present and active in the unwounded plant. Unwounded tomato plants normally express prosystemin mRNA at a low level, but without concomitant induction of proteinase inhibitor synthesis. Expression of high levels of prosystemin may saturate cellular mechanisms designed to sequester either the precursor or the processed peptide(s). An alternative explanation is that by producing more of the signal molecule, systemin, the sensitivity of the system has been enhanced, so that the plant now responds to mild environmental stimuli, such as touch.

The plants overexpressing prosystemin will be invaluable in identifying other proteins which may be regulated by biologically active polypeptides derived from prosystemin, since, like the proteinase inhibitors, they should be constitutively expressed at high levels in these plants. The resistance of these plants, and of the antisense plants, to both insects and microbial pathogens is also being investigated.

Future research

We have developed a model, describing the signal transduction pathway regulating wound-induced proteinase inhibitor synthesis, in which systemin is proposed to be the mobile wound signal that stimulates a fatty-acid-based second messenger system resulting in the intracellular release of the effector molecule, jasmonic acid, or a closely related derivative [28].

This model has been supported by experiments both from our laboratory and from others. The level of jasmonic acid has been shown to increase in plants in response to wounding [29,30], and the wound response element of the Inhibitor-II gene has been shown to be identical to the jasmonic acid-responsive element, which is conserved throughout several jasmonic acid-responsive genes [31]. Transgenic tomato plants expressing the antisense gene respond weakly to wounding but are fully responsive to systemin and to jasmonic acid when these compounds are supplied through the cut stem (McGurl, B. and Ryan, C.A. unpublished work). These results cumulatively support a wound-inducible signal transduction system that requires both systemin and jasmonic acid as key components of the pathway.

Further studies of the systemin receptor, the role of the receptor in the cascade leading to the release or production of linolenic acid and jasmonic acid, and the role of jasmonic acid in activating proteinase inhibitor genes, are in progress.

This research was supported in part by the Washington State University College of Agriculture and Home Economics Project 1791, and by Grants IBN-9244361 and IBN-9117795 from the National Science Foundation.

References

1. Bishop, P., Makus, D.J., Pearce, G. and Ryan, C.A. (1981) Proc. Natl. Acad. Sci. U.S.A. 78, 3536–3540
2. Bishop, P., Pearce, G., Bryant, J.E. and Ryan, C.A. (1984) J. Biol. Chem. **259**, 13172–13177

3. Walker-Simmons, M., Hadwiger, L. and Ryan, C.A. (1983) Biochem. Biophys. Res. Commun. **110**, 194–199
4. Peña-Cortés, H., Sánchez-Serrano, J.J., Mertens, R., Willmitzer, L. and Prat, S. (1989) Proc. Natl. Acad. Sci. U.S.A. **86**, 9851–9855
5. Farmer, E.E. and Ryan, C.A. (1990) Proc. Natl. Acad. Sci. U.S.A. **87**, 7713–7716
6. Pearce, G., Strydom, D., Johnson, S. and Ryan, C.A. (1991) Science **253**, 895–898
7. Ryan, C.A. (1989) BioEssays **10**, 20–24
8. Wingate, V.P.M., Broadway, R.M. and Ryan, C.A. (1989) J. Biol. Chem. **264**, 17734–17738
9. Wingate, V.P.M. and Ryan, C.A. (1991) Plant Physiol. **97**, 496–501
10. Ryan, C.A. and An, G. (1988) Plant Cell Environ. **11**, 345–349
11. Graham, J.S., Hall, G., Pearce, G. and Ryan, C.A. (1986) Planta **169**, 399–405.
12. Walker-Simmons, M. and Ryan, C.A. (1977) Plant Physiol. **60**, 61–63
13. Narvaez-Vasquez, J., Franceschi, V.R. and Ryan, C.A. (1994) Planta, in the press
14. Johnson, R., Narvaez, J., An, G. and Ryan, C.A. (1989) Proc. Natl. Acad. Sci. U.S.A. **86**, 9871–9875
15. Nelson, C.E., Walker-Simmons, M., Makus, D., Zuroske, G., Graham, J. and Ryan, C.A. (1983) in Plant Resistance to Insects (Hedin, P.A., ed.), pp. 105–122, ACS, Washington
16. Ryan, C.A. (1967) Anal. Biochem. **19**, 434–440
17. Trautman, R., Cowan, K. and Wagner, G. (1971) Immunochemistry **8**, 901–916
18. Ryan, C.A. (1974) Plant Physiol. **54**, 328–332
19. Baydoun, E.A.-H. and Fry, S. (1985) Planta **165**, 269–276
20. Ryan, C.A. (1988) Biochemistry **27**, 8879–8883
21. Wildon, D.C., Thain, J.F., Minchin, P.E.H., Gubb, I.R., Reilly, A.J., Skipper, Y.D., Doherty, H.M., O,Donnell, P.J. and Bowles, D.J. (1992) Nature (London) **360**, 62–65
22. Doherty, H.M. and Bowles, D.J. (1990) Plant Cell Environ. **13**, 851–855
23. Pearce, G., Johnson, S. and Ryan, C.A. (1993) J. Biol. Chem. **268**, 212–216
24. McGurl, B., Pearce, G., Orozco-Cardenas, M. and Ryan, C.A. (1992) Science **255**, 1570–1573
25. Jung, L.J. and Scheller, R.H. (1991) Science **251**, 1330–1335
26. McGrath, J.P. and Varshavsky, A. (1989) Nature (London) **340**, 400–404
27. Farmer, E.E., Johnson, R.R. and Ryan, C.A. (1991) Plant Physiol. **98**, 995–1002
28. Farmer, E.E. and Ryan, C.A. (1992) Plant Cell **4**, 129–134
29. Creelman, R.A., Tierney, M.L. and Mullet, J.E. (1992) Proc. Natl. Acad. Sci. U.S.A. **89**, 4938–4941
30. Gundlach, H., Muller, M.J., Kutchan, T.M. and Zenck, M.H. (1992) Proc. Natl. Acad. Sci. U.S.A. **89**, 2389–2393
31. Choi, J.-L. (1992) Ph.D. Thesis, Washington State University, Pullman, WA, U.S.A.

Biochem. Soc. Symp. **60**, 155–164
Printed in Great Britain

Signalling events in the wound response of tomato plants

Dianna J. Bowles*

Centre for Plant Biochemistry and Biotechnology, University of Leeds, Leeds LS2 9JT, U.K.

Synopsis

The wound response of tomato plants provides a useful experimental system to analyse local and systemic signalling events. Wounding one region on the plant leads to changes in the expression of genes at the local site of damage and elsewhere in unwounded tissues. A wound stimulus is thus converted to signal(s) that are transduced locally and signals that are involved in establishing long-distance spread of the initial stimulus. Data from studies at Leeds will be integrated into a wider discussion of the available evidence, to work towards an integrated model for understanding signalling events in the wound response.

Introduction

The wound response of plants encompasses a wide range of molecular events dependent on the plant species, the type of cells that are wounded and their developmental context [1]. The events that will be discussed in this paper concern studies initiated by Clarence Ryan's group in the 1970s. They analysed proteinase inhibitor (pin) proteins in vegetative tissues of tomato and potato plants; injury to one leaf led to an increased accumulation of pin proteins at the injury site and, significantly, in distant unwounded leaves of the plant [2]. Their results indicated that there was some mechanism of long-distance signalling in the plant such that the wound stimulus was communicated to distant sites.

Pin proteins, such as the well-characterized pin1 and pin2 of tomato and potato plants, are thought to be defence proteins that provide a general protection system against pathogens, pests and herbivores that rely on proteinases for effective invasion of a plant. This idea is supported by data on the effects of pin proteins in artificial diets, the ability of pin proteins to inhibit enzymes identified in the potential invaders, and the

*Present address: The Plant Laboratory, Department of Biology, University of York, Heslington, York YO1 5DD, U.K.

increased resistance to insect attack of transgenic plants expressing pin proteins constitutively in their foliage [3–7].

Wounding leaves induces pin proteins in other leaves: both younger and older ones, relative to the developmental age of the leaf that is wounded [8]. Pin proteins also accumulate in other aerial organs in response to leaf injury, including the stem and petioles. However, wounding the aerial regions of the plant does not induce the accumulation of pin proteins in the root system [9]. In contrast, root damage — whether mechanical or caused by a root pest, such as a cyst nematode — induces systemic pin proteins in the leaves [9,10].

Thus there is organ-to-organ signalling in the wound response, and the cell types that initiate the signal(s) may well be very different to those receiving the signal(s). In this context, the wound response can be thought to involve two distinct cell populations: the cells that are stimulated and the cells that are receiving the long-range consequences of that stimulation. There is no reason to assume that the signals and transduction pathways operating in the first set of cells will be the same as those operating in the second set of cells. The first population will be responding to the initial wound stimulus, whereas the second will be responding to the systemic signal(s) released from the wound-site.

In their early study, Green and Ryan [2] suggested that injury led to the formation of a chemical signal, and that the transport of this signal out of the wounded leaf and around the plant gave rise to the systemic response. In general, the chemicals are considered as positive inducing agents with increased levels causing the response. Over the years, a number of compounds have been put forward as candidates for the mobile signal [11]. These include oligosaccharides [12,13], jasmonic acid (JA)/methyljasmonate (MeJA) [14], abscisic acid (ABA) [15] and a peptide called systemin [16]. In one instance, a negative chemical regulator is envisaged, an auxin (1AA) [17,18], such that the wound response is triggered by decreased levels of the regulator. While it is clear that all of these different compounds have an effect on *pin* gene activation, their causal involvement in the specific events that lead to *pin* gene activation *in planta* remains an open question. It is equally possible, however, that the transport of a mobile chemical signal is not the systemic mechanism, but rather that a physical system of stimulus spread is operating. In recent studies, both hydraulic signals [19] and electrical signals have been put forward as the mechanism of systemic signalling [20–24]. While these events undoubtedly occur on wounding, their causal involvement in the events that lead to *pin* gene activation *in planta* remains an open question.

Cellular damage is also known to lead to a wide range of additional effects, beyond the activation of *pin* genes. Some of these effects involve transcriptional activation of genes, while others involve post-transcriptional regulation of gene expression, including RNA stability changes. Yet others involve changes in protein and lipid turnover and changes in the activity of preformed enzymes [25]. For example, a number of other wound-induced genes have been identified, such as those identified by Grierson and co-workers and including pTOM 13, corresponding to ethylene-forming enzyme [25–29]. Changes at the protein level encompass studies from the laboratory of Kauss, on callose synthetase activation and the consequences of 1,3-β-glucan accumulation and turnover [31–34], and studies from the laboratory of Lamb, on changes in cell wall organization induced by wounding and the 'oxidative burst' shown to accompany cellular damage [35–38].

Wound-induced changes in the oxidation state of the damaged cells is also correlated with changes in membrane lipids [39,40]. One consequence is the potential biosynthesis of a number of lipid derivatives, including the 'wound hormone' trans-2-dodecenedoioc acid, named traumatic acid, JA and its methyl ester MeJA. Wound-induced synthesis of JA in tomato and potato plants has recently been demonstrated [41], and exogenous JA to tomato and potato plants will induce *pin* gene expression [14]. JA synthesis requires the action of a number of well-defined enzymes, including β-lipoxygenase (LOX), which catalyses the oxidation of unsaturated fatty acids containing the *cis,cis*-1,4-pentadiene system, as well as linolenic and linoleic acids which are common constituents at the 2-position of plant phospholipids. LOX cannot act directly on the membrane bilayer, but rather is dependent on the prior release of the fatty acids by other lipases. In principle, these could be phospholipases A_1 or A_2, removing fatty acids from the 1- and 2-positions, respectively. LOX substrates could also arise from lipolytic acyl hydrolase action that removes fatty acids from both the 1- and 2-positions. Although all of these enzymes have been identified in plants [39,40], their direct involvement in signalling events has not been demonstrated. Likewise, their existence in leaves of tomato and potato plants and their wound-induction have not been investigated directly. Indirect evidence arises from the fact that application of exogenous linolenic acid and other metabolic precursors of JA leads to *pin* gene expression [42].

Use of bioassays to study molecular events in the wound response

A typical strategy in signalling studies is to apply chemicals to the plant and analyse their effects on a marker gene(s) or physiological response [43]. If the chemical has a positive effect there is a strong tendency to assume a role for that chemical *in planta*, even though it may only mimic an event within the endogenous signalling cascade. The method of application, however, can influence the results gained. Chemicals can be applied directly to the leaves, e.g. as volatile agents, sprayed onto the leaf surface, or infiltrated into the air spaces. Alternatively, they can be applied through the cut stem of excised plants and enter the system via the transpiration stream. The extent of access cannot readily be ascertained; leaf cuticle thickness/permeability will vary, and chemicals applied to cut surfaces may induce events locally at the cut site on the stem, or be carried in the transpiration stream to act locally in the leaves. For example, oligogalacturonides derived from pectin can induce *pin* gene expression if applied to the cut stem of excised plants, but have no effect on *pin* genes if infiltrated into the air spaces of the leaf lamina (K. Dalkin and D.J. Bowles, unpublished work). This contrasts with glycan elicitors of the hypersensitive reaction which are effective when infiltrated into air spaces [44]. ABA sprayed onto potato plants induces *pin* genes; sprayed onto tomato and tobacco plants it has no effect, and must, therefore, be applied via the transpiration stream [45].

Using bioassays involving chemical application, a number of positive inducing agents of *pin* gene expression have been identified. These include (i) an aqueous extract of autoclaved leaf material, the original 'PIIF' [46]; (ii) fragments of pectin, closely related epimers of galacturonides and other oligosaccharides, including chitosan

[12,13,44]; (iii) lipids and lipid derivatives, including linolenic acid [14,42]; (iv) JA and MeJA [14,44,45]; (v) ABA [18]; and (vi) an 18-mer peptide, called systemin [16].

Bioassays have also been used to assay the effects of potential inhibitors of *pin* gene expression induced by wounding or chemical application. Work at Leeds has shown that salicylic acid (SA), its acetylated derivative aspirin and a number of related hydroxybenzoic acids act as potent and reversible inhibitors of the response [47]. SA is now recognized to be an important endogenous signal in defence responses of plants involving systemic acquired resistance (SAR) [48]. Interestingly, the structural specificity required for inhibition of the wound response is identical to that known for the induction of pathogenesis-related proteins and SAR [23,48]. We extended our studies on aspirin and SA to show that a range of inhibitors of ion transport could equally block the response of tomato plants to wounding or chemical inducers, for example, fusicoccin at 10^{-8} M was found to completely abolish *pin* gene induction [49].

Movement of chemicals through the plant: implications for mobile systemic signals

Glycans

Ryan first suggested that the systemic signal (PIIF) was a cell wall fragment [11]. This was based on application of glycans to tomato plants in the bioassay and assumptions that fragments were released at the wound site *in planta*, and were mobile in the phloem. In 1985, Fry's group published a paper which showed that radiolabelled oligosaccharides applied to the wounded surface of a leaf did not exit the petiole, and that their mobility was restricted to a relatively local region of lamina adjacent to the injury [50]. This study led to the conclusion that cell wall fragments could not be the systemic signal. However, application of exogenous glycans to a damaged surface may not represent a good analogy for the events that occur *in planta*, particularly if the causal signalling events happen rapidly/spontaneously on damage, i.e. symplastic connections and access to long-distance transport routes, may be very different during and after the injury.

Recently, Selvendran's group have shown that oligogalacturonides with a degree of polymerization of 6 are mobile within the plant, when applied to the cut stem in the standard bioassay [51]. Mobility in this system would be the consequence of movement in the transpiration stream of the xylem. It remains to be seen whether other forms of movement of glycans are equally possible, for example, diffusion from cell to cell through cell walls and intercellular spaces, and movement in the phloem. It will also be of interest to determine whether there is a size dependence, and whether larger fragments are equally mobile. From these studies, it would seem to remain an open question, as yet, as to whether oligosaccharides can be mobile systemic signals *in planta*, whether they act by diffusion within a single lamina or whether they are transported around the plant.

ABA

Willmitzer's group working within the model of ABA involvement in the wound response did measure levels of the growth regulator in wounded leaves of potato plants and in the unwounded, systemically responding leaves [15]. From control plants, levels of 162–177 ng of ABA/g of leaf (fresh weight) were found and after wounding this level doubled to 315–366 ng of ABA/g of leaf, whether extracted from the wounded or unwounded leaves. The measurements were carried out on tissue harvested 24 h after the event, which is very much later than the wound-induced changes in *pin* gene transcript level. These can be detected in potato plants within 20 min of injury, whether local to the wound-site or systemically [9]. Since spraying one leaf of a potato plant with ABA also leads to both local and systemic *pin* gene induction, a plausible inference is that ABA is the mobile signal, and the increased levels of the growth regulator after 24 h represents an amplification of the response. Both roots and shoots can synthesize ABA, and within another context changes in ABA content of xylem sap have been recorded [52]. Willmitzer's group emphasize that although changes in the level of ABA at the local and systemic site occurred, transport of ABA from the wound site to distant regions has not been demonstrated [15]. The involvement of ABA in the wound response of tomato plants, and transgenic tobacco expressing a potato *pin* gene promoter/reporter gene fusion, is less clear. Several studies have addressed the question and different effects of ABA (including inhibition) have been shown [17,46,53].

Systemin

Most recently, Ryan and co-workers have suggested that the mobile signal in the systemic wound response is systemin [16]. This is a small 18-mer peptide, synthesized as part of a much larger protein. Application of the peptide to cut stems in the bioassay showed inducing activity at 40 fmol/plant, i.e. the specific activity was some 10^5 times greater than that of glycan fragments applied in the same manner. When radiolabelled systemin was applied to a wound-site on a tomato leaf, radiolabel was observed throughout the lamina of the wounded leaflet and other leaflets of the compound leaf. Radiolabelled peptide could also be recovered by immersing the cut petiole of the wounded leaves in 25 mM EDTA, pH 7.0, defined as 'phloem exudate'. Since application of the peptide to the wounded lamina led to recovery in material collected from the cut petiole, systemin is mobile within the plant. Whether systemin is the mobile effector systemic signal *in planta* remains as yet an open question.

Alternative mechanisms to transport of a mobile chemical signal

In collaboration with electrophysiologists, we have recently addressed whether phloem transport is indeed required for systemic spread of the wound response [24]. Using tomato seedlings and either injury to the cotyledon or to the first fully expanded leaf, we showed that the causal agent of *pin* gene induction exited the petiole rapidly, and while phloem translocation of ^{11}C-assimulate was shown to be inhibited by a cold block. These data rule out the need for transmission of a mobile chemical signal in the phloem in the experimental system that was analysed. In older tomato plants, there is some evidence to suggest that systemic spread from the wounded leaf is much slower [54]. These rates are estimated by the length of time required for the wounded leaf (or cotyledon) to remain attached to the plant to attain the full response. Petiole excision alone does not induce systemic *pin* gene induction; therefore, if a wound is applied and the wounded leaf detached at time intervals following injury, the minimum time required to result in the typically high levels of pin transcripts (pin proteins, or pin activity) in unwounded leaves can be determined. In young seedlings, the full response can be attained even if the wounded cotyledon/leaf is detached within 20 s [24]. In older tomato plants, attachment of the wounded leaf for in excess of 30 min is essential to induce the full response [54]. These differences may well represent developmental regulation of the mechanisms used by the plant for systemic spread of a wound stimulus. For example, in young tissues where symplastic domains are likely to be widespread, rapid signalling involving physical propagation of an action potential may be the preferred route for cell-to-cell communication. In older tissues, it is known that symplastic domains become restricted; for example, internodal regions each represent a separate domain along the stem of a young plant. Perhaps in larger and more mature plants, systemic spread can afford to be slower, and indeed involve a mechanism relying on the relatively slow rates of phloem transport.

Use of mutants to understand the wound response

Two groups have used mutants to investigate signalling events in the wound response. Willmitzer's group have analysed local and systemic responses in ABA-deficient mutants [15,45]. The altered phenotype of the potato *droopy* and tomato *sit* mutants arises from lower internal levels of ABA and can be reversed by exogenous application of the growth regulator. Wounding the mutant plants led only to very low levels of *pin* gene induction; application of ABA led to the full response. These data indicate that ABA is very likely to play a role *in planta* in the wound induction of *pin* genes. More recent analyses of the mutant plants have shown that ABA application also leads to the induction of four additional wound-responsive genes, encoding a cysteine proteinase inhibitor, an aspartate proteinase inhibitor, an amino peptidase and threonine dehydratase [45].

Ryan's group expressed antisense RNA of prosystemin in transgenic tomato plants to determine the effect on the systemic wound response [55]. Pin protein levels were determined in the upper unwounded leaves, 24 h after injury to the lower leaf. Of 28 F1-generation transformants, 'three-quarters of the plants inheriting the antisense construct responded more weakly to wounding than the control population', and six of the 28 produced much-decreased systemic levels of pin protein ($<15\%$ pin1, $\sim 0\%$ pin2). The data from these analyses clearly indicate that systemin plays a role at some level in the wound response *in planta*.

Integration of the available information: towards a working model for signalling events *in planta*

We are still some considerable way from understanding the causal events in the signalling cascade leading from injury to *pin* gene activation, i.e. whether the *pin* genes are induced close to the injury site or in distant unwounded leaves of the plant.

Use of the potato and tomato ABA-deficient mutants provides compelling evidence for the involvement of the growth regulator at some level in the wound response. In potato plants, the effect induced by exogenous ABA appears to be identical to that induced by wounding, given that spraying one leaf induces *pin* genes systemically in precisely the same tissues as the wound stimulus. If the applied ABA does not move from the sprayed leaf, and as yet there is no evidence that it does (nor that endogenous ABA/ABA metabolites move out of wound site), then it may be significant to analyse the local events induced by ABA in the sprayed leaf and determine which of these are causal in setting up the systemic response. Systemin is, undoubtedly, highly effective in *pin* gene induction, and the antisense experiment provides excellent evidence for the involvement of the peptide *in planta*. However, the protein sequence of the prosystemin product derived from the cDNA has not been reported to have a signal peptide [55]. If further data confirm that the prosystemin is a cytosolic protein, it is possible that the systemin peptide may exit the cell directly from the cytosol, such as is shown for a number of regulatory peptides in mammalian cells. In the absence of this mechanism, access to the long-distance transport routes of xylem and/or phloem is difficult to envisage. On wounding, steady-state levels of transcripts of prosystemin increase both in wounded and unwounded leaves of tomato plants, i.e. the gene is locally and systemically wound inducible [55]. Interestingly, as reported at the Leeds meeting, McGurl and Ryan have used prosystemin promoter/GUS fusions to analyse promoter activity in transgenic plants. These data indicated that the prosystemin promoter was active in only restricted cell types: the vascular parenchyma. The cell specificity of promoter activity was maintained on wound induction. This suggests to the author a model in which systemin provides a whole plant amplification of the wound-response, i.e. the regulatory peptide is released in the cytosol of individual cells in response to a long-range signal passing through the vascular tissue and perhaps, as is shown for electrical signals, passing directly through the vascular parenchyma. This model contrasts with that proposed by Ryan, which predicts that peptides are released from prosystemin molecules at the wound-site and that they move around the plant, binding to receptors on the individual cells to evoke the systemic response.

MeJA and/or JA are also clearly able to induce *pin* gene activation; the lipid derivatives are beginning to be recognized as key regulatory molecules capable of inducing a wide range of responses in plants, both at the level of physiology, such as tendril coiling [56], and at the level of gene expression, such as genes encoding vegetative storage proteins [57]. As yet there is no direct evidence that the molecules function during a wound response *in planta*, although wound-induced elevation of JA levels is clearly significant [41]. Equally, the fact that pretreatment of the plants with aspirin, leading to an inhibition of the wound response, could be reversed by exogenous MeJA/JA is a clear pointer to the role of jasmonates *in planta* [41]. Since JA application led to induction of wound-inducible genes in the potato ABA-deficient mutant, and did not lead to increased ABA levels in the plants, Hildemann *et al.* suggested that MeJA/JA acted at a later step than ABA in the signalling pathway triggered by mechanical wounding [45]. There is, however, at this stage no evidence to assume the two signals are causally connected in a cascade. In theory, the genes may be activated by different routes, or JA may mimic the ABA effect. Similarly, as yet there is no evidence to support the suggestion that systemin and MeJA/JA are causally related in a signalling chain. Experiments in which MeJA/JA are applied to plants expressing prosystemin antisense RNA would be informative but, again, would not be definitive in confirming that the two function within the same cascade.

Oligogalacturonides also induce *pin* gene activation — and, as with the MeJA/JA, cell wall fragments are increasingly recognized as important regulatory molecules [34]. There is no evidence, direct or indirect, for their involvement *in planta* in a wound response. Their action would necessitate cleavage and release from the structural wall polysaccharides. Pectin-degrading enzymes are known to be secreted by a wide range of pests and pathogens [34]. An endopolygalacturonase activity within leaves of tomato and potato plants has not been demonstrated, but may be a highly regulated enzyme given the bioactivity of its product. There is structural and size specificity in the glycan induction of *pin* genes, and the same specificity holds for glycan-induced phosphorylation of a tomato plasma membrane polypeptide, pp34 [58]. These two lines of evidence are suggestive of a signalling chain, but it is not known (or, at least, published) whether wounding as such induces phosphorylation of the same polypeptide. These data would clarify to some extent whether the glycan pathway operates in the wound response, or is related only to the pathogen/pest induction of *pin* genes.

The inhibitor studies with SA, fusicoccin and other ion-transport inhibitors, imply that movement of ions across the plasma membrane is in some way involved in setting up the wound response [47,49]. Using an intracellular electrode system and tomato leaf mesophyll cells, changes in membrane potential were detected when glycan fragments were applied to the leaf tissue [59]. Propagated electrical signals were also detected by surface electrodes attached to the petiole of a wounded leaf [23,24]. Although there is a strong correlation between movement of the electrical signal generated by a wound stimulus or heat stimulus and *pin* gene activation (timing and passage through a cold block [24]), it remains a correlation until the input of an electrical signal can be shown to lead to *pin* induction.

A hydraulic signal with secondary electrical activity has also been postulated as the causal mechanism *in planta* [19]. Since leaves or plants can be excised with no effect as such on *pin* gene induction, it is difficult to envisage a direct role for hydraulic signals. Experiments in which living tissue in the petiole is either destroyed or disrupted should immediately clarify any causal role for the remaining xylem vessels in the conduction of a systemic signal.

In summary, while data are accumulating on a number of effector molecules and systemic signalling mechanisms, it may be some years yet before there is a precise clarification as to how the different pieces of the molecular puzzle fit together.

Research on the wound response at Leeds has been funded by the AFRC.

References

1. Bowles, D.J. (1993) Encycl. Mol. Biol., in the press
2. Green, T.R. and Ryan, C.A. (1972) Science **175**, 776–777
3. Ryan, C.A. (1990) Annu. Rev. Phytopathol. **28**, 425–449
4. Hilder, V.A., Gatehouse, A.M.R., Sheerman, S.E., Barker, R.F. and Boulter, D. (1987) Nature (London) **330**, 160–163
5. Johnson, R., Narvaez, J., An, G. and Ryan, C.A. (1989) Proc. Natl. Acad. Sci. U.S.A. **89**, 9871–9875
6. Thornburg, R.W., Kernan, A. and Moblin, I. (1990) Plant Physiol. **92**, 500–505
7. Roby, D., Toppan, A. and Esquerre-Tugaye, M.T. (1987) Physiol. Mol. Plant Pathol. **30**, 6453–6460
8. Graham, J.S., Hall, G., Pearce, G. and Ryan, C.A. (1986) Planta **169**, 399–405

9. Pena-Cortez, H., Sanchez-Serrano, J., Rocha-Sosa, M. and Willmitzer, L. (1988) Planta **174**, 84–89
10. Bowles, D.J., Hammond-Kosack, K., Gurr, S. and Atkinson, H.J. (1991) in Biochemistry and Molecular Biology of Plant–Pathogen Interactions (Smith, C.J., ed.), pp. 225–236, Oxford University Press, Oxford
11. Ryan, C.A. (1992) Plant Mol. Biol. **19**, 123–133
12. Bishop, P., Makus, K.J., Pearce, G. and Ryan, C.A. (1981) J. Biol. Chem. **259**, 13172–13177
13. Bishop, P.D., Pearce, G., Bryant, J.E. and Ryan, C.A. (1984) J. Biol. Chem. **259**, 13172–13177
14. Farmer, E.E. and Ryan, C.A. (1990) Proc. Natl. Acad. Sci. U.S.A. **87**, 7713–7716
15. Pena-Cortez, H., Sanchez-Serrano, J.J., Mertens, R. and Willmitzer, L. (1989) Proc. Natl. Acad. Sci. U.S.A. **86**, 9851–9855
16. Pearce, G., Strydom, D., Johnson, S. and Ryan, C.A. (1991) Science **253**, 895–898
17. Kernan, A. and Thornburg, R.W. (1989) Plant Physiol. **91**, 73–78
18. Thornburg, R.W. and Li, X. (1990) Plant Physiol. **93**, 500–504
19. Malone, M. and Stankovic, B. (1991) Plant Cell Environ. **14**, 431–436
20. Pickard, B. (1973) Bot. Rev. **39**, 172–201
21. Van Sambeek, J.W. and Pickard, B. (1976) Can. J. Bot. **54**, 2642–2650
22. Robin, G. and Bonnemain, J.L. (1985) Plant Cell Physiol. **26**, 1273–1283
23. Wildon, D.C., Doherty, H.M., Eagles, G., Bowles, D.J. and Thain, J.F. (1989) Ann. Bot. **64**, 691–695
24. Wildon, D.C., Thain, J.F., Minchin, P.E.H., Gubb, I., Reilley, A., Skipper, Y., Doherty, H., O'Donnell, P. and Bowles, D.J. (1992) Nature (London) **360**, 62–65
25. Bowles, D.J. (1990) Annu. Rev. Biochem. **59**, 873–907
26. Smith, C.J.S., Slater, A. and Grierson, D. (1986) Planta **168**, 94–100
27. Hamilton, A.J., Lycett, G.W. and Grierson, D. (1990) Nature (London) **346**, 284–287
28. Hamilton, A.J., Bonzayer, M. and Grierson, D. (1991) Proc. Natl. Acad. Sci. U.S.A. **88**, 7434–7437
29. Spanu, P., Reinhart, D. and Boller, T. (1991) EMBO J. **10**, 2007–2013
30. Kauss, H. (1987) Annu. Rev. Plant Physiol. **38**, 47–72
31. Kauss, H. (1990) in The Plant Plasma Membrane (Larsson, C. and Moller, I.M., eds.), pp. 321–350, Springer-Verlag, Heidelberg
32. Delmer, D.P. (1987) Annu. Rev. Plant Physiol. **38**, 259–290
33. Robards, A.W. and Lucas, W.J.A. (1990) Annu. Rev. Plant Physiol. Plant Mol. Biol. **41**, 369–419
34. Darvill, A., Augar, C., Bergmann, C. *et al.* (1992) Glycobiology **2**, 181–198
35. Bradley, D.J., Kjellbom, P. and Lamb, C.J. (1992) Cell **70**, 21–30
36. Fry, S.C. (1986) Annu. Rev. Plant Physiol. **37**, 165–186
37. Sutherland, M.W. (1991) Physiol. Mol. Plant Pathol. **39**, 79–94
38. Alscher, R.G. (1989) Physiol. Plant **77**, 457–464
39. Anderson, J.M. (1989) in Second Messengers in Plant Growth and Development, pp. 181–212, Alan Liss, New York
40. Hamberg, M. and Gardner, W.H. (1992) Biochim. Biophys. Acta **1165**, 1–18
41. Pena-Cortes, H., Albrecht, T., Prat, S., Weiler, E.W. and Willmitzer, L. (1993) Planta **191**, 123–128
42. Farmer, E.E. and Ryan, C.A. (1992) Plant Cell **4**, 129–134
43. Bowles, D.J. (1991) Curr. Biol. **1**, 165–167
44. Walker-Simmons, M., Hadwiger, L. and Ryan, C.A. (1983) Biochem. Biophys. Res. Commun. **110**, 194–199
45. Hildemann, T., Ebneth, M., Pena-Cortez, H., Sanchez-Serrano, J.J., Willmitzer, L. and Prat, S. (1992) Plant Cell **4**, 1157–1170

46. Ryan, C.A. (1974) Plant Physiol. **54**, 328–332
47. Doherty, H.M., Selvendran, R.R. and Bowles, D.J. (1988) Physiol. Mol. Plant Pathol. **33**, 377–384
48. Enyedi, A.J., Yalpani, N., Silverman, P. and Raskin, I. (1992) Cell **70**, 879–886
49. Doherty, H.M. and Bowles, D.J. (1990) Plant Cell Environ. **13**, 851–855
50. Baydoun, E.A.-H. and Fry, S.C. (1985) Planta **165**, 269–276
51. MacDougall, A.J., Rigby, A.M., Needs, P.W. and Selvendran, R.R. (1992) Planta **188**, 566–574
52. Bowles, D.J. (1990) Nature (London) **343**, 314–315
53. Doherty, H.M. (1989) The wound response of tomato plants, Ph.D. thesis, University of Leeds
54. Nelson, C.E., Walker-Simmons, M., Makus, D., Zuroska, G., Graham, J. and Ryan, C.A. (1983) in Mechanisms of Plant Resistance to Insects (Hedin, P., ed.), pp. 103–122, American Chemical Society, Washington
55. McGurl, B., Pearce, G., Orozco-Cardenas, M. and Ryan, C.A. (1992) Science **255**, 1570–1573
56. Falkenstein, E., Groth, B., Mithöfer, A. and Weiler, E.W. (1991) Planta **185**, 316–322
57. Mason, H.S. and Mullet, J.E. (1990) Plant Cell **2**, 569–579
58. Farmer, E.E., Moloshok, T.D., Saxton, M.J. and Ryan, C.A. (1991) J. Biol. Chem. **226**, 3140–3145
59. Thain, J.F., Doherty, H.M., Bowles, D.J. and Wildon, D.C. (1990) Plant Cell Environ. **13**, 569–574

Biochem. Soc. Symp. **60**, 165–172
Printed in Great Britain

Regulation of gene expression in ripening fruits by sense and antisense genes

Don Grierson

BBSRC Research Group in Plant Gene Regulation, University of Nottingham, Department of Physiology and Environmental Science, Sutton Bonington Campus, Loughborough LE12 5RD, U.K.

Introduction

Ethylene plays an important regulatory role in many aspects of plant growth and development, including responses to pathogen attack and environmental stress, senescence of leaves and flowers, abscission and the ripening of climacteric fruits. The ethylene synthesized during these processes is believed to stimulate the production of new mRNAs and proteins that function to alter the metabolism or developmental fate of plant cells. Tomatoes, and other climacteric fruit, such as apple, pear and banana, greatly increase their production of ethylene at the onset of ripening. This stimulates new gene expression required to bring about the changes in colour, flavour, texture and aroma that are an integral part of the ripening process. The importance of ethylene can be demonstrated by supplying silver ions to fruit tissue. This suppresses ripening and is believed to function by interfering with ethylene perception or action. The process of tomato ripening affects all cell compartments. Metabolism of the cell walls leads to changes in texture; degradation of chlorophyll, dismantling of the photosynthetic apparatus and the synthesis of carotenoids are responsible for the conversion of chloroplasts to chromoplasts; there is also an increase in respiration and many other biochemical changes in response to ethylene. There is now good evidence that altered rates of synthesis of specific mRNAs and proteins are responsible for these events. Enzymes that show a clear increase in activity include polygalacturonase (PG) and pectinesterase (involved in cell wall metabolism), invertase, l-aminocyclopropane-l-carboxylate (ACC) synthase (for ethylene synthesis) and phytoene synthase (required for carotenoid production). Construction and differential screening of a ripening-related cDNA library from tomato led to the isolation of cDNAs for 19 mRNAs that accumulated during ripening [1]. Studies with transgenic plants showed that the expression of ripening genes could be inhibited specifically by introducing either antisense or sense genes into the genome via Ti plasmids [2]. This approach made it possible to test the function of specific ripening genes *in vivo* and also provided a means of altering aspects of the physiology and biochemistry of ripening in a directed manner. In this article I shall review the main features of antisense and sense gene inhibition of the PG gene, describe how antisense techniques were used to identify ACC oxidase [ethylene-forming enzyme (EFE)] and phytoene synthase genes, and discuss the over-expression and co-suppression of phytoene synthase genes in transgenic plants.

Regulation of the PG gene

Several endo- and exo-polygalacturonases have been described in ripening fruit, abscission zones and pollen; these are encoded by different genes and are regulated separately. The endopolygalacturonases expressed specifically during tomato ripening appear to be encoded by a single gene [3] whose translation product is modified post-translationally by glycosylation and/or association with a separate polypeptide (the β-subunit) to generate three distinct isoforms (Fig. 1) [4]. The accumulation of PG mRNA during ripening has been shown to be due to transcriptional activation of this ripening-specific PG gene [5]. When a 1.4 kb 5′ PG-gene-flanking region was fused to the chloramphenicol acetyltransferase (CAT) reporter gene, ripening-specific CAT activity was found in a small number of transgenic plants [3]. More recently (F. Nicholass, C.J.S. Smith, C.F. Watson, C.R. Bird, W. Schuch and D. Grierson, unpublished work), we have obtained 1000-fold higher levels of ripening-specific CAT expression employing a 5 kb 5′ PG-gene-flanking region to direct reporter gene expression. There are conflicting reports concerning the role of ethylene in stimulating PG gene expression. Treating ripening tomatoes with silver ions, which are believed to interfere with ethylene action, prevented the increase in PG mRNA [6], whereas inhibiting ethylene production by over 99% in transgenic plants containing ACC synthase antisense genes did not [7]. It remains to be established whether Ag^+ has some other effect, in addition to interfering with ethylene, or, alternatively, whether even trace levels of ethylene in the ACC synthase antisense fruit are sufficient to stimulate PG gene expression.

The accumulation of PG mRNA, and subsequent synthesis of the enzyme, was strongly inhibited by expressing antisense PG mRNA in transgenic tomato plants [8,9]. Down-regulation of the PG gene was obtained using either a full-length PG cDNA to construct the antisense gene [9] or a 730 bp (half-length) fragment from the 5′ end [8]. Interestingly, although the accumulation of PG mRNA from the endogenous gene was greatly reduced, transcription was unaffected, measured by RNA synthesis in isolated nuclei *in vitro* [9]. Furthermore, the amount of antisense RNA generated from the antisense gene driven by the CAMV-35S promoter was also reduced when the endogenous PG gene would normally have been switched on (see Fig. 3 in [8]), indicating a process involving mutual gene inactivation. These observations, and probably preconceived ideas about how antisense RNA would function, have tended to favour the idea that the sense and antisense RNA form duplexes, which render the transcripts unstable. So far, however, there has been no clear demonstration of the mechanism of action. The PG antisense genes were found to be subject to 'position effects', with the same construct being more effective in some transformants than others. In addition, inheritance studies led to five clear conclusions [10]. (i) The antisense genes were stably inherited. (ii) Inhibition of the endogenous PG gene co-segregated with the antisense

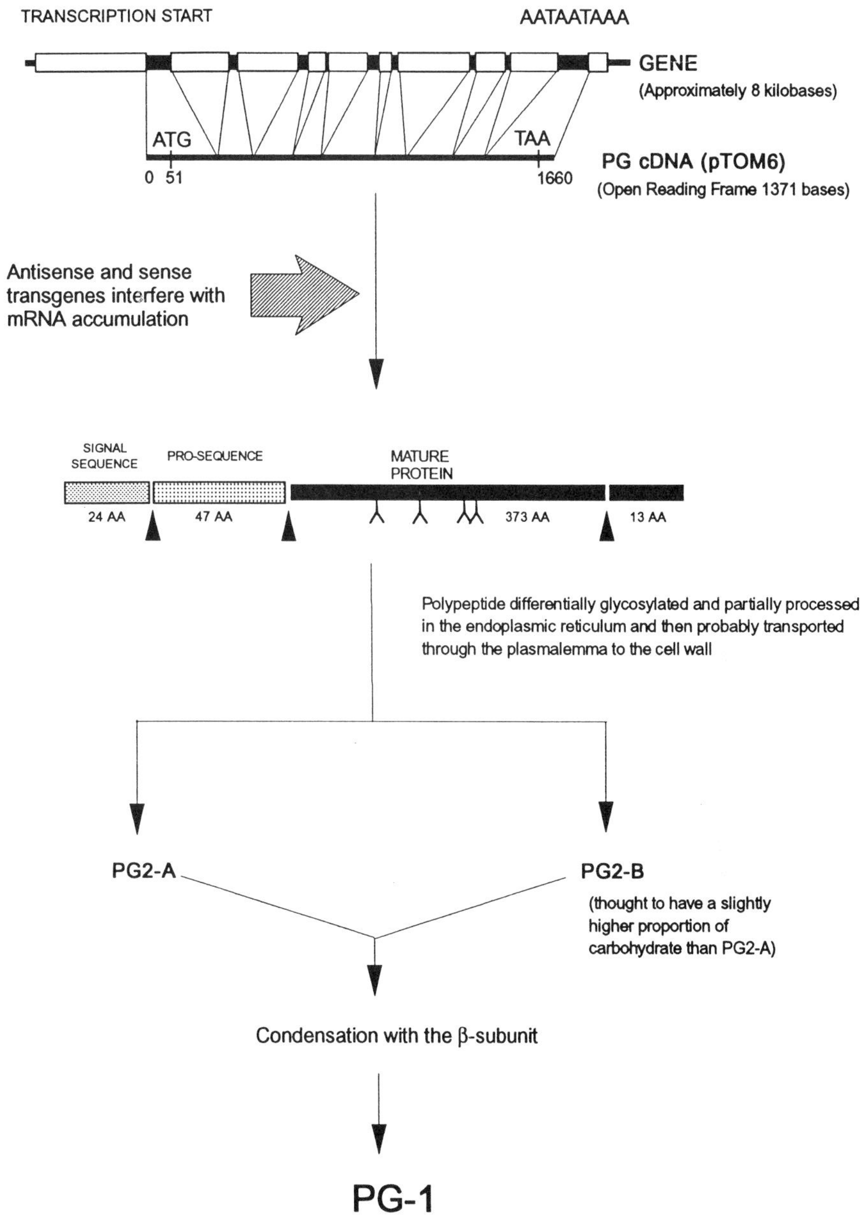

Fig. 1. Origin and control by sense and antisense genes of PG isoenzymes in ripening tomatoes.

genes. (iii) There was a gene-dosage effect, with a greater inhibition of the PG gene in progeny that were homozygous for the antisense gene compared with those that were hemizygous. (iv) Wild-type PG gene activity was recovered in progeny that did not inherit an antisense gene, indicating that no homologous recombination or permanent inactivation of the PG gene was occurring. (v) Only PG gene expression was down-regulated, and no other aspect of ripening was affected. In subsequent experiments it was found that using an identical 730 bp PG cDNA fragment in the sense orientation also caused down-regulation of the endogenous PG gene [11]. This surprising result was similar to the 'co-suppression' phenomenon reported for the inhibition of chalcone synthase in petunia flowers [12,13].

At the present time, the mechanisms of sense and antisense inhibition are not clear. It seems likely that they both involve nucleic acid base pairing, which would explain the specificity. There are, however, strong similarities between sense and antisense effects, and we have suggested that they might both involve similar mechanisms, perhaps by production of antisense RNA in both cases [14]. An alternative explanation, favoured by Jorgensen for sense effects, is that ectopic pairing between homologous DNA sequences is involved in the down-regulation [15], and it is possible that this could also apply to antisense genes.

Identification of ACC oxidase using antisense genes

The pathway for ethylene synthesis from methionine was established by Adams and Yang in 1979 [16]

$$\text{Met} \rightarrow \text{SAM} \rightarrow \text{ACC} \rightarrow C_2H_4$$

The conversion of SAM (*S*-adenosylmethionine) to ACC is catalysed by ACC synthase, and the conversion of ACC to ethylene by EFE (now renamed ACC oxidase). Despite the importance of ethylene in fundamental and applied research, it took at least 10 years for these enzymes to be purified and the genes to be cloned and identified. ACC synthase was present in relatively small amounts, and the ACC oxidase was believed to require intact membranes to function and defied all attempts at solubilization and purification. In the absence of any molecular probes for the identification of cDNAs involved in ethylene synthesis, we used a different approach. Screening a tomato-ripening-related cDNA library had identified 19 clones involved in ripening [1] and we thought these probably included clones encoding enzymes required for ethylene synthesis. We tested whether any homologous mRNA sequences were also expressed in mechanically wounded leaf tissue, where ethylene synthesis occurs rapidly. This screen showed that mRNAs related to the clone TOM13 were expressed during ripening and in response to wounding [17] and later work showed a similar mRNA also accumulated during leaf senescence [18], all situations involving ethylene synthesis. This correlation led us to suggest that the TOM13 mRNA was involved in ethylene synthesis [16]. Sequencing of TOM13 (and related genomic clones) gave no clue as to its function [19–21], so its postulated role in ethylene synthesis [17] was tested by inhibiting TOM13 mRNA accumulation in transgenic plants using an antisense gene [22]. These experiments showed that TOM13 mRNA was reduced in an antisense gene dosage-dependent manner. Ethylene production was inhibited during ripening by approximately 95% and the synthesis of ethylene in wounded leaves was also greatly reduced. Assay of tissue disks from the transgenic plants showed there was a corresponding reduction in ACC oxidase activity [22], suggesting that this enzyme was encoded by TOM13 (Fig. 2).

A direct confirmation of this was obtained by expressing the tomato TOM13 sequence in *Saccharomyces cerevisiae* [23]. Transgenic yeast were capable of converting

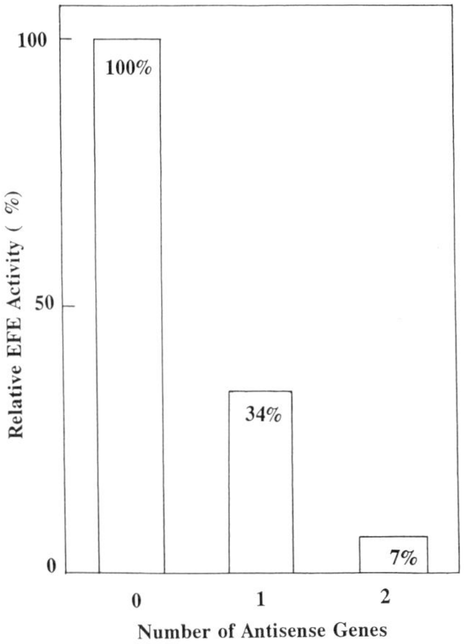

Fig. 2. ACC oxidase (EFE) activity in transgenic tomato plants with one or two TOM13 antisense genes.

ACC to ethylene, and the enzyme activity exhibited the same stereospecificity for isomers of ACC analogues as had been established for plant tissue sections [24]. These experiments showed, first, that ACC oxidase was a 35 kDa polypeptide which, on the basis of the predicted amino acid sequence, was unlikely to be a membrane-associated protein. Secondly, sequence similarity with flavanone 3-hydroxylase suggested that the two enzymes were related and that the ACC oxidase might require Fe^{2+} and ascorbate. This was confirmed in the transgenic yeast experiments [23]. Thirdly, there are at least three ACC oxidase genes in tomato, expressed during ripening, leaf senescence, wounding and probably abscission.

This was the first example of gene identification by an antisense approach. Once the properties of the enzyme were deduced from the molecular biology experiments, it proved relatively straightforward to solubilize ACC oxidase (now recognized as a dioxygenase) retaining full enzyme activity [25].

Analysis of the transgenic plants in which ACC oxidase was inhibited by antisense genes showed that the ripening of the fruit was retarded, and the shrivelling and spoilage associated with over-ripening was greatly reduced. The inhibition of ripening was increased if fruit were detached from the parent plant, indicating either that low levels of ethylene in the fruit rapidly diffused out after picking, or, alternatively, that some other factor from the plant could stimulate ripening in the presence of low levels of ethylene [26]. Interestingly, the senescence of the leaves of these transgenic plants was retarded, confirming many indirect experiments that had implicated ethylene in the control of leaf senescence. Experiments by Theologis and colleagues led to the identification of ACC synthase genes by more conventional procedures. Antisense inhibition of ACC synthase with 10 antisense genes almost completely inhibited ethylene synthesis and virtually stopped the ripening process [27].

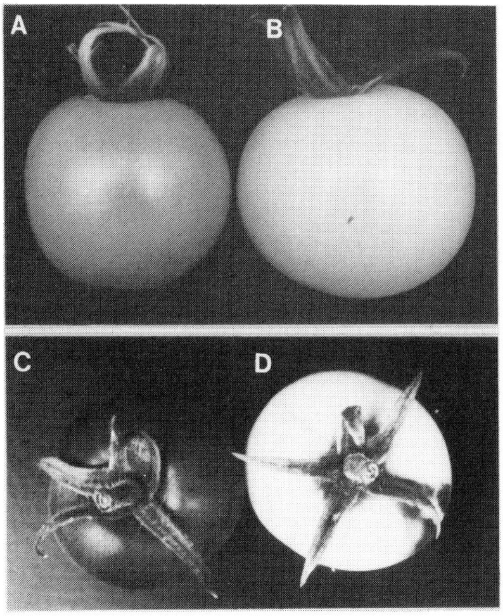

Fig. 3. Overexpression and co-suppression of phytoene synthase in transgenic tomatoes. The yellowflesh mutant turns yellow (B), not red, when it ripens, owing to the failure to accumulate lycopene and other carotenoids. The yellow pigment is naringenin chalcone, which is normally masked by carotenoids. Expressing a wild-type phytoene synthase cDNA in transgenic mutants restores lycopene production (A). Prior to ripening, wild-type and yellowflesh fruit are green (C) except in certain transgenic lines which contain a copy of a sense phytoene synthase gene capable of causing co-suppression. In such cases, green fruit can become white (D) by co-suppression of a separate phytoene synthase gene required for production of carotenoids in green tissue. The carotenoid-deficient chloroplasts are susceptible to photobleaching and turn white in the light. These white tomatoes turn yellow when they ripen, by producing naringenin chalcone.

Co-suppression and overexpression of phytoene synthase

Following the successful identification of ACC synthase using an antisense approach, TOM5 — another ripening-related gene — was shown by antisense inhibition to be involved in phytoene production and this evidence, plus sequence similarity to bacterial genes involved in carotenoid biosynthesis, indicated that it probably encoded a phytoene synthase [27]. A naturally occurring tomato mutant 'yellowflesh' was shown to be deficient in the wild-type TOM5 mRNA expressed in fruit [28]. This mutant lacks carotenoids, such as lycopene and β-carotene, when ripe and has a pale yellow colour due to the production of naringenin chalcone, which is masked by carotenoids in wild-type tomatoes. The mutant TOM5 mRNA was shown to be altered, at the 3′ end, by the insertion of an unrelated sequence into the phytoene synthase gene normally expressed in fruit. This explained why the phytoene synthase activity was absent in yellowflesh fruit. When the mutant was transformed with a wild-type TOM5 cDNA, under control of the CAMV-35S promoter, the mutation was complemented (Fig. 3). Lycopene production during ripening was restored to wild-type levels, and unscheduled pigment synthesis was also detected in callus tissue and abscission zones. These experiments confirmed that the TOM5 sequence encoded phytoene synthase and explained the basis of the yellowflesh mutation.

Interestingly, white tomatoes were obtained from some transgenic plants containing the CAMV-35S–TOM5 construct designed to overexpress phytoene synthase [28]. These transgenic yellowflesh fruit were white at the 'mature green' stage, due to photobleaching caused by lack of carotenoids. As these fruit ripened, they turned yellow by the accumulation of naringenin chalcones (Fig. 3). Green leaves from these plants were also susceptible to photobleaching [28]. These results indicated that in some transformants co-suppression of a different but related phytoene synthase gene, normally expressed in green tissues, was occurring. They also showed that the same gene construct can lead to either overexpression, when the yellowflesh mutation was complemented, or co-suppression, when the 'green' phytoene synthase gene was inhibited. Since the only known difference between the various transgenic plants was the site of insertion of the transgene, this highlights the important role of position effects in phenomena such as co-suppression [28]. These experiments show that co-suppression of homologous sequences is not an inevitable consequence of transgenic experiments, (see [11–13]) and normal overexpression can occur. The challenge is, of course, to understand the molecular basis of the position effects which govern the behaviour of the transgene.

This work was supported by grants from the AFRC, SERC and BBSRC. I wish to thank Alan Evans for producing Fig. 1 and Rupert Fray for providing the photographs for Fig. 3.

References

1. Gray, J., Picton, S., Shabbeer, J., Schuch, W. and Grierson, D. (1992) Plant Mol. Biol. **19**, 60–87
2. Smith, C.J.S., Hamilton, A.J. and Grierson, D. (1993) in Control of Plant Gene Expression (D.P.S. Verma, ed.), pp. 535–546, CRC Press, Boca Raton
3. Bird, C.R., Smith, C.J.S., Ray, J.A., Moureau, P., Bevan, M.W., Bird, A.S., Hughes, S., Morris, P.C., Grierson, D. and Schuch, W. (1988) Plant Mol. Biol. **11**, 651–662
4. Hobson, G.E. and Grierson, D. (1993) in Biochemistry of Fruit Ripening G. Seymour, J. Taylor and G. Tucker, eds.), pp. 405–442, Chapman and Hall, London
5. Dellapenna, D., Lincoln, J.E., Fischer, R.L. and Bennett, A.B. (1989) Plant Physiol. **90**, 1372–1377
6. Davies, K.M., Hobson, G.E. and Grierson, D. (1988) Plant Cell Environ. **11**, 729–738
7. Oeller, P.W., Min-Wong, L.M., Taylor, L.P., Pike, D.A. and Theologis, A. (1991) Science **254**, 437–439
8. Smith, C.J.S., Watson, C.F., Ray, J., Bird, C.R., Morris, P.C., Schuch, W. and Grierson, D. (1988) Nature (London) **334**, 724–726
9. Sheehy, R.E., Kramer, M. and Hiatt, W.R. (1988) Proc. Natl. Acad. Sci. U.S.A. **85**, 8805–8809

10. Smith, C.J.S., Watson, C.F., Morris, P.C., Bird, C.R., Seymour, G.B., Gray, J.E., Arnold, C., Tucker, G.A., Schuch, W., Harding, S. and Grierson, D. (1990) Plant Mol. Biol. **14**, 369–379
11. Smith, C.J.S., Watson, C.F., Bird, C.R., Ray, J., Schuch, W. and Grierson, D. (1990) Mol. Gen. Genet. **224**, 477–481
12. Van der Krol, A.R., Mur, L.A., Beld, M., Mol, J.N.M. and Stuitje, A.R. (1990) Plant Cell **2**, 291–299
13. Napoli, C., Lemieux, C. and Jorgensen, R. (1990) Plant Cell **2**, 279–289
14. Grierson, D., Fray, R.G., Hamilton, A.J., Smith, C.J.S. and Watson, C.F. (1991) Trends Biotechnol. **9**, 122–123
15. Jorgensen, R. (1990) Trends Biotechnol. **8**, 340–344
16. Adams, D.O., and Yang, S.F. (1979) Proc. Natl. Acad. Sci. U.S.A. **76**, 170–174
17. Smith, C.J.S., Slater, A. and Grierson, D. (1986) Planta **168**, 94–100
18. Davies, K.M. and Grierson, D. (1989) Planta **179**, 73–80
19. Holdsworth, M.J., Bird, C.R., Roy, J., Schuch, W. and Grierson, D. (1987) Nucleic Acids Res. **15**, 731–739
20. Holdsworth, M.J., Schuch, W. and Grierson, D. (1988) Plant Mol. Biol. **11**, 81–88
21. Holdsworth, M.J., Schuch, W. and Grierson, D. (1987) Nucleic Acids Res. **15**, 10600
22. Hamilton, A.J., Lycett, G.W. and Grierson, D. (1990) Nature (London) **346**, 284–287
23. Hamilton, A.J., Bouzayen, M. and Grierson, D. (1991) Proc. Natl. Acad. Sci. U.S.A. **88**, 7334–7337
24. Hoffman, N.E., Yang. S.F., Ichihara, A. and Sakamura, S. (1982) Plant Physiol. **70**, 195–199
25. John, P. (1991) Plant Mol. Biol. Rep. **9**, 192–194
26. Picton, S., Barton, S.L., Bouzayer, M., Hamilton, A.J. and Grierson, D. (1993). Plant J. **3**, 469–481
27. Bird, C.R., Ray, J.A., Fletcher, J.D., Boniwell, J.M., Bird, A.S., Teulieres, C., Blain, I., Bramley, P.M. and Schuch, W. (1991) Bio/Technology **9**, 635–639
28. Fray, R.G. and Grierson, D. (1993) Plant Mol. Biol. **22**, 589–602

Biochem. Soc. Symp. **60**, 173–182
Printed in Great Britain

Perception and transduction of an elicitor signal in cultured parsley cells

Thorsten Nürnberger, Christiane Colling, Klaus Hahlbrock, Thorsten Jabs, Annette Renelt, Wendy R. Sacks and Dierk Scheel*

Max-Planck-Institut für Züchtungsforschung, Abteilung Biochemie, Carl-von-Linné-Weg 10, D-50829 Köln, Germany

Synopsis

Treatment of cultured parsley cells or protoplasts with a purified extracellular glycoprotein from *Phytophthora megasperma* f.sp. *glycinea* induces the transcription of the same set of defence-related genes as is activated in parsley leaves upon infection. Elicitor activity was shown to reside in a specific portion of the protein moiety which was isolated, sequenced and synthesized. Partial cDNAs encoding the entire mature protein as well as other related proteins have been isolated, indicating the presence of a small gene family. The elicitor-active oligopeptide is located in the *C*-terminal portion of the deduced amino acid sequence. Binding of the elicitor to target sites on the parsley plasma membrane appears to be the initial event in defence gene activation. The subsequent intracellular transduction of the elicitor signal was shown to involve rapid and transient influxes of Ca^{2+} and H^{+}, as well as effluxes of K^{+} and Cl^{-}. Inhibition of elicitor-induced ion fluxes by channel blockers also inhibited phytoalexin synthesis, while stimulation of similar ion fluxes by treatment of cells or protoplasts with the polyene antibiotic, amphotericin B, induced the production of phytoalexins and activated the complete set of defence-related genes in the absence of elicitor.

Introduction

Most plants are resistant to the majority of potential pathogens in their environment (species or basic resistance). Relatively few true host/pathogen interactions have evolved in which the pathogen is virulent and the plant is susceptible (basic compati-

*To whom correspondence should be addressed.

bility). Specific cultivars of such susceptible species are highly resistant to certain races of the pathogen (cultivar resistance) [1,2].

Both types of resistance are believed to be caused by the initiation of a multi-component defence response which is similar or identical in cultivar and species resistance (Table 1) [3]. While many elements of the defence reactions are initiated by activation of specific genes, the oxidative burst and the accumulation of callose appear to result from specific enzyme activation [4,5]. The molecular basis of local necrosis, the so-called hypersensitive response, is not well understood, but may not be causally linked to transcriptional activation of plant defence genes [6].

The induction of defence responses requires efficient perception of appropriate signals by the plant cell and intracellular transmission to the specific sites of activation. The spatial and temporal complexity of successful resistance reactions make it difficult to investigate these processes in the infected plant. For our studies on signalling in plant defence, we have, therefore, employed a system of reduced complexity consisting of cultured parsley cells and an elicitor from *P. megasperma* f.sp. *glycinea*, a fungal soybean pathogen [7]. Parsley exhibits typical species resistance to this pathogen. Treatment of suspension-cultured parsley cells or protoplasts with fungal elicitor results in the activation of most of the defence responses that are also observed after infection of leaves with fungal zoospores [8–10]. However, cell death and accumulation of callose are not stimulated by these elicitors.

Elicitor characterization

The accumulation of furanocoumarin phytoalexins in the culture medium of elicitor-treated parsley cells or protoplasts is readily detected under u.v.-light, owing to the autofluorescence of these substances [11]. This particular defence response was, therefore, chosen as an indicator of defence-gene activation in our attempts to isolate an elicitor from *P. megasperma* f.sp. *glycinea* [12,13]. A 42 kDa glycoprotein, which stimulated the same pattern of defence responses that are activated by the crude elicitor preparation, was purified from the culture filtrate of *P. megasperma* f.sp. *glycinea*. The elicitor activity was insensitive to heating and chemical or enzymic deglycosylation of

Table 1. Typical pathogen defence responses of plants.

Defence response	Mechanism of activation
Hypersensitive response	Unknown
Oxidative burst	Enzyme activation
Phytoalexin accumulation	Gene activation
Cell wall modifications	
Callose apposition	Enzyme activation
Impregnation with phenolics	Gene activation
Incorporation of proteins	Gene activation
Accumulation of hydrolytic enzymes	Gene activation
Accumulation of pathogenesis-related proteins	Gene activation

the glycoprotein. However, digestion with Pronase or trypsin completely destroyed elicitor activity. These results suggested that elicitor activity was a property of amino acid sequence rather than native structure, and that the carbohydrate portion was not required.

Amino acid sequences of the *N*-terminus and tryptic fragments of the glycoprotein were used to design degenerate oligonucleotides which were employed as primers in the polymerase chain reaction (PCR) with fungal DNA as template (Fig. 1). The resulting PCR product was then used to detect corresponding cDNA clones in a fungal cDNA library. Three different types of cDNA were identified which displayed 80–90% sequence similarity among themselves and the PCR product. These results, as well as Southern blot experiments, supported the existence of a small gene family. The

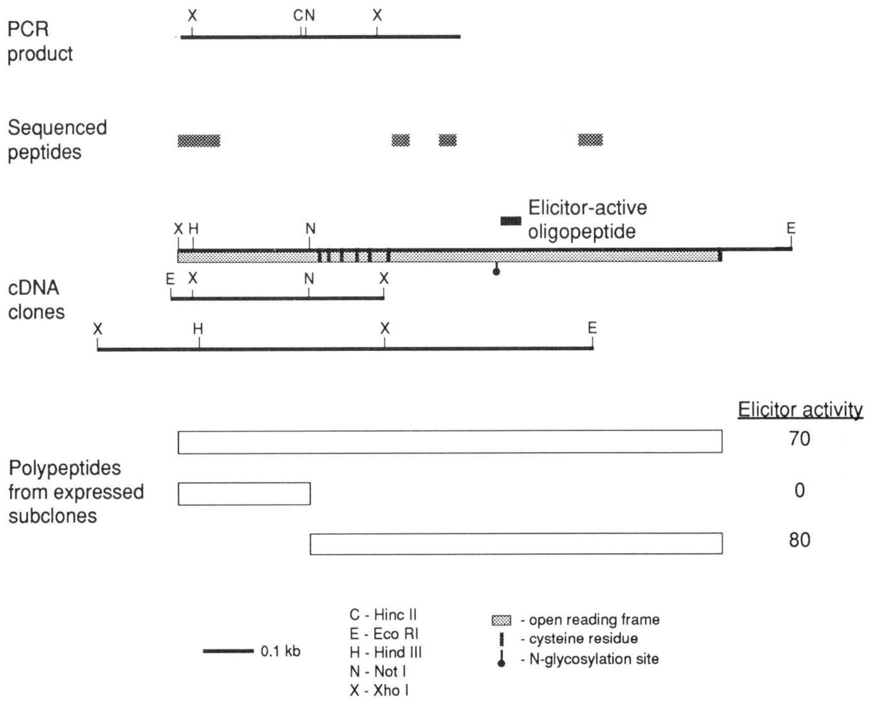

Fig. 1. Schematic representation of cDNA clones encoding isoforms of the mature 42 kDa glycoprotein elicitor. Portions of cDNAs encoding related polypeptides, with respect to sequenced regions of the purified glycoprotein, the PCR product used to isolate the clones, and the location of the elcitor-active oligopeptide are indicated. Regions expressed as fusion proteins in *E. coli*, as well as their elicitor activity, are also shown. Elicitor activity was determined by measuring the accumulation of phytoalexins in the media of parsley protoplasts 24 h after addition of the various preparations. The activity is given as a percentage amount of phytoalexins produced by protoplasts in response to treatment with equivalent concentrations of the purified glycoprotein.

largest cDNA clone was found to encode the entire mature protein, a 3′-untranslated region and a poly (A) tail, but lacked a signal sequence and 5′-untranslated region.

The complete cDNA or portions encoding the *N*-terminal 25% or the *C*-terminal 75% of the mature protein were cloned into an *Escherichia coli* expression vector [14]. The resulting proteins expressed as fusions with glutathione *S*-transferase were tested for elicitor activity in parsley protoplasts. The fusions containing the mature protein, or its *C*-terminal 75%, stimulated phytoalexin accumulation, while the *N*-terminus and the protein expressed from the vector alone were inactive (Fig. 1).

To further localize elicitor activity within the protein, we tested different proteases for their abilities to digest the glycoprotein and to release peptide fragments that retained elicitor activity [15]. Endoproteinase Glu-C was the only protease that generated an elicitor-active oligopeptide from the glycoprotein elicitor. This peptide was sequenced and located within the *C*-terminal portion of the elicitor (Fig. 1). The smallest fragment with elicitor activity consisted of 13 amino acids.

Elicitor perception

Recognition of pathogens by plants is believed to be mediated by specific binding of pathogen or plant-derived elicitors to receptor-like target sites on the plant plasma membrane [16]. Such elicitor-binding sites can be detected and characterized by incubation of membrane preparations or protoplasts with a radioactively labelled elicitor, followed by separation of bound and unbound ligand. It is, however, important to label the elicitor such that its biological activity is retained. We introduced ^{125}I into the glycoprotein elicitor by three different methods and determined the elicitor activity of the labelled protein (Table 2). Direct iodination of tyrosine residues resulted in loss of elicitor activity, whereas coupling of iodinated 4-hydroxyphenylproprionate from Bolton–Hunter reagent to primary amine functions had no adverse effect. In binding experiments with protoplasts or membrane preparations and the Bolton–Hunter-labelled elicitor, we were able to detect competable binding sites on the parsley plasma membrane [17]. However, the extreme hydrophobicity of the glycoprotein, together with the apparent low abundance of the elicitor target sites, precluded detailed analysis of the binding site with this ligand. We will now use the less hydrophobic, elicitor-active oligopeptide for binding studies.

Signal transduction

The most rapid responses of cultured parsley cells to treatment with either the crude or the peptide elicitors from *P. megasperma* f.sp. *glycinea* were changes in the ion permeability of the plasma membrane [18]. Within 2–5 min of elicitor addition, increased influxes of Ca^{2+} and H^{+} and effluxes of K^{+} and Cl^{-} were observed that lasted for 2–3 h. Omission of Ca^{2+} from the culture medium, or inhibition of the ion fluxes with ion-channel inhibitors, drastically depressed elicitor-stimulated accumulation of phytoalexins (Table 3, Fig. 2) and activation of defence-related genes [16,18]. Interestingly, the accumulation of flavonoids in response to u.v.-irradiation of parsley cells did not require extracellular Ca^{2+} (Fig. 2), although flavonoids, like the furanocou-

Table 2. Elicitor activity of radioactively labelled 42 kDa glycoprotein.

Labelling method	Specific radioactivity (Ci/mmol)	Elicitor activity (% of unlabelled glycoprotein)
Direct iodination		
Iodobeads (Pierce) (immobilized chloramine T)	280	50
Enzymobeads (BioRad) (immobilized lactoperoxidase)	770	12
Prosthetic group		
Bolton–Hunter reagent [*N*-succinimidyl-3-(4-hydroxyphenyl)proprionate]	300–1200	100

Table 3. Effects of Ca^{2+}-channel blockers on elicitor-stimulated Ca^{2+} uptake and phytoalexin accumulation in cultured parsley cells and protoplasts, respectively.

Inhibitor	Target Ca^{2+} channel	IC_{50} (μM) phytoalexin accumulation	Ca^{2+} uptake
Benzoethiazipines			
Diltiazem	L-type	No effect	No effect
Diphenylbutylpiperidines	L-type	~50	Slight inhibition
Phenylalkylamines			
Bepridil	L-type	No effect	No effect
Methoxyverapamil	L-type	No effect	No effect
Verapamil	L-type	No effect	No effect
1,4-Dihydropyridines			
Isradipine	L-type	~25	Not determined
Nifedipine	L-type	~50	No effect
Piperazines			
Cinnarizine	T-type	~20	~50
Flunarizine	T-type	~16	~50

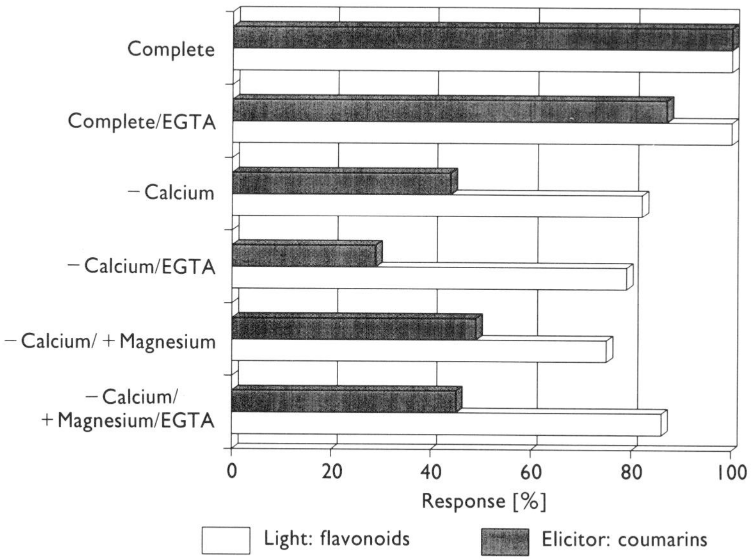

Fig. 2. Ca^{2+} requirement of the responses of cultured parsley cells to treatment with elicitor or u.v.-containing white light. Accumulation of furanocoumarin phytoalexins in response to elicitor or of flavonoids upon u.v. irradiation was measured spectrophotometrically 24 h after initiation of treatment. The responses were determined in complete media, media lacking Ca^{2+} and media in which Ca^{2+} had been replaced by Mg^{2+}. In all cases, the response was also measured in the presence of 1 mM EGTA.

marin phytoalexins, originate from the general phenylpropanoid pathway. Most known inhibitors of Ca^{2+} channels had no effect on the elicitor-stimulated ion fluxes. Only the piperazines, cinnarizine and flunarizine, at low concentrations, reduced both Ca^{2+} uptake and phytoalexin synthesis (Table 2). Stimulation of qualitatively and quantitatively similar ion fluxes by Amphotericin B in the absence of elicitor activated the same set of genes as was activated by elicitor and also resulted in the accumulation of phytoalexins [18], suggesting that elicitor-responsive ion channels in the plasma membrane of parsley cells are important elements in elicitor signal transduction.

The elicitor-stimulated, transient increase in Ca^{2+}-channel activity was shown to result in elevated levels of cytosolic Ca^{2+}, which increased from 150 nM in untreated protoplasts to 350 nM within 30–45 min of addition of elicitor [19]. Although calmodulin antagonists inhibited the accumulation of phytoalexins, suggesting the involvement of calmodulin, the level of its transcript appeared to be unaffected by elicitor treatment of parsley cells or protoplasts [15]. However, the pattern of calmodulin-binding proteins in elicitor-treated cells differed from that of untreated cells by the appearance of at least one additional band [15], suggesting the possible involvement of calmodulin-regulated proteins, such as protein kinases or phosphatases, in this signalling process.

In vivo phosphorylation experiments with elicitor-treated parsley cells indeed resulted in the transient appearance of an elicitor-specific pattern of phosphorylated proteins on two-dimensional gels [20]. This elicitor response also required the presence of Ca^{2+} in the culture medium. The appearance of the phosphorylated proteins fits well in the time frame between stimulation of ion fluxes and activation of defence-related genes. However, the protein kinase inhibitors K-252a and staurosporine did not affect phytoalexin accumulation, whereas the protein phosphatase inhibitor okadaic acid efficiently blocked this elicitor response [17]. The elicitor-stimulated changes in phosphorylation patterns of proteins may, therefore, be caused predominantly by specific elicitor/Ca^{2+}/calmodulin-regulated protein phosphatases, although the involvement of elicitor-responsive protein kinases that are insensitive to K-252a and staurosporine can not be excluded.

Since the key elements of most signalling cascades known to exist in animals also appear to be present in plants [21], we developed a method that enabled us to test their possible involvement in elicitor signal transduction in parsley. Protoplasts were electroporated under conditions that facilitated the introduction of low-molecular-mass compounds without significantly reducing their elicitor responsiveness or viability [17]. We then loaded these permeabilized protoplasts with compounds, such as stable GTP analogues, known to interfere with specific signalling events. In addition, we determined the cytosolic levels of cyclic AMP in elicitor-treated and -untreated cells. Although cyclic AMP, as well as heterotrimeric GTP-binding proteins, appeared to be present in parsley cells, we were unable to obtain convincing evidence for their involvement in elicitor signal transduction.

Inositol phosphate metabolites, however, interfere with elicitor signalling in parsley [17]. Although none of the inositol phosphate metabolites loaded into permeabilized parsley protoplasts activated defence responses in the absence of elicitor, inositol 1,4,5-trisphosphate and its metabolically stable analogue, inositol 1,4,5-trisphosphorothioate, significantly enhanced elicitor-induced phytoalexin accumulation. In addition, neomycin, an inhibitor of phospholipase C, drastically reduced elicitor-stimulated phytoalexin accumulation at concentrations that only slightly affected cell viability. Inositol 1,4,5-trisphosphate might, therefore, be involved in elicitor-stimulated release of Ca^{2+} from internal stores, or in the regulation of ion channels in the plasma membrane. However, extensive additional experimental work is required to verify any causal relationships.

Discussion

Plant cells appear to recognize potential pathogens by binding of signal molecules, the so-called elicitors, to receptor-like target sites on their plasma membrane [16]. The elicitors, which are of diverse structure, are either derived from surface components of the pathogen or are released from the plant cell wall during pathogenesis [16,22]. Pure elicitor molecules appear to be recognized by only a single (or a few closely related) plant species and, in the case of race/cultivar-specific elicitors, only by certain plant cultivars [23,24].

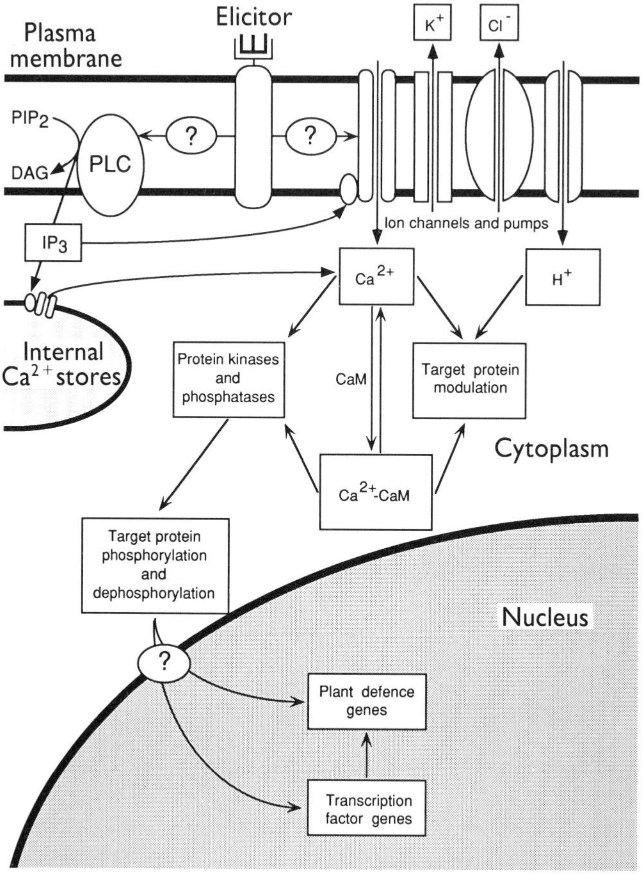

Fig. 3. Hypothetical model of elicitor signal transduction in parsley.

The existence of specific binding sites for an elicitor on the plant plasma membrane has been unequivocally demonstrated only for a glucan elicitor from *P. megasperma* f.sp. *glycinea* in soybean, the host plant of this fungal pathogen [25–29]. This elicitor binds with high affinity to a low-abundance 70 kDa protein of soybean plasma membranes [28]. The glucan elicitor is not recognized by parsley cells, which appear to specifically bind the 42 kDa glycoprotein from mycelial walls of the same fungus [17,30]. On the other hand, the glycoprotein is not recognized by soybean cells and does not compete for binding of the glucan elicitor to soybean membranes [16].

In both plants, fragments of extracellular fungal macromolecules represent the smallest elicitor-active compounds: a hepta-β-glucoside in soybean [31] and an oligopeptide in parsley. Comparable results have also been reported for oligogalacturonide elicitors [22]. The minimal structural requirements for most proteinaceous elicitors, however, are not known. One possible exception is specific glycopeptides proteolytically released from yeast invertase, which elicit the defence response in tomato [32]. While the important role of oligosaccharides as signal molecules in plants is well documented [33], examples of oligopeptide signals are still an exception in plants. Systemin, an oligopeptide that serves as a highly specific systemic signal in potato and tomato, clearly demonstrates the potential importance of peptide signals in plants [34]. In addition, a novel thyrotropin-releasing, hormone-like tripeptide with unknown function has recently been found to be present in alfalfa at high concentration, suggesting that plants, like animals, employ oligopeptides in signalling [35].

Our current knowledge of elicitor signal perception and transduction in parsley is summarized in a working hypothesis shown in Fig. 3. In soybean and parsley, the intracellular signals possibly generated by binding of the elicitors to their target sites involve transient changes in the activities of plasma membrane-located ion channels and changes in the phosphorylation status of proteins [18,20,36,37]. Elicitor-stimulated phosphorylation of specific proteins has also been observed in tomato, where results from inhibitor experiments indicate the involvement of elicitor-responsive protein kinases [38,39]. The role of inositol 1,4,5-trisphosphate, and possibly other inositol phosphates, in elicitor signalling must be investigated more thoroughly. In animal cells, these compounds are important elements in the regulation of cytosolic Ca^{2+} levels [40]. In principle, the same function could be fulfilled by these substances in plants as well [41].

We thank Magdalena Jung and Katharina Adamitza for excellent technical assistance. This work was supported by the Deutsche Forschungsgemeinschaft (Sche 235/3-1) and by doctoral and postdoctoral fellowships from Bayer Chemical Company, Leverkusen, to A.R. and T.N.

References

1. Bailey, J.A. (1983) in The Dynamics of Host Defence (Bailey, J.A. and Deverall, B.J., eds.), pp. 1–32, Academic Press, New York
2. Ellingboe, A.H. (1984) Adv. Plant Pathol. 2, 131–151
3. Hahlbrock, K. and Scheel, D. (1987) in Innovative Approaches to Plant Disease Control (Chet, I., ed.), pp. 229–254, Wiley, New York
4. Sutherland, M.W. (1991) Physiol. Mol. Plant Pathol. 39, 79–93
5. Kauss, H., Waldmann, T., Jeblick, W., Euler, G., Ranjewa, R. and Domard, A. (1989) in Signal Molecules in Plants and Plant–Microbe Interactions (Lugtenberg, B.J.J., ed.), pp. 107–116, Springer-Verlag, Berlin

6. Jakobek, J.L. and Lindgren, P.B. (1993) Plant Cell **5**, 49–56
7. Jahnen, W. and Hahlbrock, K. (1988) Planta **173**, 197–204
8. Scheel, D., Hauffe, K.D., Jahnen, W. and Hahlbrock, K. (1986) in Recognition in Microbe–Plant Symbiotic and Pathogenic Interactions (Lugtenberg, B., ed.), pp. 325–331, Springer-Verlag, Berlin
9. Dangl, J.L., Hauffe, K.D., Lipphardt, S., Hahlbrock, K. and Scheel, D. (1987) EMBO J. **6**, 2551–2556
10. Schmelzer, E., Krüger-Lebus, S. and Hahlbrock, K. (1989) Plant Cell **1**, 993–1001
11. Hauffe, K.D., Hahlbrock, K. and Scheel, D. (1986) Z. Naturforsch. **41c**, 228–239
12. Parker, J.E., Knogge, W. and Scheel, D. (1991) in Phytophthora (Lucas, J.A., Shattock, R.C., Shaw, D.S. and Cooke, L.R., eds.), pp. 90–103, Cambridge University Press, Cambridge
13. Parker, J.E., Schulte, W., Hahlbrock, K. and Scheel, D. (1991) Mol. Plant–Microbe Interact. **4**, 19–27
14. Sacks, W.R., Hahlbrock, K. and Scheel, D. (1993) in Mechanisms of Defence Responses in Plants (Fritig, B. and Legrand, M., eds.), p. 144–147, Kluwer, Dordrecht
15. Sacks, W.R., Ferreira, P., Hahlbrock, K., Jabs, T., Nürnberger, T., Renelt, A. and Scheel, D. (1993) Adv. Mol. Genet. Plant–Microbe Interact. **2**, 485–495
16. Ebel, J. and Scheel, D. (1992) in Genes Involved in Plant Defence (Boller, T. and Meins, F., eds.), pp. 183–205, Springer-Verlag, Wein
17. Renelt, A., Colling, C., Hahlbrock, K., Nürnberger, T., Parker, J.E., Sacks, W.R. and Scheel, D. (1993) J. Exp. Bot. **44** (Suppl.), 257–268
18. Scheel, D., Colling, C., Hedrich, R., Kawalleck, P., Parker, J.E., Sacks, W.R., Somssich, I.E. and Hahlbrock, K. (1991) Adv. Mol. Genet. Plant–Microbe Interact. **1** 373–380
19. Renelt, A. (1992) Ph.D. thesis, University of Köln, Germany
20. Dietrich, A., Mayer, J.E. and Hahlbrock, K. (1990) J. Biol. Chem. **265**, 6360–6368
21. Boss, W.F. and Morré, D.J., eds. (1989) Second Messengers in Plant Growth and Development, Alan R. Liss, New York
22. Darvill, A.G. and Albersheim, P. (1984) Annu. Rev. Plant. Physiol. **35**, 243–275
23. Scheel, D. and Parker, J.E. (1990) Z. Naturforsch. **45c**, 569–575
24. Knogge, W. (1991) Z. Naturforsch. **46c**, 969–981
25. Schmidt, W.E. and Ebel, J. (1987) Proc. Natl. Acad. Sci. U.S.A. **84**, 4117–4121
26. Cosio, E.G., Pöpperl, H., Schmidt, W.E. and Ebel, J. (1988) Eur. J. Biochem. **175**, 309–315
27. Cosio, E.G., Frey, T. and Ebel, J. (1990) FEBS Lett. **264**, 235–238
28. Cosio, E.G., Frey, T. and Ebel, J. (1992) Eur. J. Biochem. **204**, 1115–1123
29. Cheong, J.-J. and Hahn, M. (1991) Plant Cell **3**, 137–147
30. Parker, J.E., Hahlbrock, K. and Scheel, D. (1988) Planta **176**, 75–82
31. Sharp, J.K., McNeil, M. and Albersheim, P. (1984) J. Biol. Chem. **259**, 11321–11336
32. Basse, C.W., Bock, K. and Boller, T. (1992) J. Biol. Chem. **267**, 10258–10265
33. Ryan, C.A. (1988) Biochemistry **27**, 8879–8883
34. Pearce, G., Strydom, D., Johnson, S. and Ryan, C.A. (1991) Science **253**, 895–898
35. Lackey, D.B. (1992) J. Biol. Chem. **267**, 17508–17511
36. Ebel, J., Cosio, E.G., Feger, M., Frey, T., Kissel, U., Reinold, S. and Waldmüller, T. (1993) Adv. Mol. Gen. Plant–Microbe Interact. **2**, 477–484
37. Grab, D., Feger, M. and Ebel, J. (1989) Planta **179**, 340–348
38. Grosskopf, D.G., Felix, G. and Boller, T. (1990) FEBS Lett. **275**, 177–180
39. Felix, G., Grosskopf, D.G., Regenass, M. and Boller, T. (1991) Proc. Natl. Acad. Sci. U.S.A. **88**, 8831–8834
40. Berridge, M.J. (1993) Nature (London) **361**, 315–325
41. Drøbak, B.K. (1992) Biochem. J. **288**, 697–712

Biochem. Soc. Symp. **60**, 183–197
Printed in Great Britain

Ion channels and calcium signalling in plants: multiple pathways and cross-talk

Dale Sanders*, James M. Brosnan, Shelagh R. Muir, Gethyn Allen, Alan Crofts and Eva Johannes

Biology Department, University of York, York YO1 5DD, U.K.

Introduction

There is increasing evidence that changes in cytosolic free calcium concentration ($[Ca^{2+}]_c$) provide an essential component in stimulus–response coupling in plant cells. This evidence has been generated by the effective application of several classes of probe for measurement of $[Ca^{2+}]_c$, including fluorescent indicators, the bioluminescent indicator aequorin and ion-selective electrodes [1,2]. Work with fluorescent probes, in particular, has lent itself to digital imaging technologies and confocal microscopy, with the result that measurement of local, rather than spatially averaged, free Ca^{2+} has become possible within a single cell.

Table 1 lists a selection of environmental signals which have been shown to elicit a change in $[Ca^{2+}]_c$, together with the respective biochemical or physiological responses which ensue signal perception. These measurements, of a stimulus-evoked change in $[Ca^{2+}]_c$ preceding a cellular response, provide circumstantial evidence for a role of Ca^{2+} in stimulus–response coupling. A rigorous appraisal of the notion that $[Ca^{2+}]_c$ is an obligatory element in a given transduction pathway relies also on the demonstration that the cellular response occurs after artificial modulation of $[Ca^{2+}]_c$ (e.g. by iontophoresis), and that the pathway is blocked when the change in $[Ca^{2+}]_c$ is inhibited (e.g. with a Ca^{2+} buffer).

Exploration of these latter criteria presents technical difficulties in multicellular tissue with intercellular coupling, but where attempted (for example, on stomatal guard cells [3]), such studies have reinforced the conclusion that a rapid stimulus-evoked change in $[Ca^{2+}]_c$ plays a major role in signal transduction. In addition, a number of putative effector proteins, competent in responding to changes in free Ca^{2+} in the low or sub-micromolar range, have been identified in plants. Such targets include enzymes of intermediary metabolism [4], Ca^{2+}-dependent protein kinases [5], ion pumps [6,7] and ion channels [8].

*To whom correspondence should be addressed.

Table 1. Some examples of Ca^{2+} mediation in stimulus–response coupling.

Tissue	Stimulus	Effect on $[Ca^{2+}]_c$	Response	Reference
Charophyte	Electrical	Transient increase	Action potential	62
Charophyte	Photosynthesis	Sustained decrease	Sucrose synthesis	4, 63
Mougeotia	Blue light	Sustained increase	Chloroplast movement	64
Wheat leaf protoplast	Red light	Transient increase	Swelling	65
Guard cells	Abscissic acid	Transient increase	Stomatal closure	66
Maize coleoptile	Gravity	Sustained increase	Gravitropism	67
Tobacco cotyledons	Cold shock	Transient increase	Temperature adaptation	68
Tobacco cotyledons	Fungal elicitors	Transient increase	Phytoalexin production	68
Tobacco cotyledons	Touch	Transient increase	Growth slows	68

The information in Table 1 gives rise to two questions which are of fundamental importance for our understanding of Ca^{2+}-mediated intracellular signalling. First, how is modulation of $[Ca^{2+}]_c$ attained? Answering this question involves identifying the pathways for transmembrane Ca^{2+} transport and characterizing the controls which act upon these transport systems — ultimately *in vivo*. Secondly, how can this wide array of environmental signals each be transduced through a change in $[Ca^{2+}]_c$ to elicit a stimulus-specific response? The answer to this question is likely, by analogy with animal cells, to be multi-faceted, and will involve a detailed assessment of the magnitude of the Ca^{2+} signal and its spatio-temporal properties [9]. Furthermore, stimulus specificity might be engendered through selective expression of response elements or through permissive interactions with other signalling pathways (so-called cross-talk).

These important questions are only beginning to be addressed in plant cells. While the identities of the energized transport systems responsible for maintenance of the low resting level of $[Ca^{2+}]_c$, in the region of 200 nM, have been established for most intracellular membranes [6], we have little idea of the role of Ca^{2+}-permeable channels which are thought to open to allow rapid elevation of $[Ca^{2+}]_c$. In the one case shown in Table 1, for which a stimulus-induced lowering of $[Ca^{2+}]_c$ has been reported, light-induced uptake of Ca^{2+} across the chloroplast envelope membrane seems likely to be responsible [10,11]. Nevertheless, in most cases, this lack of information even extends to the identity of the membrane at which such channels are located.

We have argued previously [12] that the source of Ca^{2+} used for elevation of $[Ca^{2+}]_c$ is likely to be influenced by habitat. Thus, the vast majority of aquatic plants are bathed in media where Ca^{2+} is present at higher electrochemical activity than in the cytosol, and in effectively inexhaustible supply. For these plants, opening of plasma membrane Ca^{2+} channels can be anticipated to be a sufficient condition for elevating $[Ca^{2+}]_c$ during signal transduction, as exemplified by euryhaline charophyte algae in response to osmotic stress [13]. By contrast, the ready supply of extracellular Ca^{2+} cannot be guaranteed for terrestrial plants — even for their roots where Ca^{2+}-depletion zones readily form in the soil — and, therefore, Ca^{2+}-mediated signalling based on mobilization of intracellular stores might be more likely to predominate. While the generality of these statements almost invites the discovery of exceptions, there is, nevertheless, some basis for believing that detailed study of the control of Ca^{2+}-channel activity at endomembranes might provide clues to the origins and control of Ca^{2+} signals in terrestrial plants.

The large central vacuole is a dominant and distinctive feature of most mature higher plant cells. By virtue of its volume, and of the fact that Ca^{2+} is sequestered there at an activity around four orders of magnitude higher than in the cytosol, the vacuole might be regarded as a prime candidate for a store from which Ca^{2+} can be mobilized in terrestrial plants. In support of this idea, considerable localized release of Ca^{2+} from the vicinity of the vacuole has been visualized with imaging techniques during guard cell closure [14]. In the first part of this paper, we report on the properties of two Ca^{2+}-release channels from the vacuolar membrane of *Beta vulgaris* (beet) storage root, and speculate on their physiological significance. We then consider the issue of cross-talk in the context of the interactions of cytsolic free Ca^{2+}, cytosolic pH and protein phosphorylation on a Cl^--release channel at the plasma membrane of *Chara*, and the role of this channel in cytosolic pH regulation.

An inositol trisphosphate-gated Ca^{2+} channel at the vacuolar membrane

Inositol 1,4,5-trisphosphate signalling in animal and plant cells

Many types of intracellular signalling in animal cells involve intracellular mobilization of Ca^{2+} by the water-soluble ligand inositol 1,4,5-trisphosphate [Ins(1,4,5)P_3] [15]. In brief, agonist binding extracellularly results in G-protein-mediated activation of a specific phospholipase C which hydrolyses phosphatidylinositol 4,5-bisphosphate into diacylglycerol and Ins(1,4,5)P_3. Diacylglycerol then activates various isoforms of protein kinase C, while Ins(1,4,5)P_3 interacts with a specific Ca^{2+} channel at the endoplasmic reticulum (ER), or sub-fraction thereof, to mobilize Ca^{2+} intracellularly.

To date, and despite considerable experimental effort, attempts to extrapolate this simple scheme, even in the barest outlines described above, from animals to plants have met with limited success. Reports of diacylglycerol-mediated signalling have been fragmentary for *in vivo* investigations, and are not supported by the general failure of molecular–genetical investigations to produce evidence for protein kinase C regulatory domains in plants [16]. By contrast, stimulus-evoked inositol phospholipid turnover, consistent with a role of Ins(1,4,5)P_3 in Ca^{2+} release, has been demonstrated in the unicellular alga *Chlamydomonas* [17] and leaf pulvini of *Samanea* [18]. Preliminary reports are beginning to emerge from guard cells [19]. Elevation of Ins(1,4,5)P_3 levels during turgor regulation in beet has also been reported [20]. It seems likely that Ins(1,4,5)P_3 has a role in intracellular Ca^{2+} mobilization during some forms of signal transduction in plant cells, even though the details of inositol phosphate metabolism differ between animals and plants [21].

Ins(1,4,5)P_3-elicited Ca^{2+} release from membrane vesicles

Micro-injection of guard cells with caged Ins(1,4,5)P_3 [3], and studies with vacuole-enriched membrane vesicles [22] and intact vacuoles [23,24] have demonstrated that Ins(1,4,5)P_3 releases Ca^{2+} across the vacuolar membrane. Given the differences, noted above, between animals and plants in their handling of inositol lipid-derived signals, and the obvious difference in membrane location at which Ins(1,4,5)P_3-elicited release is occurring in the two systems, it is pertinent to ask how the properties of Ins(1,4,5)P_3-triggered Ca^{2+} release compare in animals and plants.

We have studied this question using red beet microsomes which are enriched in vacuolar membrane. Calcium uptake and release have been assayed using radiometric filtration. Calcium can be loaded into the vesicles with ATP, and ATP-dependent uptake is eliminated by protonophores and inhibitors of the vacuolar H^+-ATPase. This is consistent with uptake driven by H^+/Ca^{2+} antiport which, in contrast to the ER and plasma membrane, is the dominant mode of energization for Ca^{2+} transport at the vacuolar membrane [6].

In common with the findings of earlier workers on oat root vacuolar vesicles [22], the maximum Ca^{2+} release elicited by Ins(1,4,5)P_3 amounts to only around 20% of the total ATP-dependent uptake [25]. However, this fractional release is quantitatively explicable in terms of finite Ca^{2+}-channel density [26]. Thus, electrophysiological measurements have estimated the presence of about 1200 Ins(1,4,5)P_3-gated channels per vacuole of red beet [24]. Since the mean vesicle diameter is about 1% that of the vacuole, some 12% of vesicles can be anticipated to possess a channel. Limited overall

Ca^{2+} release from vesicles by saturating doses of Ins(1,4,5)P_3 is, therefore, likely to result from complete Ca^{2+} release by those vesicles possessing an Ins(1,4,5)P_3-gated channel, with the remainder of the vesicles failing to respond to Ins(1,4,5)P_3, even though they are derived from the same membrane. In support of this argument, the Ca^{2+} which is released by Ins(1,4,5)P_3 is rapidly reabsorbed — presumably by vesicles without an Ins(1,4,5)P_3-gated channel — unless a protonophore is present in which case sustained release is observed [22,25].

As implied above, Ins(1,4,5)P_3-gated Ca^{2+} release is dose-dependent. The $K_{1/2}$ for Ins(1,4,5)P_3 is 540 nM. Similar values have been previously reported for both plant and animal cells [22,27]. Release is specific for Ins(1,4,5)P_3: neither Ins(1,4)P_2 nor Ins(1,3,4,5)P_4 is effective in releasing Ca^{2+}. Potent inhibition by low-molecular-mass (5 kDa) heparin constitutes a further point of similarity between Ins(1,4,5)P_3-elicited Ca^{2+} release in plants and animals [25,28,29]. In the red beet vesicle preparation, heparin behaves as a competitive inhibitor with a K_i = 34 nM [30]. Heparin of higher molecular mass (6–20 kDa) is 10- to 20-fold less potent.

These findings are significant in two respects. First, they suggest the presence of a similar class of Ins(1,4,5)P_3-gated Ca^{2+} channel in animals and plants, despite the distinctive membrane location of the channel in the two kingdoms (ER and vacuolar membrane, respectively). Secondly, the high affinity and specificity of the plant receptor reinforces the conclusions from emerging cell biological studies [21] which indicate a firm role for Ins(1,4,5)P_3 in intracellular signalling.

Equilibrium binding of Ins(1,4,5)P_3 in solubilized membranes

More-detailed studies on the ligand-binding properties of the putative Ins(1,4,5)P_3 receptor have been performed on microsomal membrane preparations from red beet solubilized in 1% Triton X-100. The binding of [^{3}H]Ins(1,4,5)P_3 in these preparations was stopped by addition of polyethylene glycol (PEG), and radioactivity assayed in the PEG precipitate. Non-specific binding was defined as that remaining in the presence of 10 μM unlabelled Ins(1,4,5)P_3.

Specific binding of Ins(1,4,5)P_3 reaches equilibrium after 3 to 5 min and typically comprises around 40% of total binding. This rather small fraction might reflect relatively low receptor density in red beet storage root: we have recently determined that in microsomal preparations from cauliflower florets, specific binding can approach 70% of total binding. Specific binding exhibits a pH optimum of 7.5–8.0, while non-specific binding is not pH sensitive.

The dissociation constant (K_d) for the binding site can be determined by displacement of [^{3}H]Ins(1,4,5)P_3 by various concentrations of unlabelled Ins(1,4,5)P_3 providing the concentration of radiolabel is maintained constant and significantly below the K_d. Using 1.25 nM [^{3}H]Ins(1,4,5)P_3, the results indicate the presence of a single class of binding site with a K_d of 121 ± 10 nM. In the absence of unlabelled Ins(1,4,5)P_3, specific binding of [^{3}H]Ins(1,4,5)P_3 amounts to 8.6 fmol/mg of protein. From this value and the K_d, a binding site density of 0.84 nmol/mg of protein can be calculated. As anticipated, if the Ins(1,4,5)P_3 receptor is located predominantly on the vacuolar membrane of red beet, Ins(1,4,5)P_3-specific binding sites co-purify with vacuolar type H^+-ATPase activity on sucrose step gradients.

None of the other inositol phosphates tested [Ins(1,4)P_2, Ins(1,3,4,5)P_4 and InsP_6] displaces [^{3}H]Ins(1,4,5)P_3 with a K_d within an order of magnitude of that for Ins(1,4,5)P_3. Since these compounds are also ineffectual in Ca^{2+}-release assays (see above), there is circumstantial support for the notion that the [^{3}H]Ins(1,4,5)P_3 binding assays report ligand binding to the physiological receptor. This conclusion is reinforced by the finding that low-molecular-mass heparin competes effectively for the Ins(1,4,5)P_3-specific binding site. (The derived K_d for heparin is 301 ± 72 nM.)

However, one point of potential disparity between the Ca^{2+} release and binding data concerns the efficacy of Ins(1,4,5)P_3: the measured $K_{1/2}$ for Ins(1,4,5)P_3 in the Ca^{2+}-release assays is 540 nM, while the K_d for specific binding appears to be 120 nM. This disparity is neatly resolved by the finding that ATP binds — albeit with low affinity — to the Ins(1,4,5)P_3-specific sites. The K_d for ATP binding, as determined in the [^{3}H]Ins(1,4,5)P_3 displacement assay, is 0.66 mM, and 3 mM ATP is present to power vesicular Ca^{2+} accumulation in the Ca^{2+}-release assays. The anticipated $K_{1/2}$ ($K_{1/2}'$) for Ins(1,4,5)P_3-elicited Ca^{2+} release in the absence of ATP can be estimated to be 97 nM from the relationship

$$K_{1/2}' = K_{1/2}/\{1 + [\mathrm{ATP}/K_{\mathrm{d,ATP}}]\}$$

which is in excellent agreement with the K_d for Ins(1,4,5)P_3-specific binding to the solubilized binding sites. In almost all respects, the results of the equilibrium binding studies on red beet exhibit a remarkable degree of quantitative similarity with the properties of the solubilized Ins(1,4,5)P_3 receptor from cerebellum. These conclusions extend to the pH optimum (cf. [31]) and the K_ds for Ins(1,4,5)P_3 (cf. [32]), for low-molecular-mass heparin and ATP (cf. [33]). Only in the context of membrane location and binding-site abundance do the Ins(1,4,5)P_3 binding sites in cerebellum and red beet differ significantly. Typically, Ins(1,4,5)P_3-specific binding sites are present at a density of about 10 nmol/mg of protein in crude preparations from cerebellum [32–34], or about 10-fold greater than in beet. This finding is, perhaps, unsurprising since cerebellum is obviously more active in intracellular signalling than red beet. By contrast, binding-site density in red beet is in good agreement with that reported for peripheral tissues (e.g. liver, see [34]).

In summary, both Ins(1,4,5)P_3-elicited Ca^{2+} release and Ins(1,4,5)P_3-specific binding to solubilized plant membranes indicates the presence of an Ins(1,4,5)P_3 receptor with very similar properties to those in animal cells. Our results suggest that the critical elements for Ins(1,4,5)P_3-mediated signalling evolved before the division of the plant and animal kingdoms.

A voltage-gated Ca^{2+}-release channel at the vacuolar membrane

A voltage-sensitive Ca^{2+}-release pathway in vesicles

The indication that the vacuolar membrane might possess supplementary pathways for Ca^{2+} release was first obtained from radiometric assays on vacuolar-enriched vesicles from red beet [35]. A physiological, cytoplasm side-negative, membrane electrical potential can be generated across the membrane of Ca^{2+}-loaded vesicles by millimolar concentrations of the lipophilic cation triphenylmethylphosphonium ($TPMP^+$). The imposition of a membrane potential induces Ca^{2+} release, typically to a maximum at 20 mM $TPMP^+$ of about 30% of the ATP-dependent uptake [35]. Two additional characteristics of the $TPMP^+$-induced Ca^{2+} release indicate that the voltage-sensitive pathway is distinct from the Ins(1,4,5)P_3-gated Ca^{2+}-release channel. First, Ca^{2+} release by saturating doses of Ins(1,4,5)P_3 and $TPMP^+$ is essentially additive. Secondly, the two pathways exhibit distinct pharmacological profiles. Thus, while the Ins(1,4,5)P_3-gated pathway is acutely inhibited by heparin and 8-(*N*,*N*-diethylamino)-octyl 3,4,5-trimethoxybenzoate (TMB-8), the voltage-sensitive pathway is effectively unresponsive. By contrast, $TPMP^+$-gated release is inhibited by 1 mM Zn^{2+} (to 20% control) and by 0.1 mM Gd^{3+} (to 10% control), while Ins(1,4,5)P_3-gated release is largely unaffected by the presence of these cations.

Single-channel studies of a voltage-gated Ca^{2+} channel

It might justifiably be argued that the dual presence of independent pathways for Ca^{2+} release in the vacuolar membrane of red beet is open to question from this biochemical evidence, on the grounds that the vesicle preparation is contaminated by non-vacuolar membranes. Furthermore, the Ca^{2+}-release assay does not permit an assessment of the ionic specificity of the exit pathway: is this via a genuine voltage-gated Ca^{2+} channel, or does Ca^{2+} efflux occur via some non-selective $TPMP^+$-induced permeability? These questions have been addressed by application of the patch-clamp technique [36] to measure ionic currents in inside-out patches obtained from intact vacuoles of the storage root of beet.

In conditions where Ca^{2+} is the sole inorganic cation on the lumenal (bathing medium) side of inside-out patches, activity of single channels passing inward currents is observed. (Note that in the discussion below, we adopt the convention recently proposed by Bertl *et al.* [37] regarding the reference point for voltage and current measurements across the vacuolar membrane. Thus, voltage is referenced to the vacuolar lumen and inward currents denote the passage of positive charge into the cytosol.) Five distinctive features emerge [30,35]. (i) The unitary currents increase, then saturate at 0.6 pA, as a function of membrane voltage over the range +30 to −40 mV. The saturating current is independent of the lumenal Ca^{2+} concentration over the range 5–20 mM. The single-channel conductance at non-saturating voltages is 12 pS. (ii) The zero-current potential (reversal potential) of the channel shifts with lumenal Ca^{2+} concentration in a manner which implies a 15- to 20-fold selectivity for Ca^{2+} over K^+ (iii) As the holding potential swings increasingly negative of zero, the open probability of the channel increases exponentially over the physiological range of membrane potentials, thought to be −20 to −40 mV with respect to the lumen. Channel openings at positive membrane potentials are rare. In other words, the channel exhibits genuine voltage-gating characteristics. (iv) The open probability of the channel is also very sensitive to the lumenal Ca^{2+} concentration. At −30 mV, for example, the open state probability increases from about 0.01 to 0.15 as lumenal Ca^{2+} is raised from 5 to 20 mM. (v) Channel activity is potently inhibited by Gd^{3+}. The inhibition constant (K_i) for the effect of Gd^{3+} on open-state probability (35 μM) is in close agreement with the K_i for Gd^{3+} on $TPMP^+$-elicited Ca^{2+} release in membrane vesicles (21 μM). This finding constitutes, at first sight, evidence that the voltage-sensitive release pathway assayed in membrane vesicles is the same as that observed at single-channel level in the patch-clamp recordings.

The ionic selectivity of the channel [see (ii) above] permits its description as a Ca^{2+} channel. Even so, as long as sufficient lumenal Ca^{2+} is provided to facilitate channel opening [cf. (iv) above], the channel will effectively catalyse K^+ transport at a considerable rate. In practice, 30 μM lumenal Ca^{2+} is required to observe channel opening. With the supplementary presence of 50 mM K^+, the unitary conductance is about 16-fold larger than that observed with 5–20 mM Ca^{2+} as a charge carrier. As lumenal Ca^{2+} is progressively increased to 10 mM, currents at negative membrane potentials decrease to values closer to those observed in the absence of lumenal K^+. This phenomenon is adequately explained by the concept of a Ca^{2+}-binding site within the channel, with Ca^{2+} competing more effectively for the binding site. Calcium, therefore, lingers longer at the binding site, and ionic selectivity is engendered at the expense of catalytic efficiency. A similar phenomenon occurs in L-type Ca^{2+} channels at the plasma membrane of animal cells [38]. The competition between K^+ and Ca^{2+} is voltage dependent, and this can be used to estimate the intramembrane location of the binding site. Our recent analysis indicates that the Ca^{2+}-binding site lies within the membrane di-electric, 9% of the distance from the lumenal side. This does not preclude the existence of an additional intramembrane binding site nearer the cytosolic side.

This ability of the channel to conduct K^+ currents can be used to elucidate the nature of activation by lumenal Ca^{2+}. In the complete absence of bivalent cations, the channel fails to open at all. Although addition of lumenal Mg^{2+} facilitates opening, channel activity is considerably less than in the presence of comparable concentrations of lumenal Ca^{2+}. This implies the presence of a Ca^{2+}-specific gating site at the lumenal surface, and eliminates mechanisms in which Ca^{2+} has a relatively non-specific, charge-screening role.

Besides the effects of voltage and Ca^{2+} on channel gating, additional regulatory mechanisms must also be present. Channel activity is very high indeed in the presence of physiological concentrations of K^+ and Ca^{2+}, and at pH 7.5 on each side of the membrane. Indeed, we calculate that, were this level of activity sustained *in vivo*, the channel would rapidly dissipate the transmembrane Ca^{2+} gradient, with presumably fatal consequences resulting from elevation of $[Ca^{2+}]_c$ [35]. However, at the more physiological lumenal pH 5.5, channel openings become very rare, and there is a marked, time-dependent decrease in activity [30]. Furthermore, many recordings display seemingly spontaneous transition — from a low-conductance state with low open probability, to a state with higher conductance and high open probability. Such observations are suggestive of the dissociation of a regulatory compound which serves to repress activity *in vivo*, and we are currently investigating the biochemical identity of this regulator.

Relationship to other vacuolar Ca^{2+} channels

Several reports have indicated the presence of at least one class of additional voltage-gated Ca^{2+} channel at the vacuolar membrane [39–41]. In all cases, the channel behaves in a fashion which would mediate Ca^{2+} uptake into the vacuole, with significant opening only in the positive range of membrane potentials. It has been proposed [39] that the physiological role of this channel might be to catalyse vacuolar uptake of Ca^{2+} after Ca^{2+} release via an Ins(1,4,5)P_3-gated channel, if opening of the Ins(1,4,5)P_3-gated channel results in a sufficiently positive swing of the membrane potential. However, the argument essentially results in a perpetual motion machine, and contravenes the Laws of Thermodynamics. Thus, net Ca^{2+} uptake by the vacuole would require the generation of a membrane potential more positive than the equilibrium potential for Ca^{2+}, and opening of an Ins(1,4,5)P_3-gated Ca^{2+} channel could not, by definition, achieve this. Thus, the physiological role of these Ca^{2+} channels — gated open by positive voltages — remains questionable.

Vacuolar calcium release by two channels: physiological implications

Two independently controlled Ca^{2+} channels reside in the vacuolar membrane of plants: an Ins(1,4,5)P_3-gated channel, and a Ca^{2+} channel gated open by voltages in the physiologically negative range. What can this tell us about Ca^{2+}-signalling mechanisms? The answers at present remain speculative, but intriguing.

First, it is possible that the two channels reside in separate, cell-specific sub-types of vacuoles. This arrangement would endow a degree of cell specificity on environmental signals which require elevation of $[Ca^{2+}]_c$ for their transduction. Thus, signals working via the Ins(1,4,5)P_3 pathway would have the capacity to elevate $[Ca^{2+}]_c$ only in those cells possessing a vacuolar Ins(1,4,5)P_3-gated Ca^{2+} channel, while the converse would hold for cells selectively transducing signals via an as yet unknown pathway which activates the voltage-gated Ca^{2+} channel.

Secondly, if the two channels reside in the same vacuoles, it remains possible that independent control of the two channels endows the Ca^{2+} signal with a degree of flexibility consistent with its role as a multifarious messenger. Thus, the amplitude and dynamics of the elevation of $[Ca^{2+}]_c$ which ensues channel activation might well be expected to differ, depending on which channel type is activated. This might then lead to responses specific to a given stimulus, even though all transduction is ultimately through elevation of $[Ca^{2+}]_c$.

Thirdly, the presence of two Ca^{2+}-release channels might simply indicate the evolutionary convergence of independent signalling pathways upon a common intermediate: elevation of $[Ca^{2+}]_c$. In this case, the response would not be stimulus-specific, but would enable a variety of input pathways to elicit a common response.

It is noteworthy that in animal cells at least two pathways are present for intracellular Ca^{2+} mobilization: an Ins(1,4,5)P_3-gated Ca^{2+} channel and one sensitive to the plant alkaloid ryanodine. These pathways can co-reside in the same cell [42]. Currently, we favour the possibility that the presence of two independently controlled vacuolar Ca^{2+} channels might provide a starting point for an explanation of the diverse roles of Ca^{2+} in cell signalling (Table 1).

Cross-talk between signalling pathways at a plasma membrane anion channel

Cross-talk and anion-channel function in plants

As pointed out in the Introduction, specificity in stimulus–response coupling can also be attained through cross-talk of signalling pathways. Thus, the ability of a given intracellular signal to evoke a response might depend on the extent to which the effector protein is already primed by the presence of a second intracellular signal. Documentation of cross-talk during signalling in animal cells is now widespread — for example, between cyclic AMP-mediated signals and Ca^{2+}, between Ca^{2+} and protein kinases, and between lysophospholipids and diacylglycerol in activating protein kinase C [43]. Analogous studies in plant cells are in their infancy. However, plasma membrane anion channels provide one point at which signalling pathways might be anticipated to interact, because of the many physiological functions attributed to these channels.

Although plasma membrane anion channels in plants generally exhibit rather poor selectivity for Cl^- over other anions, such as NO_3^- and even malate [44], it is generally believed that in physiological conditions Cl^- comprises a principal ionic component of currents passing through these channels when open. At the plasma membrane of glycophytic plants, at least, the equilibrium potential for Cl^- normally lies more than 200 mV positive of the resting membrane potential and, therefore, plasma membrane anion channels invariably catalyse Cl^- efflux from the cell [45].

Studies of cellular Cl^- release in general, and of anion channels in particular, have led to elucidation of a number of functions for anion channels in plant cell biology. One class of plasma membrane anion channel is clearly gated open by cytosolic Ca^{2+}. Such channels participate in net salt loss, during both hypo-osmotic regulation of cell turgor around a constant set-point (as exemplified in euryhaline charophyte algae [13]), and volume and turgor changes which entail a shift to a new set-point (e.g. closure of stomatal guard cells [8,46]). Membrane depolarization, resulting from the inward Cl^- current, would in all cases activate an outwardly rectifying K^+ channel, which is probably ubiquitously distributed in the plasma membranes of plants [47]. These functions of anion-conducting channels are, of course, unsurprising given the major role of Cl^- as a cellular osmoticum.

However, other functions for anion channels are also apparent. Electrical signalling in plants might well involve opening of anion channels through which a major component of the depolarizing current can flow. The best-defined example is in the charophyte algae where, again, the Ca^{2+}-activated anion channel opens in response to a depolarizing stimulus which elevates cytosolic free Ca^{2+}. Even though the function of the charophyte action potential remains obscure [48], it remains possible, by analogy, that Cl^- currents are involved in electrical signalling in the sensitive plant *Mimosa* [49], rapid movements associated with prey capture in the insectivorous plants *Drosera* and *Dionaea* [50,51] and general wounding responses [52], all of which are thought to involve signalling via action potentials.

Cl^- release across the plasma membrane also plays an indirect, but important, role in cytosolic pH regulation. Cytosolic acidosis in plants is normally countered by stimulation of the primary H^+-ATPase at the plasma membrane [53]. This enzyme is electrogenic, and the membrane hyperpolarization which could result from its enhanced activity is potentially counter-productive: the elevated electrical driving force on H^+ across the plasma membrane would not only tend to inhibit the electrogenic H^+ pump [54,55], but would also increase H^+ influx back across the plasma membrane. However, this tendency for membrane hyperpolarization is off-set by opening of anion channels which facilitate Cl^- release — and hence membrane depolarization — during cytosolic acidosis, both in charophyte algae [56] and higher plants (I.C. Sejbl and D. Sanders, unpublished work).

The potentially conflicting cellular demands on anion channels imply sophisticated modes of control ensuring, for example, that cytosolic pH regulation does not trigger Ca^{2+} signalling. In part, flexibility might be achieved by the presence of more than one type of channel, each independently regulated, as in the case of vacuolar Ca^{2+} channels. Indeed, recent evidence points strongly to the existence of more than one class of anion channel at the plasma membrane of guard cells, although it is not known whether these channels are differentially regulated by intracellular messengers [46,57]. Additionally, disparate signals might converge and interact at a single class of anion channel.

Control of anion-channel activity in *Chara*

The giant internodal cells of charophyte algae represent ideal material for the study of membrane-interacting intracellular control mechanisms, since cytosolic composition can be easily modulated by intracellular perfusion [58,59]. Briefly, an internodal cell several centimetres in length is mounted in a three-compartment chamber, and the ends of the cell which project into the two outermost compartments are excised (Fig. 1). An intracellular perfusion medium of defined composition is then allowed to flow through the cell under the influence of a small hydrostatic pressure gradient, but at a rate sufficient to wash away the vacuolar membrane and the contents of the streaming cytosol. After ligation of the section between the two outer compartments, pond water is added to the central compartment. Efflux of Cl^- into the central compartment can then be determined if radiolabelled Cl^- has been incorporated into the perfusion medium. To eliminate secondary effects of membrane potential on efflux, conditions were chosen such that the membrane potential was clamped approximately at the equilibrium potential for K^+. Direct measurements of membrane potential for a range of the experimental treatments described below showed it to remain constant and in the range of -100 ± 10 mV.

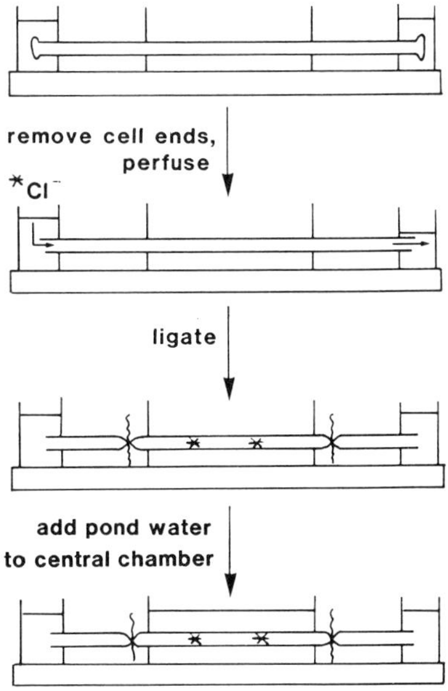

Fig. 1. Method for intracellular perfusion of charophyte internodal cell.

In the absence of intracellular Ca^{2+}, Cl^- efflux from the cell fragment in the central compartment is sensitive to internal pH in a direction suggestive of the role of Cl^- channels in cytosolic pH homoeostasis. Thus, efflux is small at the normal value of cytosolic pH of 7.8 [60], and rises almost as a first-order function of H^+ activity as the pH is lowered by one unit. The projected pK_a of the relevant H^+-binding site is about 6.2.

At a constant internal pH of 7.8, Cl^- efflux is activated by Ca^{2+}, as has been shown previously by others [61]. Efflux shows a Michaelian dependence on Ca^{2+} activity, with a $K_{1/2}$ for Ca^{2+} of about 3 μM. This value implies that the channel is well primed for activation by Ca^{2+} during an action potential, during which $[Ca^{2+}]_c$ rises from a value in the region of 200 nM to between 6 and 10 μM [62]. The saturating efflux, which can be observed at 100 μM Ca^{2+}, is about 50 $pmol \cdot cm^{-2} \cdot s^{-1}$ and is considerably in excess of the flux measured at even the lowest value of internal pH tested in the absence of Ca^{2+} (25 $pmol \cdot cm^{-2} \cdot s^{-1}$, pH 6.8).

Two notable features emerge if the Ca^{2+} titration is carried out over a range of lowered internal pHs in the range 7.8–7.2. First, the saturating efflux at 100 μM Ca^{2+} is completely unaffected, and remains at 50 $pmol \cdot m^{-2} \cdot s^{-1}$. Since the effect of pH alone on efflux is not additive with that of Ca^{2+}, this implies that the effects of pH and Ca^{2+} are on a single class of anion channel. This conclusion is broadly supported by the finding that the anion-channel blocker 5-nitro-2-(phenylpropylamino) benzoic acid inhibits 90% of Cl^- efflux with kinetics displaying a single K_i. The second finding from the Ca^{2+} titration studies is that the $K_{1/2}$ for Ca^{2+} activation of efflux decreases dramatically from 3 μM to 0.4 μM, 0.1 μM and 0.07 μM, respectively, at internal pH values of 7.8, 7.6, 7.4 and 7.2. The pH dependence of the $K_{1/2}$ for Ca^{2+} is so steep that the gating mechanism must involve binding of at least three H^+.

The physiological significance of these observations seems clear. The Ca^{2+}-activated anion channel is competent to respond to pH by virtue of the pH-sensitivity of Ca^{2+} gating, even at constant $[Ca^{2+}]_c$. Thus, during cytosolic acidosis, the Ca^{2+}-activated channel can undergo activation by a factor of 10, as cytosolic pH falls from 7.8 to 7.2, without the requirement for an increase in Ca^{2+} which would have potentially far-reaching consequences. The charophyte plasma membrane anion channel, therefore, comprises a significant example in which intracellular H^+ and Ca^{2+} signalling pathways interact at the level of a single effector. However, the significance of anion channel regulation as a crossroads in cellular signalling is likely to extend even further than indicated by the direct effects of H^+ and Ca^{2+}. After electrical excitation, which involves a large increase in plasma membrane anion channel activity, a significant refractory period ensues which can last several minutes [48]. The purpose of this refractory period is, presumably, to prevent excessive salt loss, but the mechanism has never been adequately explained.

The intracellular perfusion of charophyte cells with ATP and the catalytic subunit of protein kinase A results in a remarkable inhibition of Cl^- efflux. Indeed, efflux is barely detectable even at the most permissive Ca^{2+} activity (100 μM) or at the most permissive pH (6.8). This effect is protein-kinase-dependent. By contrast, perfusion with alkaline phosphatase elicits efflux, but only to moderate levels of 30–35 $pmol \cdot cm^{-2} \cdot s^{-1}$. Moreover, in the presence of phosphatase, the channel is refractory to changes in both internal pH and Ca^{2+}.

It seems reasonable to propose that the refractory period which follows an action potential is engendered by phosphorylation (probably Ca^{2+}-dependent) of the Ca^{2+}-

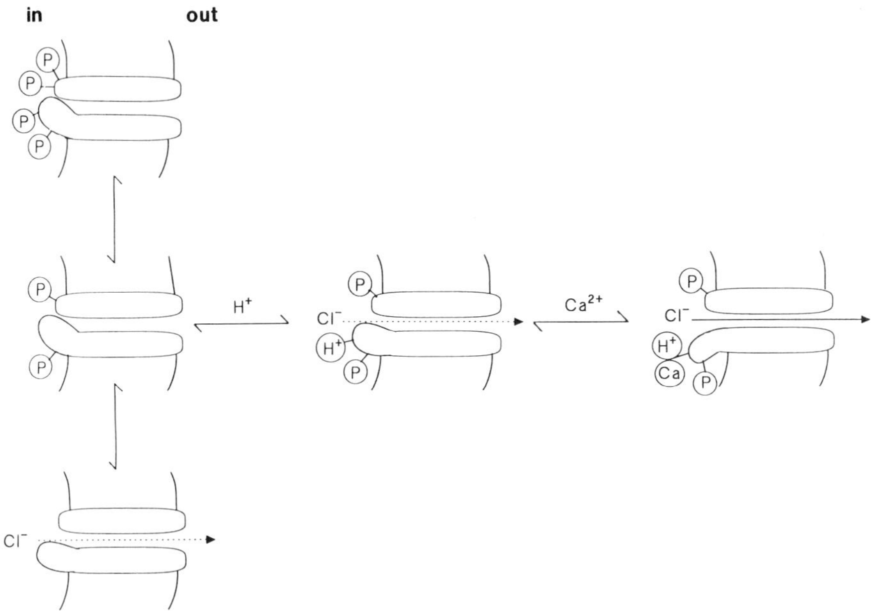

Fig. 2. Scheme for regulation of plasma membrane anion-channel activity by phosphorylation, cytosolic H^+ and Ca^{2+} in *Chara*.

dependent anion channel. If so, then the cross-talk between the Ca^{2+} signalling and protein phosphorylation pathways could provide an important negative feedback control on channel activity, thereby facilitating action potentials at a frequency which does not lead to excessive loss of cell turgor.

Fig. 2 shows a scheme for control of anion-channel activity which embodies the principal features of the results discussed above. Phosphorylation of the channel results in complete closure, while complete dephosphorylation results in a moderate frequency of opening. In both states, the channel is not competent to respond to changes in the cytosolic activities of either H^+ or Ca^{2+}. The channel is, presumably, partially phosphorylated in the state in which its activity is normally assayed in perfused cells. There, the channel is also closed in the (ideal) condition of perfusion with H^+- and Ca^{2+}-free medium. The presence of H^+ in the physiological range of cytosolic pH elicits partial activity. The full capacity for Cl^- efflux via this channel results only if the channel is both primed by H^+ and then activated by Ca^{2+}. It is envisaged that the binding of H^+ might expose Ca^{2+}-binding sites which then allow full channel activity.

Concluding remarks

The studies reported above relate to separate membranes — the vacuolar and plasma membranes — yet, independently, provide some insights into the manner in which the widespread observations of intracellular Ca^{2+} signalling can be coupled to provide stimulus-specific responses. First, the presence of multiple pathways for elevation of $[Ca^{2+}]_c$ enables flexibility in the amplitude and dynamics of the signal.

Secondly, interaction of Ca^{2+} signals with others involving cytosolic H^+ and protein kinases can provide a finely tuned control which enables more than one cellular function to reside on a single effector, and, at the same time, negative feedback control on effector activity.

Unpublished work referred to in this paper has been supported by grants from the Agricultural and Food Research Council and studentships from the AFRC and the Science and Engineering Research Council.

References

1. Felle, H. (1989) Plant Physiol. **91**, 1239–1242
2. Read, N.D., Shacklock, P.S., Knight, M.R. and Trewavas, A.J. (1993) Cell Biol. Int. **17**, 111–125
3. Gilroy, S., Read, N.D. and Trewavas, A.J. (1990) Nature (London) **346**, 769–771
4. Brauer, M., Sanders, D. and Stitt, M. (1990) Planta **182**, 236–243
5. Harper, J.F., Sussman, M.R., Schaller, G.Z., Putnam-Evans, C., Charbonneau, H. and Harmon, A.C. (1991) Science **252**, 951–954
6. Evans, D.E., Briars, S.A. and Williams, L.E. (1991) J. Exp. Bot. **42**, 285–303
7. Rea, P.A., Britten, C.J., Jennings, I.R., Calvert, C.M., Skiera, L.A., Leigh, R.A. and Sanders, D. (1992) Plant Physiol. **100**, 1706–1715
8. Schroeder, J.I. and Hagiwara, S. (1989) Nature (London) **338**, 427–430
9. Berridge, M.J. (1993) Nature (London) **361**, 315–332
10. Kreimer, G., Melkonian, M., Holtum, J.A.M. and Latzko, E. (1985) Planta **166**, 515–523
11. Kreimer, G., Melkonian, M. and Latzko, E. (1985) FEBS Lett. **180**, 253–258
12. Johannes, E., Brosnan, J.M. and Sanders, D. (1991) BioEssays **13**, 331–336
13. Okazaki, Y. and Tazawa, M. (1990) J. Membr. Biol. **114**, 189–194
14. Gilroy, S., Fricker, M.D., Read, N.D. and Trewavas, A.J. (1991) Plant Cell **3**, 333–344
15. Berridge, M.J. and Irvine, R.F. (1989) Nature (London) **341**, 197–205
16. Lawton, M.A., Yamamoto, R.T., Hanks, S.K. and Lamb, C.J. (1989) Proc. Natl. Acad. Sci. U.S.A. **86**, 3140–3144
17. Musgrave, A., Kuin, H., Jongen, M., de Wildt, P., Schuring, F., Klerk, H. and Van den Ende, H. (1992) Planta **186**, 442–449
18. Morse, M.J., Crain, R.C. and Satter, R.L. (1987) Proc. Natl. Acad. Sci. U.S.A. **84**, 7075–7078
19. MacRobbie, E.A.C. (1992) Phil. Trans. R. Soc. London B **338**, 5–18
20. Srivastava, A., Pines, M. and Jacoby B. (1989) Physiol. Plant 77, 320–325
21. Droback, B.K. (1992) Biochem. J. **288**, 697–712
22. Schumaker, K.S. and Sze, H. (1987) J. Biol. Chem. **262**, 3944–3946
23. Ranjeva, R., Carrasco, A. and Boudet, A.M. (1988) FEBS Lett. **230**, 137–141
24. Alexandre, J., Lassalles, J.-P. and Kado, R.T. (1990) Nature (London) **343**, 567–570
25. Brosnan, J.M. and Sanders, D. (1989) FEBS Lett. **260**, 70–72
26. Brosnan, J.M. (1990) Nature (London) **344**, 593
27. Berridge, M.J. and Irvine, R.F. (1984) Nature (London) **312**, 315–321
28. Alexandre, J. and Lassalles, J.-P. (1992) Phil. Trans. R. Soc. London B **338**, 53–61
29. Chopra, L.C., Twort, C.H., Ward, J.F. and Cameron, I.R. (1989) Biochem. Biophys. Res. Commun. **163**, 262–268
30. Johannes, E., Brosnan, J.M. and Sanders, D. (1992b) Phil. Trans. R. Soc. London B **338**, 105–112

31. Rossier, M.F., Capponi, A.M. and Vallotton, M.B. (1989) J. Biol. Chem. **264**, 14078–14084
32. Suppatapone, S., Worley, P.F., Baraban, J.M. and Snyder, S.H. (1988) J. Biol. Chem. **263**, 1530–1534
33. Challis, J., Willcocks, A.L., Mulloy, B., Potter, B.V. and Nahorski, S.R. (1991) Biochem. J. **274**, 861–867
34. Nunn, D.L., Potter, B.V. and Taylor, C.W. (1990) Biochem. J. **270**, 227–232
35. Johannes, E., Brosnan, J.M. and Sanders, D. (1992) Plant J. **2**, 97–102
36. Hamill, O.P., Marty, A., Neher, E., Sakmann, B. and Sigworth, F.J. (1981) Pfluegers Arch. **391**, 85–100
37. Bertl, A., Blumwald, E., Coronado, R., Eisenberg, R., Findlay, G., Gradmann, D., Hille, B., Köhler, K., Kolb, HA., MacRobbie, E., *et al.* (1992) Science **258**, 873–874
38. Tsien, R.W. and Tsien, R.Y. (1990) Annu. Rev. Cell Biol. **6**, 715–760
39. Pantoja, O., Gelli, A. and Blumwald, E. (1992) Science **255**, 1567–1570
40. Ping, Z., Yabe, I. and Muto, S. (1992) Biochim. Biophys. Acta **1112**, 287–290
41. Ping, Z., Yabe, I and Muto, S. (1992) Protoplasma **171**, 7–18
42. Walton, P.O., Airey, J.A., Sutko, G.A., Sudhof, T.C., Deerinck, T.J. and Ellisman, M.H. (1991) J. Cell. Biol. **113**, 1145–1157
43. Nishizuka, Y. (1992) Trends Biochem. Sci. **17**, 367
44. Hedrich, R. and Jeromin, A. (1992) Phil. Trans. R. Soc. London B **338**, 31–38
45. Tyerman, S.D. (1992) Annu. Rev. Plant Physiol. Plant Mol. Biol. **43**, 351–373
46. Schroeder, J.I. and Keller, B.U. (1992) Proc. Natl. Acad. Sci. U.S.A. **89**, 5025–5029
47. Blatt, M.R. (1991) J. Membr. Biol. **124**, 95–112
48. Beilby, M.J. (1984) Plant Cell Environ. **7**, 415–421
49. Samejima M. and Sibaoka, T. (1980) Plant Cell Physiol. **21**, 467–479
50. Williams, S.E. and Pickard, B.G. (1972) Planta **103**, 222–240
51. Hodick, D. and Sievers, A. (1988) Planta **174**, 8–18
52. Davies, E. (1987) Plant Cell Environ. **10**, 623–631
53. Kurkdjian, A. and Guern, J. (1989) Annu. Rev. Plant Physiol. Plant Mol. Biol. **40**, 271–303
54. Sanders, D., Hansen, U.-P. and Slayman, C.L. (1981) Proc. Natl. Acad. Sci. U.S.A. **78**, 5903–5907
55. Blatt, M.R., Beilby, M.J. and Tester, M. (1990) J. Membr. Biol. **114**, 205–223
56. Smith, F.A. and Reid, R.J. (1991) J. Exp. Bot. **42**, 173–182
57. Schroeder, J.I. (1992) Phil. Trans. R. Soc. London B **338**, 83–89
58. Tazawa, M., Kikuyama, M. and Shimmen, T. (1976) Cell Struct. Funct. **1**, 165–176
59. Sanders, D. (1980) J. Membr. Biol. **52**, 51–60
60. Smith, F.A. (1984) J. Exp. Bot. **35**, 1525–1536
61. Shiina, T. and Tazawa, M. (1987) J. Membr. Biol. **99**, 137–146
62. Williamson, R. and Ashley, C.C. (1982) Nature (London) **296**, 647–651
63. Miller, A.J. and Sanders, D. (1987) Nature (London) **326**, 397–400
64. Russ, U., Grolig, F. and Wagner, G. (1991) Planta **184**, 105–112
65. Shacklock, P.S., Read, N.D. and Trewavas, A.J. (1992) Nature (London) **358**, 753–755
66. McAinsh, M.R., Brownlee, C. and Hetherington, A.M. (1990) Nature (London) **343**, 186–188
67. Gehring, C.A., Williams, D.A., Cody, S.H. and Parish, R.W. (1990) Nature (London) **345**, 528–530
68. Knight, M.R., Campbell, A.K., Smith, S.M. and Trewavas, A.J. (1991) Nature (London) **352**, 524–526

Biochem. Soc. Symp. **60**, 199–206
Printed in Great Britain

Using T-DNA tagging to search for genes involved in the mechanism of phytohormone action

Richard Walden, Hiroaki Hayashi, Klaus Fritze and Jeff Schell*

Max Planck Institut für Züchtungsforschung, Carl von Linné Weg 10, 50829 Köln, Germany

Introduction

Genetic analysis of a biological system depends on the generation and analysis of mutations effecting the phenomenon in question. Such an analysis, in contrast, for example, to a biochemical analysis, can be carried out in the absence of detailed knowledge of the process under study. Moreover, a genetic analysis can be initiated without making assumptions of how a particular process operates. Traditionally, in plant research, genetic analysis has involved the study of mutants which have arisen spontaneously, or have been generated by treatment with mutagenic agents. The majority of mutations that have been studied in plants are recessive, with the loss of gene product activity resulting from the insertion of DNA, sequence modification or deletion [1,2]. Dominant mutations are observed less frequently using standard methods of mutagenesis. However, dominant mutations can be advantageous, not least because they ease selection.

Gene tagging has emerged as a powerful tool to isolate genes from plants, with the idea being that insertion of a characterized sequence of DNA within or near a target gene allows its subsequent isolation from the genome by the use of hybridization probes representing the gene tag. In the first instance, this strategy of gene isolation in plants proved successful using transposable elements [3]. However, with the development of convenient methods for plant transformation, strategies of gene tagging have been extended to include not only transposable elements [4], but also the T-DNA of *Agrobacterium tumefaciens* [5,6]. Gene tagging by the use of transposable elements, in either homologous or heterologous hosts, takes advantage of the movement through

*To whom correspondence should be addressed.

the genome of well-characterized, mobile elements — with the idea that, from many insertion events, one might occur in a gene of interest. T-DNA tagging relies on the insertion of a single, or low copy, element into the genome as a result of transformation carried out naturally by *Agrobacterium* [7]. In both instances, gene disruption generally results in a recessive mutation requiring selfing of the primary mutated population to identify the phenotype by loss of function. With current methods of both transposon and T-DNA tagging, the isolation of a mutated gene involved in a specific biochemical or developmental pathway is a matter of chance.

Clearly, to construct mutations which can be used to address specific biological questions, it would be advantageous to carry out mutagenesis and select specifically for a defined phenotype. This has, indeed, been carried out after treatment with u.v.-light, X-rays or ethyl methyl sulphate (reviewed in [8]), However, while effective in producing a variety of mutations, isolation of the mutated locus requires map-based cloning which is at best tedious, and currently only feasible in *Arabidopsis* which has a small genome and a relative lack of repetitive sequences [9].

We decided to devise a new form of T-DNA tagging which would provide the means of isolating genes involved in a specific biological process. To do this, a T-DNA tagging vector was constructed in such a way that after insertion of the T-DNA into the target genome, dominant, 'gain of function' mutations would be generated. This allows not only direct selection for a specific phenotype from among the primary transformants, but also provides a means to isolate the tagged sequence by using the sequences contained on the T-DNA as a hybridization probe. We decided to apply this technique in an attempt to construct mutants which might yield information concerning two areas of plant biology of particular interest to us: the molecular basis of auxin action and the analysis of the role played by polyamines in plant growth regulation. Both auxins and polyamines are considered to be important in plant biology; however, their action at the molecular level is poorly understood. In this review, we will describe the development of this approach in gene tagging and use the examples of work from our laboratory to illustrate the general application of what has come to be known as 'activation tagging'.

The concept of activation tagging

The basis of T-DNA tagging rests on the observation that during the process of transformation, the T-DNA of *A. tumefaciens* preferentially integrates into potentially transcribed regions of the plant genome [10]. Simply, this means that the T-DNA is, in effect, targeted to coding regions of the genome. Operationally, if one assumes that the proportion of the genomic DNA that can be transcribed is essentially the same regardless of genome size, roughly the same number of tagging events will be observed in different plant species [10]. None of the internal regions of the T-DNA is required for transfer to the plant cell [11], hence the T-DNA can be engineered to contain any sequence of foreign DNA. Activation tagging capitalizes on the action of transcriptional enhancers cloned within the T-DNA — the idea being that, following insertion of such a tag into the plant genome, the expression of flanking plant genes will be stimulated. The best-characterized plant transcriptional enhancer is that derived from the 35S RNA promoter of cauliflower mosaic virus [12]. The −90 to −420 region of the promoter, once subcloned, acts to enhance transcription in an orientation-independent manner [13], and cloning it as a multimer serves to multiply the level of enhancement activity [14]. While it is rather difficult to assess the effect of the enhancer over long distances within the plant genome, it has been observed that dimers of the 35S RNA promoter-enhancer region act to enhance expression of a marker gene at a distance of up to at least 2000 bp [14]. The tagging vector we have used is based on a general purpose plant transformation vector pPCV002 [15]; it contains, at the left T-DNA border sequence, a hygromycin gene driven by the nopaline synthase promoter, a region of a high copy number plasmid vector encoding a bacterial ampicillin-resistance gene and bacterial origin of replication. The enhancer sequences from the 35S RNA promoter, cloned in tandem as a tetramer, were inserted near the right border sequence. Based on the observation that the right border sequence is probably integrated first into the plant genome [16], selection for growth of transgenic material on hygromycin will select for transformants containing the complete T-DNA.

Theoretically, following insertion into the plant genome, flanking plant sequences will become expressed as a result of the action of the enhancer sequence contained within the T-DNA. Based on experiments where the 35S RNA enhancer sequence has been fused to promoters in different positions, two types of overexpression might be obtained in tagging experiments depending on the site of T-DNA insertion. On the one hand, overexpression could be essentially constitutive, occurring in all tissues and cells of the tagged individual; on the other, overexpression might be developmentally regulated, occurring in specific tissue or cells.

Once tagging has taken place, genetic analysis is essential to demonstrate that the observed phenotype is genetically linked to the tag before the isolation of the tagged plant sequences can proceed. Isolation of the tagged sequences takes advantage of using the sequences of the tag, either as a hybridization probe (to isolate the tagged plant sequence from a genomic library constructed using DNA from the tagged individual) or as the base of a plasmid vector. In the latter instance, plant genomic DNA is cut with a restriction enzyme that cuts outside the T-DNA tag and then religated. The plasmid sequences contained within the T-DNA allow rescue of the plant sequences as a plasmid in transformed *Escherichia coli*.

The successful application of T-DNA tagging requires the ability to produce large numbers of independent transgenic individuals. Several methods have been devised to carry out *Agrobacterium*-mediated transformation [17]. Protoplast *Agrobacterium* co-cultivation [18] allows the generation of very large populations of transgenic individuals with ease. To give an idea of the numbers that are involved in such an experiment, we routinely carry out a tagging experiment with 30 million tobacco protoplasts, and obtain a transformation frequency of roughly 20–30%. This means that we select for specific mutants from a population of about 10 million primary transformants. Typically, from such an experiment, we obtain 10–15 individual mutants. *In vitro* culture of transformed protoplasts offers the possibility of applying a selection, or screen, at the level of callus growth for specific mutants. Although this greatly simplifies selection, it limits the types of selection available, and has the disadvantage that tagging genes which only exert a phenotypic effect in organized tissue may not be feasible with this approach.

Tagging genes implicated in auxin action

Auxin has been implicated in a wide variety of effects associated with plant growth and development [19,20]. However, little is known of the molecular basis of auxin action. Genetic experiments to address this have focused on two approaches: the transfer and expression of bacterial genes encoding proteins involved in activating the accumulation of growth substances in transformed plant cells [21,22], or the generation of mutants by E.M.S. or u.v.-light treatment followed by selection either for the requirement of auxin for growth, or for growth in normally toxic levels of auxin [23,24]. With the first approach, the view has emerged that phenotypic changes (that have been observed in transgenics bearing the bacterial enzymes) resemble changes that have been observed following, or could be predicted by, the external application of growth substances. In the second approach, the rationale has often been that a mutation leading to resistance to high levels of auxin might reflect changes in the pathway of growth substance perception, or turnover. A wide variety of mutant plant lines have been produced in this manner; however, definition of the mutated locus has been either a matter of guesswork, or has had to rely on map-based cloning.

We decided to design a T-DNA tagging experiment that could be used to isolate genes involved in auxin action. Isolated protoplasts from untransformed tobacco plants display an absolute requirement for both auxins and cytokinins to be applied exogenously in the culture medium, in order that cell growth and division might occur [25,26]. We carried out a tagging experiment by protoplast co-transformation with *Agrobacterium* containing a T-DNA tag bearing the transcriptional enhancers and selected from primary transformants for the ability to grow in the absence of auxin in the culture medium. In experiments carried out as described above, on average 12 independent transgenic calli grew under these conditions. From the majority of the mutant cell lines, it was possible to regenerate plants which fall into two categories: those that display obvious phenotypic changes and those that appear normal. The phenotypic changes that have been observed among the regenerated plants include increased side branching, decreased root induction and growth, as well as a tendency of the leaves to senesce earlier than normal. These phenotypes can be used to infer that changes in auxin metabolism have occurred in the plant. However, the key phenotype shared by all plants studied to date is that the selected phenotype is maintained, i.e. protoplasts isolated from leaf tissue of regenerated individuals grow in culture in the absence of auxin. Thus, we have been able to use segregation analysis to demonstrate that this phenotype is genetically linked with the hygromycin-resistance gene encoded by the T-DNA insert and is inherited both somatically and sexually.

Currently, six of these so-called 'auxin-independent' mutants are under detailed study. All contain single T-DNA inserts and restriction mapping and Southern analysis suggest that, in each, the T-DNA has inserted into a different site in the plant genome.

*Axi*159 is the tagged plant line with which our analysis has progressed the furthest [27]. This plant line displays no overt phenotypic changes other than the ability of its mesophyll protoplasts to divide in the absence of auxin *in vitro*. Genetic and Southern analysis indicates that *axi*159 contains a single, unrearranged T-DNA insert. Rescue of the T-DNA and flanking plant sequences from a partial genomic DNA library resulted in a plasmid, pHH159, which, upon re-introduction into untransformed protoplasts, resulted in the ability of the transfected protoplasts to form a callus in the absence of auxin. Protoplast transfection of deletion derivatives of pHH159, followed by screening for growth in the absence of auxin, allowed the localization of the gene responsible to a 6.2 kb sequence of plant genomic DNA flanking the right T-DNA border. Using this sequence as a hybridization probe allowed us to demonstrate that the gene responsible is overexpressed in *axi*159 when compared with untransformed control tissue, confirming the rationale of the tagging experiment. A cDNA library was constructed, using RNA isolated from *axi*159, and cDNAs corresponding to the rescued plant sequence were isolated. When subcloned into a plant expression cassette, these cDNAs also conferred protoplast growth and division in the absence of auxin. Comparison of the cDNA sequence with that of the genomic DNA indicates that the gene responsible, *axi*1, is approximately 4100 bp in size, is interrupted by nine introns, and that the T-DNA is inserted downstream from the gene, approximately 5.5 kb from the transcriptional start site (R. Walden, H. Hayashi and J. Schell, unpublished work).

Naturally, our interest now is focused on how overexpression of *axi*1 results in the ability of protoplasts to divide in culture in the absence of auxin. Strictly speaking, *axi*1 is the equivalent of an activated cellular proto-oncogene and, by analogy with what has been found in animal systems, a wide variety of functions are conceivable [28]. However, dealing with auxin, mechanisms unique to plant cells arise, such as the activation of auxin synthesis, or the release of active auxin from inactive conjugates. With this in mind, it is worth noting that it has recently been demonstrated that protoplasts isolated from plants overexpressing the *rolB* gene of *A. rhizogenes* are able to divide and form calli *in vitro* in the absence of auxin in the culture medium [29]. The rolB protein encodes a β-glucosidase [30] that is able to hydrolyse an indoxyl-β-D-glucoside and, thereby, apparently stimulate the endogenous formation of active auxin [31].

The coding region of *axi*1 reveals no obvious similarities to previously characterized genes, although two sequences resembling those sufficient to target protein to the nucleus are present. Currently, we are using gene fusions with the β-glucuronidase (GUS) reporter gene to investigate the cellular location of the AXI1 protein with the expectation that, once its localization is determined, a working hypothesis can be constructed to test its action.

Tagging genes involved in polyamine metabolism

For some time, polyamines have been implicated as being important in a variety of developmental processes, although evidence for this has largely been correlative [32]. The most exhaustive genetic analysis of mutants affected in polyamine metabolism has been carried out in tobacco, where u.v.-mutagenized cells were selected for their ability to grow on selective levels of methylglyoxal bis(guanylhydrazone) (MGBG) an inhibitor of *S*-adenosylmethionine decarboxylase (SAM-dc), a key enzyme in the biosynthesis of the polyamines, spermidine and spermine [33]. All the plants regenerated from this study displayed severe morphological changes in flower development and in general this resulted in severely reduced fertility which limited a detailed genetic analysis. Nevertheless, one regenerant, fertilized by wild-type pollen, was able to produce seed which appeared to co-segregate male sterility and MGBG resistance [34]. In addition, floral mutants of petunia [35] and tomato [36] have been shown to have increased levels of polyamines. In the latter case, it was shown that external application of polyamines induced abnormal stamen development [37] and application of an inhibitor of polyamine biosynthesis causes production of normal pollen in the mutants.

The previously described experiments with tobacco suggested that it might be feasible to carry out a tagging experiment and select for resistance to MGBG. Hence we embarked on a programme of activation tagging with tobacco protoplasts, selecting for the ability to grow on selective levels of the inhibitor (K. Fritze and R. Walden, unpublished work).

From a typical experiment, eight cell lines were recovered which were all regenerated into plants. Each of the regenerated plant lines displays a variety of phenotypic changes, including leaf wrinkling, or deformation of the flowers. This latter point has hindered the genetic analysis of progeny of single mutant lines. However, in two cases we have been able to demonstrate that the phenotypic changes, including growth in the presence of selective levels of MGBG, co-segregate with the T-DNA insert.

Preliminary biochemical analysis of two mutants indicates changes in the relative amounts of free and conjugated polyamines accumulating in the plant. Plant genomic DNA flanking the T-DNA insert has been rescued from one of the plant lines and, upon reintroduction into protoplasts, this confers the ability of calli to grow in the presence of selective levels of MGBG. Currently, work is underway to screen for cDNAs corresponding to this region of the plant genome.

Discussion

T-DNA tagging has proven to be a useful means of gene isolation, especially in isolating genes important in developmental pathways that could not be cloned by other means [38,39]. Previously, gene tagging has relied on chance to produce a mutant whose phenotype could be seen following selfing of the mutagenized population of primary transformants. Here we have described the construction and the use of a novel T-DNA tagging vector designed to produce a dominant phenotype, allowing direct selection for a desired phenotype to be applied to primary transformants.

One might imagine that in designing a tagging vector to produce dominant mutants, two types of transcriptional activating element might be used: one, such as described here, containing transcriptional enhancers alone and another containing the enhancer elements linked to a transcriptional start site. The decision to use enhancers only in the tagging vector was taken with the aim of isolating genes located some distance from the site of integration, regardless of the orientation of the T-DNA insert to the tagged gene. This has, indeed, occurred in the case of *axi*159, where the T-DNA has inserted downstream from the *axi*1. This, however, opens an element of risk in the experimental scheme because rescued plant genomic DNA directly flanking the T-DNA insert might not necessarily contain the gene responsible for the observed phenotype. This limitation could be overcome by using a vector that also contains a transcriptional start site linked to the transcriptional enhancers. Such a tagging vector would generate an mRNA, or an antisense RNA, which would require the target gene to be located relatively near the T-DNA to be functional. The use of a transcriptional start site in the construct has the additional advantage that knowledge of the sequence at the start of the transcript allows a rapid isolation of a complementary cDNA by polymerase chain reaction.

Regardless of whether a transcription start site is used or not, one of the strengths of the dominant selection scheme is the relative ease and speed with which complementation or, more correctly, expression of the dominant phenotype, can be scored in transfected wild-type protoplasts following transfection with tagged rescued DNA. Currently, one of the major limitations of tagging experiments generating recessive mutations is the rescue of function of a mutant by a specific fragment of DNA. Generally this requires subcloning of genomic DNA rescued from a wild-type individual into a T-DNA-based vector system, followed by transformation of the original mutant plant line. This can be tedious and, at times, not easy when the mutant plant is difficult to regenerate. Simple transfection of protoplasts followed by selection for growth under the appropriate culture conditions eases the process of functional rescue dramatically, allowing scoring for a specific phenotype to be carried out in a matter of days after transfection.

The coupling of dominant selection to T-DNA tagging raises the potential of tagging any gene for which overexpression in tissue culture allows a biochemical selection or an easily screenable phenotype. As described here, in the case of mutants able to grow either in the absence of auxins, or in the presence of selective levels of MGBG, we have been able to isolate genes which are implicated in the molecular basis of auxin action or polyamine metabolism. Such genes could not readily be isolated by other means. Currently, experiments are underway to apply this tagging strategy to isolate genes acting not only in response to environmental change but also other developmental pathways.

The authors would like to thank the undergraduate students who, as part of practical courses, were involved in the establishment of activation tagging: Ursula Uwer, Michaela Dehio, Cristel Schipman and Helge Lubenow. Our thanks to Csaba Koncz for his constant enthusiasm and insight into the mechanism of *Agrobacterium*-based transformation. This manuscript is dedicated to the memory of Annelli Tallberg.

References

1. Briggs, F.N. and Knowles, P.F. (1967) Introduction to Plant Breeding, Van Nostrand Reinhold, New York
2. Poehlman, J.M. (1987) Breeding Field Crops, Van Nostrand Reinhold, New York
3. Shepherd, N.S. (1988) in Plant Molecular Biology: a Practical Approach (Shaw, C.H., ed.) pp. 187–220, IRL Press, Oxford
4. Balcells, L., Swinbourne, J. and Coupland, G. (1991) Trends Biotechnol. **9**, 31–37
5. Walden, R., Hayashi, H. and Schell, J. (1991) Plant J. **1**, 281–288
6. Koncz, C., Nemeth, K., Redei, G. and Schell, J. (1992) Plant Mol. Biol. **210**, 963–976
7. Walbot, V. (1992) Annu. Rev. Plant Mol. Biol. Physiol. **43**, 49–82
8. Thomas, H. and Grierson, D. (1987) Developmental Mutants in Higher Plants, Cambridge University Press, Cambridge
9. Arondel, V., Lemieux, B., Hwang, I., Gibson, S., Goodman, H. and Somerville, C. (1992) Science **258**, 1353–1355
10. Koncz, C., Martinin, N., Mayerhofer, K., Koncz-Kamen, Zs, Körber, H., Redei, G. and Schell, J. (1990) Proc. Natl. Acad. Sci. U.S.A. **86**, 8467–8471
11. Zambryski, P., Tempe, J. and Schell, J. (1989) Cell **56**, 193–201
12. Odell, J. T., Nagy, F. and Chua, N-H. (1985) Nature (London) **313**, 810–812
13. Fluhr, R., Kuhlemeyer, C., Nagy, F. and Chua, N-H. (1986) Science **232**, 1106–1112
14. Kay, R., Chan, A., Daly, M. and McPherson, J. (1987) Science **236**, 1299–1302

15. Koncz, C. and Schell, J. (1986) Mol. Gen. Genet. **204**, 383–396
16. Zambryski, P. (1988) Annu. Rev. Genet. **22**, 1–30
17. Walden, R. and Schell, J. (1990) Eur. J. Biochem. **192**, 563–576
18. Marton, L., Wullems, G.J., Molendijk, L. and Schilperoot, R.A. (1979) Nature (London) **227**, 129–131
19. Jacobs, W.P. (1979) Plant Hormones and Development, Cambridge University Press, Cambridge
20. Davies, P. J. (ed.) (1989) Plant Hormones and their Role in Plant Growth and Development, Martin Nijhoff Dordrecht
21. Klee, H. and Estelle, M. (1991) Annu. Rev. Plant Physiol. **42**, 529–551
22. Spena, A., Estruch, J.J. and Schell, J. (1992) Curr. Opin. Biotechnol. **3**, 159–163
23. King, P. (1988) Trends Genet. **4**, 181–188
24. Estelle, M. (1992) Bioessays **14**, 439–444
25. Murashige, T. and Skoog, F. (1962) Physiol. Plant. **15**, 473–497
26. Linsmaier, E.L. and Skoog, F. (1965) Physiol. Plant **18**, 100–125
27. Hayashi, H., Czaja, I., Schell, J. and Walden, R. (1992) Science **258**, 1351–1353
28. Bishop, J. M. (1991) Cell **64**, 235–248
29. Walden, R., Czaja, I. and Schell, J. (1993) Plant Cell Rep. **12**, 551–554
30. Estruch, J. J., Schell, J. and Spena, A. (1991) EMBO J. **10**, 3125–3128
31. Spena, A., Estruch, J.J., Prinsen, E., Nacken, W., Van Onckelen, H. and Sommer, H. (1992) Theory Appl. Genet. **84**, 520–527
32. Evans, P.T. and Malmberg, R. (1989) Annu. Rev. Plant Physiol. Mol. Biol. **40**, 235–269
33. Malmberg, R. L., McIndoo, J., Hiatt, A.C. and Lowe, B.A. (1985) Cold Spring Harbor Symp. **50**, 475–482
34. Malmberg, R.L. and McIndoo, J. (1984) Mol. Gen. Genet. **196**, 28–34
35. Gerats, A.G.M., Kaye, C., Collins, C. and Malmberg, R. L. (1988) Plant Physiol. **86**, 390–393
36. Rastogy, R. and Kaur-Sawhey, V. (1990) Plant Physiol. **93**, 439–445
37. Rastogy, R. and Kaur-Sawhey, V. (1990) Plant Physiol. **93**, 446–452
38. Yanofsky, M., Ma, H., Bowman, J. L., Drews, G., Feldman, K. and Meyerowitz, E. (1990) Nature (London) **346**, 35–38
39. Deng, X-W., Matsui, M., Wei, N., Wagner, D., Chu, A. M., Feldman, K. and Quail, P. (1992) Cell **71**, 791–802

Biochem. Soc. Symp. **60**, 207–218
Printed in Great Britain

Molecular biology of resistance to potato virus X in potato

David Baulcombe*, Julie Gilbert, Matthew Goulden, Bärbel Köhm and Simon Santa Cruz

The Sainsbury Laboratory, Norwich Research Park, Colney, Norwich NR4 7UH, U.K.

Synopsis

It has been proposed that plants express resistance to pathogens when the product of a resistance gene interacts with an elicitor molecule produced by the pathogen. Although there is one instance with tobacco mosaic virus (TMV) in which virus resistance is known to act through the same type of mechanism, it is not known whether this model accounts generally for resistance interactions with plant viruses. To address this issue the interactions of resistance genes in potato with potato virus X (PVX) have been analysed at the molecular level. PVX is an RNA virus that is affected by three different types of resistance locus in various potato cultivars. By using recombinant isolates of PVX, incorporating components of strains or mutant viruses able to overcome or avoid the effects of the resistance loci, we have identified different regions of the viral genome that determine the outcome of the resistance interaction. This information has allowed us to investigate the resistance in detail. For example, with the resistance specified by the *Rx* locus, it has been shown that the coat protein is an avirulence determinant and elicitor of an induced resistance. This resistance acts by reducing virus accumulation in the inoculated cell. Although the recognition component of the resistance is highly specific, the induced response is apparently non-specific and is effective against viruses unrelated to PVX in cells doubly inoculated with PVX and a second virus. The recognition function of *Rx* is also expressed in *Gomphrena globosa* which is a non-host plant of PVX. Based on these data, we propose that virus resistance fits the paradigm of resistance to fungal and bacterial pathogens and that there are similarities between the mechanism of cultivar specific resistance and non-host resistance to pathogen attack. Further analysis of the mechanism of the non-specific response phase may ultimately allow genetic engineering of broad-spectrum virus resistance in crop plants.

*To whom correspondence should be addressed.

Resistance gene interactions in plants

The 'gene-for-gene' hypothesis is a proposed genetical explanation of how plants resist pathogen attack [1,2]. According to the hypothesis, resistance is a consequence of the genetical interaction of the resistance (*R*) gene in the host and the avirulence (*avr*) gene in the pathogen. A widespread molecular interpretation of the hypothesis is that the *R* gene encodes a recognition determinant that interacts either with the product of the *avr* gene or with a factor that is produced by the product of the *avr* gene. It is thought that in gene-for-gene interactions the resistance is the consequence of the interaction and is an induced process in the plant. This interpretation is substantiated by the molecular characterization of avirulence determinants in bacterial and fungal pathosystems [3–5], and by the analysis of induced changes in the resistant plant following challenge with an avirulent pathogen [6]; there are physical changes in the cell wall that would hinder spread of fungal and bacterial pathogens, production of enzymes with direct antifungal activity and of secondary metabolites that inhibit growth of phytopathogenic bacteria. In many but not all instances, the induced resistance is associated with a hypersensitive response resulting in death of the tissue at, or around, the site of inoculation. Interestingly, there are similar changes in different plants inoculated with different pathogens [7] and it is possible that the products of different resistance genes carry out related functions: the domain encoding the recognition specificity would vary but the domain responsible for activation of the common response could be conserved in many types of resistance gene.

It not clear to what extent resistance to virus attack fits into this view of disease resistance in plants. Some lines of evidence suggest that virus defence is part of the generalized defence system in plants. For example, it is known that viral pathogens induce expression of the same pathogenesis-related proteins as are induced in plants expressing resistance to fungal and bacterial proteins [8]. However, none of these proteins is known to have antiviral activity and they are expressed at a high level even in tissue that is not expressing resistance. In this paper we apply a gene-for-gene perspective on the analysis of resistance to plant viruses generally, and PVX in particular.

PVX and PVX-resistance genes in potato

PVX, like most plant viruses, has an RNA genome that is single stranded and that has the same polarity as the mRNA of the virus-encoded proteins, of which there are five (Fig. 1) [9–12]. The large product of open reading frame (ORF) 1 is identified, based on sequence similarity and mutation analysis, as a virus-encoded component of the replicase. The ORFs 2–4 are overlapping genes and encode proteins that are not required for virus replication in protoplasts but are essential for spread of the infection from the site of inoculation [13,14]. These proteins could be functional homologues of the movement protein of TMV [15,16], although it is also possible that they have a regulatory role affecting stages of the infection cycle not directly involved in cell-to-cell movement of the virus. The PVX coat protein encoded in ORF5 may also play a role in cell-to-cell movement of the virus, as well as being essential for long-distance movement and passage of the virus between plants [17,18].

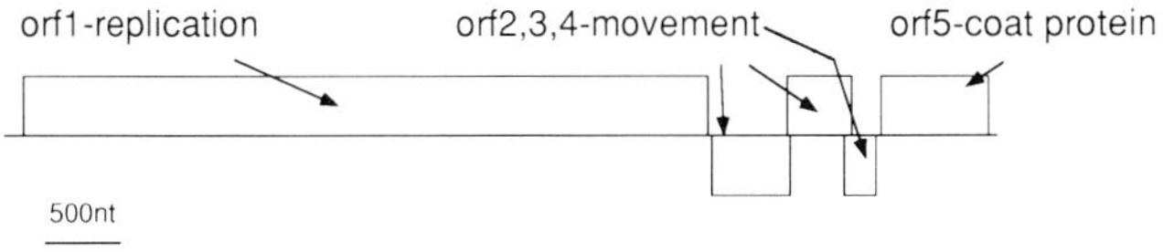

Fig. 1. The genome organization of PVX. The diagram indicates the relative size and position of the five major ORF in the viral genome. The size bar shows the length of 500 nucleotides on the diagram.

Table 1. The response of strains of PVX on potato varieties with different resistance genotypes.

	Potato genotype			
	Nx, Nb, rx	*nx, Nb, rx*	*Nx, nb, rx*	*nx, nb, Rx*
PVX group 1	HR	HR	HR	ER
PVX group 2	HR	HR	S	ER
PVX group 3	HR	S	HR	ER
PVX group 4	S	S	S	ER
PVX strain HB	S	S	S	S

Abbreviations used: HR, hypersensitive response; ER, extreme resistance; S, susceptible.

There are three types of PVX-resistance locus in potato [19]. The *Nx* and the *Nb* resistance loci both confer resistance associated with a hypersensitive response of the plant in which there may be some accumulation of the virus in the infected plant. *Rx* confers extreme resistance that prevents detectable PVX accumulation and the expression of visible symptoms even in the inoculated leaf [20,21]. The strains of PVX are categorized according to whether they are affected by the different resistance loci in potato [19] (Table 1): the group 1 strains are affected by all three resistance loci; the group 2 and group 3 strains by the *Rx* locus and either the *Nb* or the *Nx* locus; and the group 4 strains are only affected by the *Rx* locus. On the plants in which the virus is not otherwise affected by a resistance locus, the inoculation gives rise to a systemic infection (S in Table 1).

Resistance interactions can be influenced by various features of the pathogen

Resistance to fungal and bacterial pathogens

An approach to analysis of the resistance interactions is to identify components of the pathogen that are involved in, or that can affect the outcome of, the resistance interaction. The approach is normally quite straightforward due to the genetical simplicity of the pathogen and, in analysis of fungal and bacterial pathosystems, two major types of avirulence gene have been discovered. In one type the avirulence gene may encode directly the elicitor of resistance responses. An example of this situation is in the fungus *Cladosporium fulvum* in which the *avr9* gene encodes a peptide elicitor of the resistance response in tomato carrying the *Cf9* resistance gene [3]. A second type of avirulence gene encodes an enzyme that catalyses synthesis of the elicitor molecule. This alternative situation is exemplified by the *avrD* gene of *Pseudomonas syringae* pv. *tomato* that encodes an enzyme synthesizing a lactone elicitor of resistance on soybean with the *Rpg4* resistance gene [22].

Resistance to TMV

The analysis of resistance interactions in viral systems is most complete with TMV. Each of the three genes of TMV is implicated in at least one type of resistance interaction. Sequence identity in the ORF1 gene affects the resistance of the *N* gene from *Nicotiana glutinosa* [23] and *Tm1* in tomato [24,25]; the 30 kDa protein product of ORF2 affects the resistance in tomato of *Tm2* [26] and the coat protein affects the outcome of the resistance mediated by *N' in N. sylvestris* [27–33]. In each of these instances, the feature of the virus has been identified by functional analysis of strains of the virus able to overcome or avoid the effects of the resistance gene; hybrid viral isolates have been generated incorporating elements of the resistance-breaking strain and the phenotype of the hybrid isolate determined by inoculation to resistant and susceptible plants.

In principle the features identified in such an approach could be the avirulence determinant and elicitor or virulence determinants, as described above in the discussion of TMV. In this situation the property of resistance breaking would result from the absence of the molecular features necessary for avirulence. An alternative interpretation of the results from the functional analysis of viral genomes proposes that the domain implicated in the ability to overcome the resistance is an antagonist of the resistance interaction: in effect a determinant of virulence. These alternatives have been resolved in the analysis of the interaction of TMV with the *N'* gene by transgenic expression of the TMV coat protein in an *N'* plant [31,34]. The plants expressing the coat protein of an avirulent isolate of the virus displayed signs of systemic hypersensitivity: necrosis that was more pronounced on the older leaves. Expression of the coat protein of a resistance-breaking isolate had no obvious effect on the growth and development of the plant. Taking the hypersensitivity as an indicator of elicitation of resistance, these observations were interpreted as an indication that the coat protein is an elicitor and avirulence determinant [31,34]. This interpretation was reinforced by the finding that coat protein of isolates with intermediate resistance-breaking capability elicited mild necrosis in the transformed plants [31].

Resistance to PVX

Two genes of PVX have been implicated in resistance gene interactions: sequence identity in ORF1 affects the resistance mediated by *Nb* (A. Forsyth and D. Baulcombe, unpublished work), whereas the *Rx*- and *Nx*-mediated resistance is influenced by the coat protein gene [35].

A more precise identification of the sequence domain affecting the interaction of PVX with *Nx* has been carried out by exploiting the genetical instability of PVX. Jones and co-workers had carried out an analysis of PVX_{DX}, which is an isolate showing a hypersensitive response on potato carrying *Nx* [36]. They reported that after repeated passage on *Nx* potato there was the occasional production of a mutant isolate able to overcome the *Nx*-mediated resistance. One of the resistance-breaking isolates described was designated PVX_{DX4}. The reverse transition to generate isolates affected by the *Nx*-mediated resistance was achieved by repeated passage of PVX_{DX4} but in the absence of selection by *Nx*.

The phenotypic difference of PVX_{DX} and PVX_{DX4} was shown to result from mutation of the viral coat protein gene by analysis of hybrid isolates of PVX_{UK3} and PVX_{DX4}: substitution of the PVX_{DX4} coat protein into PVX_{UK3} caused this isolate to become resistance breaking [37]. There is only a single nucleotide polymorphism in the coat protein genes of PVX_{DX} and PVX_{DX4} at codon 76 and it is formally possible that *Nx*-mediated resistance was affected directly by the RNA sequence of PVX. However, this nucleotide polymorphism affects the sequence of the coat protein, and a more plausible possibility is that the resistance interaction was influenced by the protein sequence encoded around codon 76. This interpretation is supported by the further observation that *Nx*-mediated resistance is also affected by coding sequence mutations at codons 62 and 78 [37].

The viral factor affecting *Rx*-mediated resistance is also the coat protein [35], but not in the domain affecting the *Nx*-mediated resistance [37,38]. The feature of the coat protein affecting the *Rx* interaction was identified by analysis of hybrid isolates of PVX incorporating the coat protein of PVX_{CP4} and PVX_{HB}: PVX_{CP4} is affected by *Rx*-mediated resistance, whereas PVX_{HB} is the only natural isolate of PVX with an *Rx*-resistance-breaking phenotype. There are seven coding sequence differences between these isolates, and a functional analysis showed that change in codon 121 was the crucial difference between these two strains that affects the interaction with *Rx* [38]. The domains affecting the interactions with either *Nx* or *Rx* are spatially separate on a structural model [39,40] of the PVX coat protein (Fig. 2).

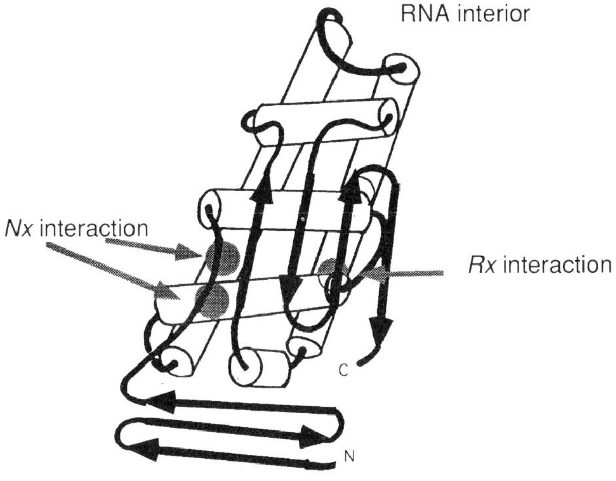

Fig. 2. A secondary structure model of the coat protein of PVX. The diagram taken from the work of Sᴗarma and colleagues [39,40] shows the *N*- and *C*-termini of the viral protein. Regions of β-sheet are shown as arrows and α-helix as cylinders. The domain of the *Nx* interaction includes residues 62 and 76 [37] and the *Rx* domain is at residue 121 [38].

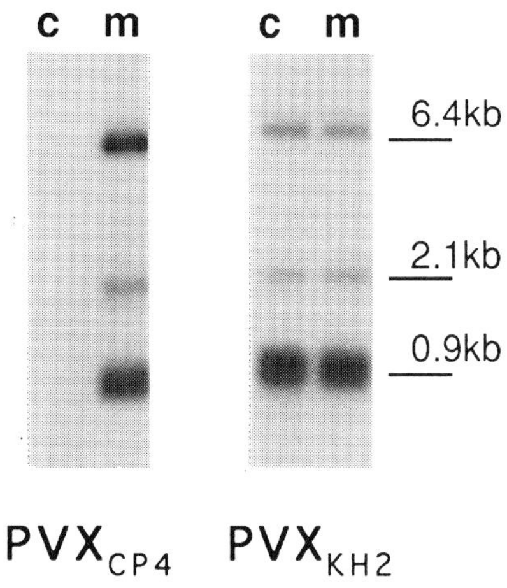

Fig. 3. The expression of *Rx*-mediated resistance in protoplasts. Protoplasts of potato varieties Cara (carries *Rx*) or Maris Bard (does not carry *Rx*) were inoculated with RNA of either PVX_{CP4}, which is affected by *Rx*-mediated resistance, or PVX_{KH2}, which is able to evade the effects of *Rx*. Total RNA was isolated from the protoplasts at 24 h post-inoculation and analysed by Northern blotting with a probe specific for the positive strand RNA of PVX. The probe detected the 6.4 kb genomic RNA and viral mRNAs of 2.1 and 0.9 kb. The RNA of Cara is in tracks marked c whereas that of Maris Bard is in tracks marked m.

An unusual property of *Rx*-mediated resistance is that it is expressed in protoplasts of the plants carrying *Rx* [20,41,42]. This result is illustrated in Fig. 3, which shows a Northern analysis of PVX RNA from infected protoplasts of either resistant or susceptible cultivars of potato. The differential accumulation of the viral RNA was seen only with the viral isolate affected by the *Rx* resistance when inoculated to plants. Most other types of virus resistance, including that specified by *Nx*, *Nb*, *N* and *N'* in tobacco and *Tm2* in tomato, are expressed only at the whole plant level (A. Forsyth, B Köhm, S. Santa Cruz and D. Baulcombe, unpublished work) [20,41,42]. These whole-plant resistance mechanisms probably involve suppression of spread of the virus away from the site of inoculation or require the presence of a cellular component that is absent from protoplasts. The *Rx*-mediated resistance is thought to involve mechanisms that suppress virus replication or promote degradation of the viral RNA.

A practical consequence of the expression of *Rx*-mediated resistance in protoplasts is that the effect of *Rx* may be tested on mutant derivatives of PVX that cannot establish full infection of plants. We have shown, for example, that various frameshift mutations in the coat protein gene of PVX_{UK3} or PVX_{CP4} caused these avirulent isolates to become resistance breaking, at least in protoplasts [42]. It was not possible to test these isolates on whole plants because the coat protein is required to be substantially intact for efficient accumulation of viral RNA and for spread of the virus between cells [18]. There was no effect of the frameshift mutations on the effect of *Rx* in protoplasts on PVX_{HB}, and this isolate was able to accumulate equally well in resistant and susceptible protoplasts as either the wild-type virus or the frameshift mutant [42]. These results lead to two conclusions. First, that the viral component affecting the outcome of the resistance gene interaction is a protein rather than RNA, and, secondly, that the active form of the coat protein, with respect to the *Rx* interaction, is from isolates affected by *Rx* rather than the resistance-breaking forms. Thus, in the terminology of gene-for-gene interactions, the coat protein is an avirulence determinant rather than a virulence determinant with respect to the interaction with *Rx* [38,42].

The mechanisms of *Rx*- and *Nx*-mediated resistance

Nx-mediated resistance

A widespread view of resistance associated with a hypersensitive response is that the necrosis is a component of the resistance mechanism: virus cannot propagate or spread from a dead cell. However, there are now several lines of evidence to suggest that this may be an oversimplified view. For example, from the analysis of the *Nx*-mediated resistance we can conclude that there may be other components to the resistance mechanism: in the *Nx* plants the expression of the hypersensitive response can be inhibited by propagation of the inoculated plants at relatively high temperature without compromising expression of resistance [37,43]. Furthermore, in the *Nx* cultivars there is several-fold less accumulation of viral RNA than in plants of the *Nx* genotype after inoculation with PVX. This restriction in viral RNA accumulation is evident before the onset of the hypersensitive response [37]. We interpret this observation and the observation that *Nx* is not active in protoplasts [44] to indicate that *Nx* affects virus spread independently of a role in the hypersensitive response. None of the changes identified in plants showing a hypersensitive response has an obvious association with cell-to-cell movement of viruses, although it is conceivable that the additional cross-linking in the cell wall of a plant showing hypersensitive response [45] may have a pleiotropic effect on the modification of plasmodesmata. This plasmodesmatal modification is thought to be necessary for cell-to-cell movement of TMV and possibly other viruses [46]. It is likely that the *Nx*-mediated response in PVX-infected plants is a multi-component process with the development of necrosis being the most conspicuous effect, but probably secondary to other processes that restrict propagation and spread of the virus.

Rx-mediated resistance

The protoplast experiments described above indicate that the primary mechanism of *Rx*-mediated resistance is in the inoculated cell. This mechanism apparently reduces the rate of virus accumulation below the level necessary to sustain the infection process. In different lines of experimentation we have started to ask whether the inhibition of virus accumulation is a direct consequence of the interaction of *Rx* and the PVX coat protein, or whether the interaction initiates a signal transduction in the plant leading eventually and indirectly to expression of the resistance process. We have also asked whether the resistance observed in protoplasts is the only component of *Rx*-mediated resistance or whether, as proposed for the *Nx*-mediated resistance, there are multiple components to the resistance mechanism.

Several lines of evidence are not consistent with the idea that the resistance is direct. For example, if the resistance is direct and an immediate consequence of the interaction of the product of *Rx* and the coat protein of PVX it would be expected that the pattern of viral accumulation in resistant protoplasts would be the same as that of an isolate of PVX with a mutation in the coat protein inoculated to susceptible protoplasts. That clearly was not the situation: the most extreme phenotype of a coat protein mutant inoculated to susceptible protoplasts was reduction of genomic RNA accumulation, and there was no effect of the mutation on the accumulation of sub-genomic mRNAs or negative-strand RNA [18]. In contrast, the *Rx*-mediated resistance affected the accumulation of all species of viral RNA [42]. Clearly there is a process in the resistant protoplasts which amounts to more than simply interaction with, and neutralization of, the coat protein as a pathogenicity factor.

The idea that this unknown process might include induced effects in the inoculated cells is consistent with the analysis of the time-course of PVX accumulation in protoplasts of resistant and susceptible protoplasts. This analysis showed that the resistance did not affect virus accumulation until between 8 h and 16 h post-inoculation [42]. Presumably the lag is accounted for partly by the time necessary for accumulation of the coat protein to a high enough level for the interaction to take place with the product of *Rx*. However, the coat protein mRNA had begun to accumulate well before 8 h and it is likely that the lag must also be accounted for — by cellular processes that are initiated by the interaction of *Rx* and the coat protein and that are necessary for the expression of the resistance.

An analysis of protoplasts of a cultivar carrying *Rx* co-inoculated with two viral isolates is also consistent with there being induced changes associated with the resistance mechanism [42]. These co-inoculation experiments showed that the *Rx*-mediated resistance was effective against an *Rx*-breaking isolate of PVX in the co-inoculated cells with virulent and avirulent isolates of PVX. Similarly, in *Rx* protoplasts co-inoculated with an avirulent isolate of PVX and either cucumber mosaic virus (CMV) or TMV, there was a lower level of accumulation of the CMV or TMV than with a co-inoculum of a virulent isolate of PVX or when the experiments were carried out in protoplasts of a PVX-susceptible cultivar of potato [42]. We suggest that these data reflect a change in the protoplasts of potato carrying the *Rx* gene so that the accumulation of the formerly virulent isolate of PVX or of an isolate of a virus not closely related to PVX is not now able to occur. This change could involve the PVX coat protein, the resistance gene product or both. Alternatively, there could be a sequence of induced changes in the cell leading ultimately to expression of the resistance.

A model of virus resistance mediated by *Rx*

A model for the *Rx*-mediated resistance is presented in Fig. 4. It proposes that there are several categories of PVX-induced response in the plants carrying *Rx*. The primary response would be the suppression of virus accumulation seen in the protoplast experiments. Under normal conditions the expression of this primary resistance would prevent the expression of the secondary components of the resistance response. However, under certain conditions, the primary response may be expressed less strongly and the secondary resistance effects would be more evident. None of the plant components of the model has been identified but there are several sets of data for which the model provides a reasonable explanation. For example, the properties of a mutant

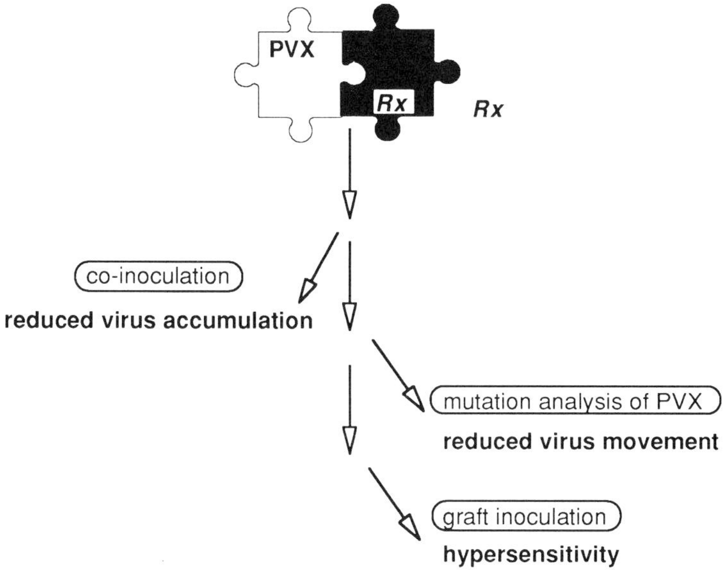

Fig. 4. A model of *Rx*-mediated resistance. It is proposed that the resistance is induced when the product of the *Rx* locus interacts with a component of PVX that is known to include the amino acid residue 121 of the coat protein of isolates that are affected by *Rx* [38]. The induced processes include an inhibition of virus accumulation [42]. When this primary inhibition is not fully expressed there may be an arrest of virus movement [38] and a limited HR [50,51]. The boxed text indicates the type of evidence for each of the stages in the response phase of the resistance interaction.

form of PVX described by Goulden *et al.* [38] indicate that the resistance response is not restricted to an effect on virus accumulation. This mutant PVX was able to overcome the *Rx*-mediated resistance when inoculated into protoplasts, but not when inoculated into plants. It is suggested that, with this mutant, the interaction between the PVX coat protein and the product of *Rx* was not strong or quick enough to elicit the primary response.

There is also an indication of complexity in the *Rx*-mediated resistance from experiments in which PVX was graft-inoculated to an *Rx* cultivar. When the graft-inoculated plants were maintained at 10°C there was low-level accumulation of PVX in the systemic leaves of an *Rx* cultivar and necrotic flecking reminiscent of the hypersensitive response [38]. These observations could be reconciled with the model by proposing that the processes leading to the primary resistance are cold sensitive, to a limited extent, and can be overcome when the inoculation pressure is intense, as is the situation with graft-inoculated plants. A link of *Rx*-mediated resistance is also provided by the report of an allele of *Rx* that controls resistance associated with hypersensitive response rather than the extreme resistance described for the alleles of *Rx* used in potato breeding programmes [19].

Further observations that fit in with the model presented above derive from the analysis of lesion formation by PVX on *G. globosa*. This ornamental plant is a member of the *Amaranthaceae* and is strictly a non-host for PVX. There is only limited accumulation of the virus on the inoculated leaf and, in nearly all instances, local lesions develop at the sites of inoculation. The one exception, as with the *Rx* interactions, is with PVX_{HB}. Inoculation of PVX_{HB} to *G. globosa* fails to induce lesion formation although there is localized accumulation of the virus at the inoculation site in the inoculated leaf: there is no systemic movement of the PVX on *G. globosa* with any of the known isolates of PVX. Surprisingly, the *G. globosa* response to PVX is affected by the codon identity at position 121 of the coat protein [47] and, as with the *Rx* interaction, the form of the coat protein that is able to elicit the response on the plant has threonine rather than lysine at this position. One interpretation of this result may well be that it is merely coincidence. However, it may also be proposed that there is a homologue of *Rx*, even in the non-host plant, with the capability to recognize the same feature of the coat protein as *Rx*. This proposed homologue of *Rx* would have the same property as the Rx_{acl}^{n} allele of *Rx* with respect to the primary and secondary resistance responses. Perhaps, as with the mutant derivative of the PVX coat protein described above, the interaction of the *Rx* homologue in *G. globosa* and the coat protein of PVX does not result in rapid initiation of the primary response so that the secondary effects, including hypersensitivity, are evident.

There is an interesting parallel of this *G. globosa* result from bacterial pathosystems. In several instances it has been demonstrated that bacterial avirulence genes can elicit a resistance response on non-host plants, and from this finding it has been suggested that gene-for-gene interactions may contribute to non-host resistance [48,49]. If the interpretation of the *G. globosa* analysis presented above is correct it would be possible to make the same hypothesis about non-host resistance to viruses.

Perspectives and applications

The main questions about resistance gene interactions and viruses are essentially the same as for other types of pathogen: what are the different types of resistance mechanism and what is the identity of the host component involved in these interactions? In the analysis of *Rx*-mediated resistance described in this paper the evidence indicates that even in the extreme resistance conferred by *Rx* the resistance process can be considered in the same way as the well-characterized examples of resistance to fungi and bacteria in plants: an interaction of the pathogen and the product of the resistance gene triggers a response in the plant that leads to resistance. The interesting finding that the recognition function of the *Rx* gene is also found in a non-host plant of PVX may indicate that the diversity of recognition of resistance genes is not maintained by direct selection for the individual specificities. One explanation of this observation is that the natural selection is for the profile of specificities in the different resistance genes of the plant genome. Alternatively there may be structural domains shared in different pathogens so that the domain of PVX recognized by *Rx* may be present in a second virus which is a severe pathogen of the non-host *G. globosa*.

The induction of a broad spectrum resistance mechanism in the protoplasts expressing the *Rx*-mediated resistance is not only of academic interest. This finding also indicates how there is the potential to manipulate a non-specific virus resistance by genetic engineering of the expression or activity of the plant components involved in the resistance mechanism. Unfortunately, none of these components has yet been identified at the molecular level although it is our hope that the identification and cloning of the genes for these plant components will be facilitated by the types of analysis described in this paper.

We are grateful to the Gatsby Charitable Foundation for support, and to Alec Forsyth for permission to cite unpublished work.

References

1. Keen, N.T. (1992) Plant Mol. Biol. **19**, 109–122
2. Keen, N.T. (1990) Annu. Rev. Genet. **24**, 447–463
3. Van den Ackerveken, G.F.J.M., Van Kan, J.A.L. and De Wit, P.J.G.M. (1992) Plant J. **2**, 359–366
4. Bonas, U., Stall, R.E. and Staskawicz, B. (1989) Mol. Gen. Genet. **218**, 127–136
5. Staskawicz, B.J., Dahlbeck, D. and Keen, N.T. (1984) Proc. Natl. Acad. Sci. U.S.A. **81**, 6024–6028
6. Lamb, C.J., Ryals, J.A., Ward, E.R. and Dixon, R.A. (1992) Biotechnology **10**, 1436–1445
7. Bowles, D.J. (1990) Annu. Rev. Biochem **59**, 873–907
8. Linthorst, H.J., Cornelissen, B.J.C., Van Kan, J.A.L., Van de Rhee, M., Meuwissen, R.J.L., Gonzalez Jaen, M.T. and Bol, J.F. (1990) in Recognition and Response in Plant–Virus Interactions (Fraser, R.S.S., ed.), pp. 361–375, Springer–Verlag, Berlin
9. Skryabin, K.G., Kraev, A.S., Morozov, S.Y., Rozanov, M.N., Chernov, B.K., Lukasheva, L.I. and Atabekov, J.G. (1988) Nucleic Acids Res. **16**, 10929–10930
10. Huisman, M.J., Linthorst, H.J.M., Bol, J.F. and Cornellissen, B.J.C. (1988) J. Gen. Virol. **69**, 1789–1798
11. Orman, B.E., Celnik, R.M., Mandel, A.M., Torres, H.N. and Mentaberry, A.N. (1990) Virus Res. **16**, 293–306
12. Querci, M., Van der Vlugt, R.A.A., Goldbach, R.W. and Salazar, L.F. (1993) J. Gen. Virol. **74**, 2251–2255
13. Beck, D.L., Guilford, P.J., Voot, D.M., Andersen, M.T. and Forster, R.L.S. (1991) Virology **183**, 695–702
14. Davies, C., Hills, G.J. and Baulcombe, D.C. (1993) Virology, **197**, 166–175
15. Deom, C.M., Oliver, M.J. and Beachy, R.N. (1987) Science **237**, 389–394
16. Meshi, T., Watanabe, Y., Saito, T., Sugimoto, A., Maeda, T. and Okada, Y. (1987) EMBO J. **6**, 2557–2563
17. Forster, R.L.S., Beck, D.L., Guilford, P.J., Voot, D.M., Van Dolleweerd, C.J. and Andersen, M.T. (1992) Virology **191**, 480–484
18. Chapman, S., Hills, G.J., Watts, J. and Baulcombe, D.C. (1992) Virology **191**, 223–230
19. Cockerham, G. (1970) Heredity **25**, 309–348
20. Saladrigas, M.V., Ceriani, M.F., Tozzini, A.C., Arese, A.I. and Hopp, H.E. (1990) Plant Cell Physiol. **31**, 749–755
21. Tozzini, A.C., Ceriani, M.F., Saladrigas, M.V. and Opp, H.E. (1991) Potato Res. **34**, 317–324
22. Keen, N.T., Sims, J.J., Midland, S., Yoder, M., Jurnak, F., Shen, H., Boyd, C., Yucel, I., Lorang, J. and Murillo, J. (1993) in Advances in Molecular Genetics of Plant–Microbe Interactions (Nester, E.W. and Verma, D.P.S., eds.), pp. 211–220, Kluwer, The Netherlands
23. Padgett, J.S. and Beachy, R.N. (1993) Plant Cell **5**, 577–586

24. Meshi, T., Motoyoshi, F., Adachi, A., Watanabe, Y., Takamatsu, N. and Okada, Y. (1988) EMBO J. 7, 1575–1581
25. Yamafuji, R., Watanabe, Y., Meshi, T. and Okada, Y. (1991) Virology **183**, 99–105
26. Meshi, T., Motoyoshi, F., Maeda, T., Yoshiwoka, S., Watanabe, H. and Okada, Y. (1989) Plant Cell **1**, 515–522
27. Knorr, D.A. and Dawson, W.O. (1988) Proc. Natl. Acad. Sci. U.S.A. **85**, 170–174
28. Saito, T., Meshi, T., Takamatsu, N. and Okada, Y. (1987) Proc. Natl. Acad. Sci. U.S.A. **84**, 6074–6077
29. Saito, T., Yamanaka, K., Watanabe, Y., Takamatsu, N., Meshi, T. and Okada, Y. (1989) Virology **173**, 11–20
30. Culver, J.N. and Dawson, W.O. (1989) Mol. Plant–Microbe Interact. **2**, 209–213
31. Culver, J.N. and Dawson, W.O. (1989) Virology **173**, 755–758
32. Culver, J.N. and Dawson, W.O. (1991) Mol. Plant–Microbe Interact. **4**, 458–463
33. Dawson, W.O., Bubrick, P.B. and Grantham, G.L. (1988) Phytopathology **78**, 783–789
34. Pfitzner, U.M. and Pfitzner, A.J. (1992) Mol. Plant–Microbe Interact. **5**, 318–321
35. Kavanagh, T., Goulden, M., Santa-Cruz, S., Chapman, S., Barker, I. and Baulcombe, D.C. (1992) Virology **189**, 609–617
36. Jones, R.A.C. (1982) Plant Pathol. **31**, 325–331
37. Santa Cruz, S. and Baulcombe, D.C. (1993) Mol. Plant–Microbe Interact. **6**, 707–714
38. Goulden, M.G., Köhm, B.A., Santa Cruz, S., Kavanagh, T.A. and Baulcombe, D.C. (1993) Virology, **197**, 293–302
39. Baratova, L.A., Grebenshchikov, N.I., Shishkov, A.V., Kasharin, I.A., Radavsky, J.L., Järvekulg, L. and Saarma, M. (1992) J. Gen. Virol. **73**, 229–235
40. Baratova, L.A., Grebenshchikov, N.I., Dobrov, E.N., Gedrovich, A.V., Kasharin, I.A., Shishkov, A.V., Efimov, A.V., Järvekulg, L., Radavsky, Y.L. and Saarma, M. (1992) Virology **188**, 175–180
41. Adams, S.E., Jones, R.A.C. and Coutts, R.H.A. (1986) J. Gen. Virol. **67**, 2341–2345
42. Köhm, B.A., Goulden, M.G., Gilbert, J.E., Kavanagh, T.A. and Baulcombe, D.C. (1993) Plant Cell **5**, 913–920
43. Adams, S.E., Jones, R.A.C. and Coutts, R.H.A. (1986) Plant Pathol. **35**, 517–526
44. Adams, S.E., Jones, R.A.C. and Coutts, R.H.A. (1985) J. Gen. Virol. **66**, 1341–1346
45. Bradley, D.J., Kjellbom, P. and Lamb, C.J. (1992) Cell **70**, 21–30
46. Wolf, S., Deom, C.M., Beachy, R.N. and Lucas, W.J. (1989) Science **246**, 377–379
47. Goulden, M.G. and Baulcombe, D.C. (1993) Plant Cell **5**, 921–930
48. Minsavage, G.V., Dahlbeck, D., Whalen, M.C., Kearney, B., Bonas, U., Staskawicz, B.J. and Stall, R.E. (1990) Mol. Plant–Microbe Interact. **3**, 41–47
49. Whalen, M.C., Stall, R.E. and Staskawicz, B.J. (1988) Proc. Natl. Acad. Sci. U.S.A. **85**, 6743–6747
50. Benson, A.P. and Hooker, W.J. (1960) Phytopathology **50**, 231–234
51. Bagnall, R.H. (1961) Phytopathology **51**, 338–340

Biochem. Soc. Symp. 60, 219–229
Printed in Great Britain

Induction, modification, and perception of the salicylic acid signal in plant defence

Daniel F. Klessig*†, Jocelyn Malamy†, Jacek Hennig†‡, Paloma Sanchez-Casas†, Janusz Indulski§, Grzegorz Grynkiewicz§ and Zhixiang Chen†

†Waksman Institute, Rutgers, The State University of New Jersey, P.O. Box 759, Piscataway, NJ 08855, U.S.A. and §Pharmaceutical Research Institute, Rydygiera Street 8, 01-793 Warsaw, Poland

Synopsis

Endogenous salicylic acid (SA) levels increase and several families of pathogenesis-related genes (including PR-1 and PR-2) are induced during the resistance response of tobacco to tobacco mosaic virus (TMV) infection. We have found that at a temperature (32°C) that prevents the induction of PR genes and resistance, the increases in SA levels were eliminated. However, when the resistance response was restored by shifting inoculated plants to lower temperatures, SA levels increased dramatically and preceded PR-1 gene expression and necrotic lesion formation associated with resistance. SA was also found in a conjugated form whose levels increased in parallel with the free SA levels. This SA β-glucoside (SAG) was as active as SA in inducing PR-1 gene expression. PR-1 gene induction by SAG was preceded by a transient release of SA. The existence of a mechanism that releases SA from SAG suggests a possible role for SAG in the maintenance of systemic acquired resistance.

Previously, we identified a soluble salicylic acid-binding protein (SABP) in tobacco whose properties suggest that it may play a role in transmitting the SA signal during plant defence responses. This SABP has been purified 250-fold by sequential chromatography on DEAE-Sephacel, Sephacryl S-300, Blue Dextran-Agarose and Superose 6. Several monoclonal antibodies (mAbs) raised against the highly purified SABP immunoprecipitated the SA-binding activity and a 280 kDa protein. This 280 kDa protein also co-purified with the SA-binding activity during the various chromatography steps, suggesting that it was responsible for binding SA. Immunoblot analysis

*To whom correspondence should be addressed.
‡Present address: Institute of Biochemistry and Biophysics, Polish Academy of Sciences, 02-532 Warsaw, Rakowiecka Str. 36, Poland.

with the SABP-specific mAbs also detected the 280 kDa protein in highly purified preparations of SABP. However, in crude homogenates these mAbs only recognized a 57 kDa protein. These and other results suggest that SABP is a multimeric complex which contains, at least, a 57 kDa protein and whose components are readily cross-linked during purification.

Introduction

We are studying the molecular and cellular responses of tobacco to TMV infection as a model system for plant–pathogen interactions. Infection of tobacco with TMV results in one of two distinct responses that depend upon the genetic background of the host, the viral strain and environmental conditions. In a susceptible tobacco cultivar, TMV infection results in the systemic spread of virus from the site of infection to distal parts of the host, with the potential for causing widespread damage. In contrast, tobacco cultivars that carry a dominant resistance locus (N) are able to restrict the spread of TMV to a small zone of tissue around the site of infection, where a necrotic lesion later forms. This resistance phenotype is called the 'hypersensitive response' and is subsequently accompanied by the induction throughout the plant of systemic acquired resistance. Development of systemic acquired resistance results in enhanced resistance to a secondary challenge by the same (e.g. TMV) or even unrelated pathogens. Both the hypersensitive response and systemic acquired resistance correlate with the synthesis of abundant amounts of host-encoded, pathogenesis-related (PR) proteins. The synthesis of these proteins is likely to be part of a general defence system against pathogenic attack, since their synthesis can also be induced by certain bacteria and fungi [1,2].

Expression of plant defence genes, such as the PR genes, is regulated by one or more poorly defined signal transduction pathways, initiated upon recognition of the pathogen by the plant host [3]. During the past few years, several new plant-encoded signalling molecules involved in plant–pathogen interaction have been described, including jasmonic acid and systemin (for review, see [4]). In addition, we have shown recently that the endogenous level of SA (a chemical inducer of the PR genes and disease resistance [4,5]) dramatically increases during TMV infection [6]. The discovery, that this increase is seen only in resistant cultivars, precedes PR-1 gene induction and occurs throughout the plant, led us to hypothesize that SA is a natural signal in the induction of the PR genes and other defence responses. Complementary studies by Métraux *et al.* [7] and Rasmussen *et al.* [8] demonstrated that rises in SA levels also correlated with the induction of systemic acquired resistance in cucumber. Here we briefly report the results of our recent studies on the induction, modification and reception of the SA signal.

Materials and methods

Plant material and growth conditions

Tobacco plants (*Nicotiana tabacum* cv. Xanthi nc) were grown at 22°C in growth chambers programmed for a 14 h light cycle and a 10 h dark cycle. For high-temperature experiments, plants were transferred to 32°C Conviron (Asheville, NC, U.S.A.) chambers 2–3 days before inoculation. TMV strain U1 was used at a concentration of 1 μg/ml in 50 mM phosphate buffer at pH 7.5 in all experiments.

Quantification and characterization of SA and SAG

Free SA was extracted and quantified essentially as reported by Raskin *et al.* [9], with minor modifications, as described by Malamy *et al.* [10] and Hennig *et al.* [11]. The SAG was characterized and quantified, as described by Malamy *et al.* [10].

Synthesis of SAG

The SAG was synthesized by condensation of acetobromoglucose with methyl salicylate, as described by Grynkiewicz *et al.* [12].

Determination of the biological activity of SAG

The biological activity of SAG was determined by injecting SAG into tobacco leaves and assaying for the accumulation of PR-1 proteins. Details can be found in Hennig *et al.* [11].

Purification of SABP

The SA-binding and competition assays were described by Chen and Klessig [13]. SABP was purified by sequential chromatography on DEAE-Sephacel, Sephacryl S-300, Blue Dextran–Agarose and Superose 6 in 20 mM citrate (pH 6.5), 5 mM MgCl, 1 mM EDTA and 30 μg/ml phenylmethanesulphonyl fluoride in the presence of various concentrations of KCl.

Immunoprecipitation of SABP

Mouse mAbs were raised against fractions from the Superose 6 column which contained the highest levels of SA-binding activity. Aliquots of 100–500 μl of hybridoma culture media were incubated with 40 μl of Protein A–Sepharose (50% slurry) at 4°C for 2 h. The antibody–Protein A–Sepharose complexes were pelleted and washed three times with RIPA buffer (150 mM NaCl, 5 mM EDTA, 0.1% SDS, 1% sodium deoxycholate, 10 mM Tris, pH 7.4) and once with a buffer containing 20 mM citrate, 20% glycerol, 150 mM KCl and 0.1% NP-40 (Nonidet P-40). The complexes were then incubated at 4°C for 2 h with 100 μl of the Blue Dextran-Agarose fraction containing approximately 15 pmol of SA-binding sites. The antigen–antibody–Protein A complexes were then pelleted and the supernatants were collected and assayed for remaining SA-binding activity. The pellets were then washed three times with RIPA buffer, resuspended in protein sample buffer, and subjected to SDS/PAGE (7.5–15%) analysis.

Immunoblot analysis

Aliquots from the Blue Dextran chromatography fraction containing the highest level of SA-binding activity, or from freshly prepared crude homogenates made in the presence of 15 mM β-mercaptoethanol (to decrease the activity of polyphenol oxidases), were fractionated on SDS/PAGE (7.5–15%) and transferred to nitrocellulose filters. The filters were subjected to immunoblot analysis using hybridoma culture media and sheep anti-(mouse IgG) antibodies conjugated to horseradish peroxidase. Immunoblots were developed using the ECL (Enhanced Chemiluminescence) detection kit from Amersham.

Results

Increases in SA levels correlate with the hypersensitive response and PR-1 gene expression

To further analyse the relationship between SA and PR gene induction and resistance, we took advantage of the observation that resistance to TMV can be reversibly blocked at elevated temperatures. When genetically resistant (e.g. *N. tabacum* cv. Xanthi nc, genotype NN) plants are inoculated with TMV and incubated at temperatures above 28°C, the replication and spread of TMV are not restricted, necrotic lesions are not formed and the PR genes are not induced. However, when the infected plants are subsequently moved to lower temperatures (22°C), PR proteins accumulate and the hypersensitive response is rapidly activated [14].

If SA is in the pathway leading to the induction of defence responses, temperatures that prevent the defence responses might also block increases in endogenous SA levels. This result would indicate that the temperature-sensitive step precedes the induction of SA. Indeed, at elevated temperatures (32°C) the characteristic increase in SA levels was completely abolished. However, when the resistance response was restored by shifting infected plants from 32°C to 22°C, there was a rapid and dramatic increase in SA levels. This rise in SA levels preceded the induction of PR-1 genes by 2 h and appearance of the necrotic lesions associated with resistance by 3–5 h (Fig. 1). These findings provide a further correlation between increases in SA levels and induction of the hypersensitive response and the PR-1 genes. (For more details, see Malamy *et al.* [10].)

Identification and characterization of SAG

In addition to the free SA analysed in previous experiments, we found that another pool of SA was released by subjecting samples to acid hydrolysis. The levels of hydrolysable SA conjugates were very low in non-inoculated or mock-inoculated plants, but rose dramatically (15–20-fold) between 24 h and 48 h after inoculation. This increase paralleled the initial rise in free SA. In contrast, both the conjugated and free forms of SA were present only at very low levels in inoculated plants maintained at 32°C. However, after shifting these plants to 22°C the levels of both forms of SA rose dramatically (data not shown). Thus, the SA conjugate, like free SA, was produced *de novo* after TMV infection, and its induction was reversibly blocked by elevated temperatures.

Many phenolic acids in plants exist as sugar conjugates [15]. Since glucosides are particularly common, it seemed likely that the SA conjugate detected after TMV infection was SAG. To test this hypothesis, plant extracts containing both free and conjugated SA were subjected to β-glucosidase digestion, which specifically releases β-linked terminal D-glucose residues. This enzymic treatment released free SA to almost the same extent as acid hydrolysis, strongly implying that the major conjugate was SAG (Fig. 2). (For more details, see Malamy *et al.* [10].)

The existence of SAG suggests additional complexity in the modulation of the SA signal. Many biologically active compounds in plants (e.g. auxin, cytokinin and

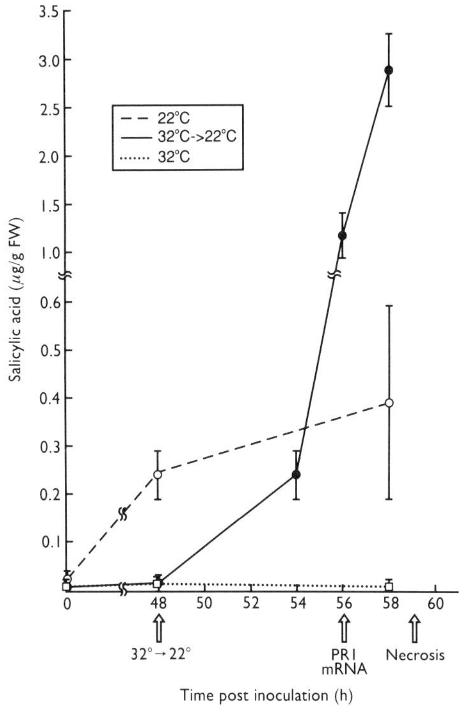

Fig. 1. SA levels after inoculation with TMV at elevated temperatures. Plants were inoculated with TMV and maintained at 22°C (dashed line), 32°C (dotted line), or transferred from 32°C to 22°C at 48 h post inoculation (solid line). Error bars reflect standard deviations. Increases in PR-1 mRNA were detected 8 h after the shift from 32°C to 22°C, while necrotic lesions which are indicative of the hypersensitive response appeared approx. 11 h after the shift. In plants inoculated and maintained at 22°C, lesions appeared at 42–43 h post inoculation under our conditions. The experiment was repeated three times. FW, fresh weight. Reproduced with permission from [10].

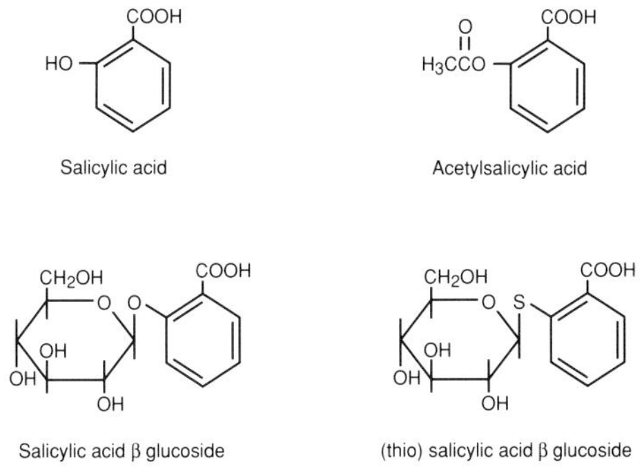

Fig. 2. Structures of salicylic acid and its analogues.

giberellins) are inactivated via conjugation with glucose. To determine whether conjugation also inactivates SA, we assayed for induction of PR-1 protein synthesis after injection of chemically synthesized SAG into tobacco plants. Unexpectedly, the SAG was biologically active and induced PR-1 gene expression at concentrations similar to those at which SA is active (Fig. 3).

This result suggested that either SAG itself was active, or it was rapidly, and perhaps transiently, hydrolysed to release free SA. To distinguish between these possibilities, SA and SAG levels were monitored after injection of leaves with SAG. Within 2 h of injection, there was a rapid and transient reduction in the level of SAG and a concomitant rapid and transient rise in the level of free SA (data not shown). This suggested that SAG was enzymically hydrolysed to free SA, which was then reconjugated to glucose. Further analysis showed that the transient increase in SA occurred in the extracellular spaces. These results suggest that SAG is hydrolysed in the spaces and the free SA is imported into the cells where it is conjugated to glucose. This model is supported by the finding that SA enters plant tissue culture cells from the surrounding medium but SAG does not. (For details, see Hennig *et al.* [11].)

To test whether cleavage of SAG is necessary for PR-1 gene induction, the thio analogue of SAG (TSAG) was chemically synthesized. TSAG is very similar in structure to SAG (Fig. 2) but, unlike SAG, it cannot be readily hydrolysed to release SA. TSAG treatment did not induce PR-1 expression, suggesting that release of free SA is required for induction (data not shown).

SABP

To elucidate the mechanism(s) of action of SA in plant signal transduction, we have attempted to identify cellular component(s) which directly interact with SA. In the search for such component(s), we have detected and partially characterized a soluble SABP in tobacco leaves [13]. The SABP has an apparent K_d of 14 μM for SA, which is consistent with the range of physiological concentrations of SA observed during the induction of plant disease responses. Moreover, the ability of analogues of benzoic acid to compete with SA for binding to the partially purified SABP is quantitatively, as well

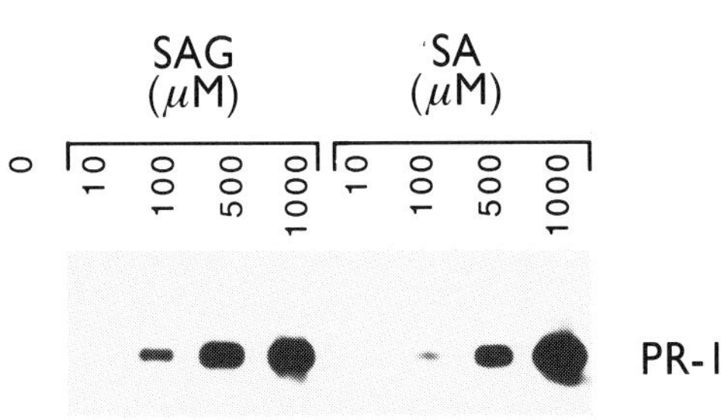

Fig. 3. Bioactivity of SAG. Different concentrations of SAG or SA in 5 mM phosphate buffer, pH 7.0, were injected into tobacco leaves. Control leaves were injected with buffer alone. Leaves were harvested 2 days later, proteins extracted, and equal amounts of proteins were separated by SDS/PAGE. PR-1 proteins were detected by immunoblot analysis using anti-PR-1 antibodies. Reproduced with permission from [11].

as qualitatively, correlated with their ability to induce genes associated with disease resistance. Highly active analogues, such as 2,6-dihydroxybenzoic acid or acetylsalicylic acid, compete very efficiently, while less biologically active analogues, like benzoic acid and 2,3-dihydroxybenzoic acid, compete less well, and inactive compounds fail to compete (Table 1). Thus, the binding affinity and specificity of SABP suggest that it might play a role in perceiving and transducing the SA signal to the appropriate response elements, ultimately activating one or more of the defence responses.

To investigate the possible involvement of SABP in signal transduction, a purification strategy for SABP was developed. Through sequential chromatography on DEAE-Sephacel, Sephacryl S-300, Blue Dextran-Agarose and Superose 6, partial purification was achieved. Several mABs (mAbs 3B6 and 1F5) raised against this partially purified SABP immunoprecipitated the SA-binding activity (Fig. 4A). These mAbs also specifically immunoprecipitated a 280 kDa protein (Fig. 4B). In contrast,

Table 1. Inhibition of [^{14}C]SA binding by benzoic acid analogues.

Benzoic acid analogues	Biological activity[a]	Inhibition (%)[b] 2×	10×
2-Hydroxybenzoic acid (SA)	++	50	91
2,6-Dihydroxybenzoic acid	++	51	91
Acetylsalicylic acid	++	25	57
Benzoic acid	+	14	25
2,3-Dihydroxybenzoic acid	±	1	10
3-Hydroxybenzoic acid	−	−2	0
4-Hydroxybenzoic acid	−	1	−2
2,4-Dihydroxybenzoic acid	−	−1	−2
2,5-Dihydroxybenzoic acid	−	1	−1
2,3,4-Trihydroxybenzoic acid	−	1	2
2,4,6-Trihydroxybenzoic acid	−	−1	3
3,4,6-Trihydroxybenzoic acid	−	1	−1
3-Aminosalicylic acid	−	1	5
4-Aminosalicylic acid	−	0	−1
5-Aminosalicylic acid	−	1	−2
Thiosalicylic acid	−	−5	1
2-Chlorobenzoic acid	−	0	2
2-Ethoxybenzoic acid	−	−3	2
o-Coumaric acid	−	1	—

[a]Biological activity is based on Abad *et al.* [16]; Van Loon [17]; White [5]; White *et al.* [18]; and Z. Chen and D.F. Klessig (unpublished work).
[b][^{14}C]SA (20 μM) binding was assayed in the presence of 40 μM (2×) or 200 μM (10×) unlabelled competitor.

A

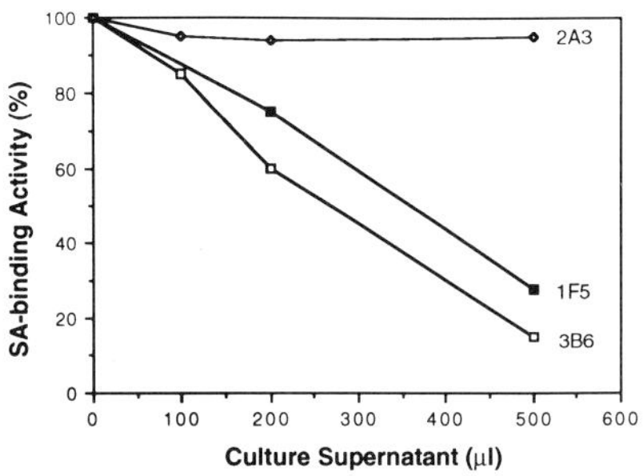

B

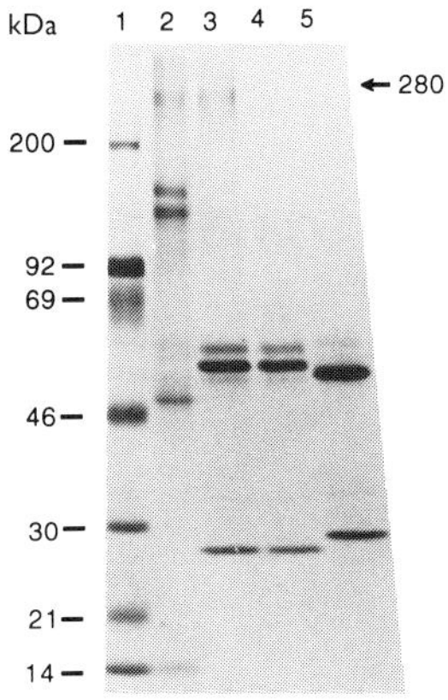

Fig. 4. Immunoprecipitation of SA-binding activity (A) and a 280 kDa protein (B) by mAbs. Increasing amounts of hybridoma culture media (supernatants) were incubated with Protein A–Sepharose. The mAb–Protein A–Sepharose complexes were reacted with the partially purified SABP obtained after the third purification step (Blue Dextran-Agarose). The antigen–mAb–Protein A–Sepharose complexes were pelleted and the supernatants assayed for remaining SA-binding activity (A). The pellets were washed and analysed by SDS/PAGE and silver staining. Lane 1, size marker proteins; lane 2, antigen only (a Blue Dextran-Agarose fraction); lane 3, antigen plus mAb 3B6; lane 4, mAb 3B6 only; and lane 5, antigen plus mAb 2A3. Note that mAbs 3B6 and 1F5 both immunoprecipitated the 280 kDa protein (data not shown for 1F5) and the SA-binding activity, while mAb 2A3 immunoprecipitated neither.

mAb 2A3, which did not immunoprecipitate the SA-binding activity, also failed to recognize the 280 kDa protein. This 280 kDa protein also co-purified with the SA-binding activity through the various chromatography steps (data not shown), arguing that it was responsible for SA binding.

The 280 kDa species was also recognized by the SABP-specific mAbs during immunoblot analysis of partially purified SABP. In contrast, these mAbs detected only a 57 kDa protein during immunoblot analysis of crude homogenates freshly prepared in the presence of a reducing agent to decrease the activity of polyphenol oxidases (data not shown). Since the native molecular mass of SABP is approx. 240 kDa, this suggests that SABP is a complex which contains the 57 kDa protein and, perhaps, one or more additional proteins. (For further details, see Chen *et al.* [18a].)

Discussion

The experiments described above provide additional evidence that SA is involved in activating defence responses, including the induction of the PR-1 genes. When plants were infected at 32°C, the increases in SA levels characteristic of the resistance response to TMV infection were abolished and both the hypersensitive response and the induction of the PR-1 genes were blocked. Upon releasing the block by shifting plants to 22°C, there was a rapid (6 h) and dramatic (100–200-fold) rise in SA levels. This rapid rise in SA levels after the temperature shift correlates with the observations of Weststeijn [19]. By transferring inoculated tobacco plants from 32°C to 22°C for short periods of time, and then returning them to the higher temperature, he discovered that necrotic lesion formation associated with resistance occurred only when the 22°C incubation period exceeded 6 h. These results suggest that an elevated level of SA is involved in lesion formation. However, it is probably one of several essential signals, since SA treatment of inoculated plants at elevated temperatures does not induce the formation of lesions [10].

We [10] and Enyedi *et al.* [20] have discovered that SA exists in a conjugated, as well as free, form in TMV-inoculated tobacco plants. The majority of these conjugates were shown to be SAG. The very low levels of both SA and SAG in non-inoculated plants or in inoculated plants held at 32°C suggest that both forms were produced *de novo* during defence responses. These observations also suggest that their synthesis is blocked by elevated temperatures.

The presence of SAG, as well as SA, in TMV-infected tobacco suggests that there may be greater complexity in the regulation of the SA signal than previously thought. For example, conjugation to glucose may serve to inactivate the SA signal or produce a new signal that has different physiological consequences. Here we have shown that SAG is physiologically active, as it induces the expression of the defence-related PR-1 genes. However, we strongly suspect that SAG itself is not active but acts via its conversion to free, active SA. After injection of SAG into tobacco leaves, the level of SAG rapidly declined with a concomitant rise in SA. This was followed by a decrease in SA to almost basal levels, with a concomitant increase in the level of SAG. It appears that hydrolysis of SAG occurred in the extracellular spaces, where the compound was injected. The released SA was then imported into the cells, where a glucosyl transferase converted it to SAG. The physical separation of the conjugating (intracellular) and deconjugating (extracellular) enzymes may play an important role in regulating the levels of the SA signal by these two opposing reactions.

Although these results do not exclude the possibility that SAG is itself an active molecule, given the precedence of other plant compounds and their glucoside conjugates [21–25], we propose that SAG serves as an inactive storage form that can be converted to free, active SA. Precedence for glucosides serving as storage forms of bioactive compounds is provided by recent studies on phytohormone deconjugation [23–25]. If SAG serves as a storage form for active SA, it may play a role in the maintenance of systemic acquired resistance. It is generally believed that a plant's ability to resist a second infection more rapidly than an initial infection (systemic acquired resistance) is due to the persistence of defence-related proteins and factors produced in both infected and uninfected tissues during the initial infection [26]. However, accumulation of SAG may represent another mechanism that allows rapid mobilization of defences during a second infection. For example, the early events at an infected site, such as changes in cell membrane permeability and cell damage, might release cellular stores of SAG into the extracellular spaces. Hydrolysis of the released SAG would provide a rapid increase in SA levels at the infection site, which could further induce defence responses.

We previously reported the identification of a soluble SABP in tobacco leaves [13]. Its binding affinity for SA and its binding specificity for biologically active SA analogues suggest that it may be involved in the induction of plant disease resistance. In order to further characterize this interesting protein and to eventually establish its functional relevance to plant signal transduction, we have purified it and are attempting to clone its gene(s) for further genetic and biochemical analysis.

During the initial stages of purification, we had difficulty separating SABP from the majority of the other proteins. It now appears that this was due to extensive aggregation that occurred during the initial ammonium sulphate precipitation step. By eliminating this step and employing a series of ion-exchange and gel-filtration chromatographic steps, the SABP has been purified 250-fold. Several mAbs, which were raised against the partially purified SABP, immunoprecipitated the SA-binding activity and a 280 kDa protein. Since the 280 kDa protein co-purified with the binding activity, and its size was consistent with the previously determined molecular mass of native SABP (approx. 240 kDa), we conclude that it is responsible for SA binding. However, the results of immunoblot analysis suggest that SABP is a homomeric or heteromeric complex containing a 57 kDa protein. Components of this complex appear to be readily cross-linked during purification to form the 280 kDa entity.

The successful identification and purification of SABP opens the way to addressing the critical question of its function. Such analysis should provide further insights into the mechanism(s) of action of SA-mediated physiological changes and perhaps the mode(s) of signal transduction in plants. Current data suggest that SA plays an important role in the induction of the disease defence responses. Multiple cellular factors appear to be involved in the synthesis, modification and reception of this SA signal. Several of these factors may be regulated by SA and other signals. Thus there may be a complex circuitry that controls the SA signal. Since SA is known to affect a diverse and extensive set of plant processes, at least when applied exogenously [27], an intricate control system may be necessary. The discovery of the mechanism(s) by which SA affects these processes and is itself modulated as a signal, should provide significant insights into signal transduction and disease defence responses in plants.

This work was supported in part by grants from the National Science Foundation (DCB-9003711) and U. S. Department of Agriculture (92-37301-5799). J.M. has been supported by Busch and Benedict-Michael Fellowships.

References

1. Cutt, J.R. and Klessig, D.F. (1992) in Plant Gene Research (T. Boller and F. Meins, Jr., eds.), pp. 181–216, Springer-Verlag, Wein
2. Linthorst, H.J.M. (1991) Crit. Rev. Plant Sci. **10**, 123–150
3. Lamb, C.J., Lawton, M.A., Dron, M. and Dixon, R.A. (1989) Cell **56**, 215–224
4. Malamy, J. and Klessig, D.F. (1992) Plant J. **2**, 643–654
5. White, R.F. (1979) Virology **99**, 410–412
6. Malamy, J., Carr, J.P., Klessig, D.F. and Raskin, I. (1990) Science **250**, 1001–1004
7. Métraux, J.P., Signer, H., Ryals, J., Ward, E., Wyss-Benz, M., Gaudin, J., Raschodf, K., Schmid, E., Blum, W. and Inverardi, B. (1990) Science **250**, 1004–1006
8. Rasmussen, J.B., Hammerschmidt, R. and Zook, M.N. (1991) Plant Physiol. **97**, 1342–1347
9. Raskin, I., Turner, I.M. and Melander, W.R. (1989) Proc. Natl. Acad. Sci. U.S.A. **86**, 2214–2218
10. Malamy, J., Hennig, J. and Klessig, D.F. (1992) Plant Cell **4**, 359–366
11. Hennig, J., Malamy, J., Grynkiewicz, G., Indulski, J. and Klessig, D. F. (1993) Plant J. **4**, 593–600
12. Grynkiewicz, G., Achmatowicz, O., Hennig, J., Indulski, J. and Klessig, D.F. (1993) Pol. J. Chem. **67**, 1251–1254
13. Chen, Z. and Klessig, D.F. (1991) Proc. Natl. Acad. Sci. U.S.A. **88**, 8179–8183
14. Kassanis, B. (1952) Ann. App. Biol. **39**, 358–369
15. Harborne, J.B. (1980) Encycl. Plant Physiol. **8**, 329–402
16. Abad, P., Marais, A., Cardin, L., Poupet, A. and Ponchet, M. (1988) Antiviral Res. **9**, 315–327
17. Van Loon, L.C. (1983) Neth. J. Plant Path. **89**, 265–273
18. White, R.F., Dumas, E., Shaw, P. and Antoniw, J.F. (1986) Antiviral Res. **6**, 177–185

18a. Chen. Z., Ricigliano, J.W. and Klessig, D.F. (1993) Proc. Natl. Acid. Sci. U.S.A. **90**, 9533–9537

19. Weststeijn, E.A. (1981) Physiol. Plant Path. **23**, 357–368
20. Enyedi, A., Yalpani, N., Silverman, P. and Raskin, I. (1992) Proc. Natl. Acad. Sci., U.S.A. **89**, 2480–2484
21. Letham, D.S. and Plani, L.M.S. (1983) Annu. Rev. Plant Physiol. **34**, 163–197
22. Reineche, D.M. and Bandurski, R.S. (1988) in Plant Hormones and Their Role in Plant Growth and Development (P. J. Davis, ed.), pp. 24–42, Martinus Nijhoff, Dordrecht
23. Estruch, J.J., Chriqui, D., Grossmann, K., Schell, J. and Spena, A. (1991) EMBO J. **10**, 2889–2895
24. Estruch, J.J., Schell, J. and Spena, A. (1991). EMBO J. **10**, 3125–3128
25. Campos, N., Bako, L., Feldwisch, J., Schell, J. and Palme, K. (1992) Plant J. **2**, 675–684
26. Carr, J.P. and Klessig, D.F. (1989) Genet. Eng. (N.Y.) **11**, 65–109
27. Cutt, J.R. and Klessig, D.F. (1992) Pharm. Technol. **16**, 26–34

Biochem. Soc. Symp. **60**, 231–240
Printed in Great Britain

Regulation of gene expression in bacterial pathogens

M.J. Daniels*, J.M. Dow, T.J.G. Wilson, S.D. Soby†, J.L. Tang‡, B. Han§ and S.A. Liddle

The Sainsbury Laboratory, John Innes Centre, Norwich Research Park, Norwich NR4 7UH, U.K.

Introduction

Bacteria which can cause disease in plants belong to a small number of genera, both Gram-positive and Gram negative [1]. The latter, including the genera *Agrobacterium*, *Erwinia*, *Pseudomonas* and *Xanthomonas*, have been much more extensively studied by molecular plant pathologists. The reasons for this are that the organisms are generally easier to grow and handle and, more importantly, tools required for genetic analysis have been available for more than 10 years. In contrast, comparable systems for Gram-positive pathogens are still being developed. The Gram-negative pathogens together can infect a wide range of plants, including important crop plants and favourite experimental 'model' plants, such as *Arabidopsis*. Some of the most damaging crop diseases, such as bacterial blight of rice and bacterial wilt of solanaceous and other crops, are caused by Gram-negative bacteria *Xanthomonas oryzae* pv. *oryzae* and *Pseudomonas solanacearum*, respectively. Bacteria tend to thrive in warm, humid environments, which means that bacterial diseases are generally more serious in tropical to warm temperate climates.

Several strategies have been used to identify and clone genes essential for pathogenicity, including (i) isolation of mutants altered in pathogenicity (by direct screening of mutagenized colonies on plants), followed by isolation of clones which restore pathogenicity from a wild-type DNA library; (ii) identification of cloned DNA which directs the synthesis of a specific detectable product, believed to be involved in pathogenesis, by expression in a suitable host bacterium; (iii) identification of genes showing differential expression when the bacteria are in plants (compared with pure cultures) [2]; and (iv) isolation of genes from a library by hybridization to specific sequences suspected to be contained in pathogenicity genes [3,4].

*To whom correspondence should be addressed.
Present addresses: †Department of Plant Pathology, University of Arizona, Tucson, AZ 85721, U.S.A.; ‡Laboratory of Molecular Genetics, Guangxi Agricultural University, Nanning, China; and §Department of Plant Sciences, University of Cambridge, Cambridge CB2 3EA, U.K.

Having isolated cloned genes, it is usual to evaluate their role in pathogenesis by creating specific mutations in the endogenous copy of the gene [5] and then to observe the effect of the mutation on plant interaction phenotype. These strategies have been applied to all the genera listed above, in many laboratories. Genes affecting the following processes have been analysed in detail and shown to be involved in pathogenicity: (i) structural genes for extracellular enzymes (*Erwinia*, *Pseudomonas*, *Xanthomonas*); (ii) enzyme secretion systems (*Erwinia*, *Xanthomonas*); (iii) avirulence genes (controlling host specificity) (*Pseudomonas*, *Xanthomonas*); (iv) extracellular polysaccharide biosynthesis (*Erwinia*, *Pseudomonas*, *Xanthomonas*); (v) auxin/cytokinin synthesis (*Agrobacterium*, *Pseudomonas*); (iv) toxin biosynthesis (*Pseudomonas*); (vii) *hrp* genes (affecting pathogenicity to susceptible plants and induction of hypersensitive response on resistant plants) (*Erwinia*, *Pseudomonas*, *Xanthomonas*); (viii) regulatory genes (*Agrobacterium*, *Erwinia*, *Pseudomonas*, *Xanthomonas*).

In addition, genes have been cloned from several bacteria whose role in pathogenesis is clear-cut but not understood (for example, see [6]).

The rest of this discussion will be confined to regulatory genes and to *hrp* genes, some of which are believed to have regulatory functions. Although our understanding of these genes and their functions is far from complete, common themes are beginning to emerge. This is particularly true of *Erwinia*, *Pseudomonas* and *Xanthomonas*.

Agrobacterium

Agrobacterium tumefaciens stands apart from the other pathogens because of its unusual mode of pathogenesis. This involves transfer of T-DNA from the bacterial Ti plasmid into plant cells, and its incorporation into the plant genome to give transformed, tumerous tissue. The transformation phenomenon has been heavily exploited to yield procedures for simple production of transgenic plants. The virulence system of *Agrobacterium* has been studied for several years [7]. Transfer of T-DNA into plants requires a set of *vir* (virulence) genes located on the Ti plasmid, together with some chromosomal loci. Two of the *vir* genes, *virA* and *virG*, regulate the expression of the other *vir* genes in response to plant-derived signals [8]. The latter include phenolics, sugars and pH changes produced as a consequence of wounding of plants. The VirA and VirG proteins belong to a large class of prokaryotic regulatory systems originally called 'two-component' regulators, but also 'sensor-regulator' and 'histidine protein kinase-response regulator' systems. Many such systems have been described in bacteria and, although all have certain features in common, there are a number of sub-classes, defined in terms of arrangement of conserved functional domains [9].

The first component is the sensor (VirA in the present instance). The *N*-terminus of the protein has transmembrane domains and a region within the periplasm. These regions are believed to recognize the external signal molecules, whereupon the conserved cytoplasmic *C*-terminal 'transmitter module' region undergoes autophosphorylation of a histidine residue. Phosphorylated VirA in turn transfers the phosphate to an aspartate residue in the conserved *N*-terminal 'receiver module' region of VirG [10,11]. Phosphorylated VirG acts as a transcriptional regulator and can bind to the promoters of *vir* genes [12,13]. Although the *virA*–*virG* system is the most extensively studied regulatory system in plant pathogens, there are still many gaps in our knowledge of the biochemical mechanisms involved, particularly concerning signal perception [14].

Erwinia

Genetic research on *Erwinia* has been greatly helped by the relatedness of the genus to *Escherichia*. Many of the genetic tricks and tools developed for *Escherichia coli* can be readily adapted for *Erwinia*. *Erwinia carotovora* and *Erwinia chrysanthemi* produce a range of extracellular enzymes (pectinases, cellulases and proteases) which can degrade plant tissues and give soft rots and related pathological conditions.

The pathways of pectin degradation and incorporation of the breakdown products into the central metabolic pathways of the cell have been elucidated in *E. chrysanthemi*. Twelve genes, which are independent transcription units, are involved, of which six encode extracellular enzymes (pectin methylesterase, *pem*, and five polygalacturonate lyases, *pelA–E*). All these genes are repressed by the product of the *kdgR* gene. The KdgR protein binds to a KdgR 'box' sequence in the upstream regions of the genes, and the formation of this repressor–operator complex is inhibited by the inducer 2-keto-3-deoxygluconate, one of the intermediates in the intracellular pectin degradation pathway [15]. In addition to *kdgR*, other regulatory genes affecting the pectin pathway are known: *pecI*, *pecS*, *pecY*, *pecL* and *pecN*. Thus the regulation system is complex; the interaction between the regulatory genes is not understood. Studies of expression of *pel* genes (using fusions to β-glucuronidase) in bacteria growing in potato tubers suggest that plant factors modulate expression [16]. In *E. carotovora*, KdgR box sequences are found in regulatory regions of some pectinase genes, but there is no direct evidence for the presence of a *kdgR* gene [17]. Pectate lyase, polygalacturonase and protease are co-ordinately regulated at the transcriptional level in *E. carotovora* by two activator genes, *aepA* and *aepB*, and a negative regulatory gene *repN* [17]. Mutation of the *expB* gene also gives a drastic reduction in synthesis of all extracellular enzymes [18] but, on the other hand, *pehR* regulates only one polygalacturonase gene [19]. *E. carotovora* also produces a pectin lyase encoded by *pnlA*. Interestingly, this gene is regulated differently to the *pme*, *pel* and *peh* genes; it is induced by DNA-damaging agents and requires *recA* and *rdg* genes [20].

Certain strains of *E. carotovora* produce an antibiotic, carbapenem, late in exponential growth. This is regulated by production of a low-molecular-mass inducer *N*-(3-oxohexanoyl)homoserine lactone, so that regulation is cell-density dependent. Similar regulatory substances are believed to be made by a number of prokaryotes [21]. Conjugal transfer of the *A. tumefaciens* Ti plasmid has recently been shown to be promoted by a similar lactone [22,23].

In addition to extracellular enzymes, virulence of *Erwinia* depends on a source of iron and on production of polysaccharide. Iron uptake depends on a siderophore (chrysobactin) and a transport protein. Expression of the genes for chrysobactin synthesis and transport (*cbsC*, *E*, *B*, *A* and *fct*) is regulated by *cbrA* and *B* in response to iron availability [24]. The latter genes also have some effect on *pel* gene expression.

Regulation of *Erwinia* extracellular polysaccharide synthesis is very similar to *E. coli*. Both *Erwinia amylovora* and *Erwinia stewartii* have *rcsA* genes which encode a regulatory protein sensitive to degradation by the *lon* protease [25–27]. In addition, *E. coli* possesses a two-component system, *rcsBC*, and it will be of interest to see what role equivalent *Erwinia* genes may have.

Pseudomonas

For the purpose of this discussion, it is convenient to consider two groups of pseudomonads: the *P. syringae* pathovars and *P. solanacearum*.

P. syringae pathogenesis seems to be less dependent on extracellular enzymes than *Erwinia*, *P. solanacearum* or *Xanthomonas campestris*. On the other hand, many strains produce low-molecular-mass phytoxins, such as coronatine, tabtoxin, phaseolotoxin, syringomycin and syringotoxin.

A gene designated *lemA* was isolated because transposon mutations in it abolished pathogenicity. Sequencing revealed that the gene product belongs to the class of two-component regulators [28]. However, it is a member of a minority sub-class (ITRO in the terminology adopted by Parkinson and Kofoid [9]) in which both transmitter and receiver modules are present in the same protein. The *Xanthomonas* regulator RpfC [29] and *A. tumefaciens* VirA share this property, the biochemical significance of which is not known. The *lemA* gene is expressed at a low level in *P. syringae* but is autoregulated [28]. *P. syringae* pv. *coronafaciens lemA* mutants fail to produce protease or tabtoxin [30]. The *lemA* gene is found in many *P. syringae* pathovars, although its function is not clear because *lemA* mutants of *P. syringae* pv. *phaseolicola* retain pathogenicity and still produce phaseolotoxin [31].

P. solanacearum is a soilborne bacterium which infects roots of susceptible plants. It then rapidly colonizes the xylem and causes 'bacterial wilt'. It is believed that the wilting symptoms result from interference with water relations in the plant [32]. *P. solanacearum* is probably the most serious bacterial plant pathogen in terms of crop losses worldwide. Virulent strains produce extracellular polysaccharide and various extracellular proteins, including pectinases and endoglucanases. These are believed to be virulence factors, although, as with most other bacterial virulence factors, their precise function is not understood. It is becoming apparent that regulation of synthesis of the factors is complex. The locus *pehR* is a positive regulator of polygalacturonase synthesis. One of the genes in this locus is a sensor member of the 'two-component' family [33]. Denny, Schell and colleagues [34] have studied the regulation of polysaccharide synthesis in *P. solanacearum* and have found a number of interacting loci. The polysaccharide synthesis (*eps*) genes are organized into two adjacent clusters, encoding at least 11 polypeptides. One of these (XpsR) is a regulator, removal of which by mutation reduces transcription of the *eps* genes by 90%. Other distinct loci which are required for *eps* expression have been identified: *phcA*, *phcB*, *vsrA*, *vsrB*, *phcA* and *vsrB* mutants grow in plants and give some symptoms, but *xpsR* and *vsrA* mutants show no growth or symptoms. PhcA may be a LysR family regulator, and VsrA and VsrB may be secreted or periplasmic proteins [34]. *phcB* may be involved in production of a diffusible and volatile regulatory effector substance.

Xanthomonas

Xanthomonas strains produce extracellular polysaccharide (xanthan), and some pathovars produce several extracellular enzyme activities. *X. campestris* pv. *campestris* produces pectinases (pectin methylesterase and polygalacturonate lyase), proteases, endoglucanase, lipase and amylase. Although the role of these factors individually in pathogenesis is not understood, together they are clearly necessary. Regulatory mutants which fail to produce enzymes and xanthan, and secretion mutants which produce enzymes but fail to export them into the medium are severely reduced in virulence and in ability to grow in plants [35].

The synthesis of enzymes and xanthan is regulated in a complex manner, at individual and global levels. Proteases are induced by oligo- and polypeptides and repressed by amino acids; polygalacturonate lyases are induced by polygalacturonate, but endoglucanase production is constitutive.

The first evidence for global, pleiotropic regulation came from the isolation of a mutant of *X. campestris* pv. *campestris* which was non-pathogenic and failed to produce enzymes and xanthan. A recombinant cosmid containing wild-type *X. campestris* pv. *campestris* DNA concomitantly restored pathogenicity and factor production when transferred into the mutant [36]. The DNA fragment was found to contain at least seven genes, designated *rpfA–G*, all of which were required for enzyme and xanthan synthesis. Sequencing revealed that *rpfC* encodes a two-component regulator, with both receiver and transmitter domains in the same protein, similar to *lemA* of *P. syringae* [29]. RpfG is also a two-component regulator protein (with a receiver domain). It is probable that RpfC transfers the phosphate group from its transmitter domain to the RpfG receiver domain, although this has yet to be proved. The RpfC protein has been overexpressed in *E. coli* and purified, and work is in progress to reconstruct the regulatory system *in vitro* (T.J.G. Wilson and M.J. Daniels, unpublished work).

Enzyme production by an *rpfF* mutant can be restored by growth in proximity to an $rpfF^+$ strain, suggesting that a diffusible substance is produced by *X. campestris* pv. *campestris* which is lacking in the *rpfF* mutant (Fig. 1). It is possible that the diffusible factor is related to the homoserine lactone derivatives discussed above in connection with *Erwinia*.

It is not clear why at least seven *rpf* genes are needed to regulate the set of enzymes and xanthan biosynthetic genes. There is some evidence that some of the regulators may regulate the synthesis of others, and there are also hints that the regulators may act ‘in series’ to form a regulatory pathway. A pathway might serve to amplify the initial signal, and would also make possible the modulation of end-product synthesis by multiple inputs.

The *rpfA–G* provides positive regulation of enzyme and xanthan synthesis. Balancing negative regulation is provided by another locus *rpfN* [37]. Mutation of the latter results in overproduction of enzymes and probably xanthan. A protein has been detected in *X. campestris* pv. *campestris* extracts which binds to a conserved sequence upstream of the protease and cellulase genes. The binding is *rpfN*-dependent, but it is not yet known whether the RpfN protein itself binds. It is supposed that the protein binding is part of the down-regulation mechanism (S. Soby, B. Han and M.J. Daniels, unpublished work). The promoter of the *rpfN* gene is probably dependent on a σ^{54} (*rpoN*, *ntrA*)-RNA polymerase for transcription, and adjacent to the promoter is a

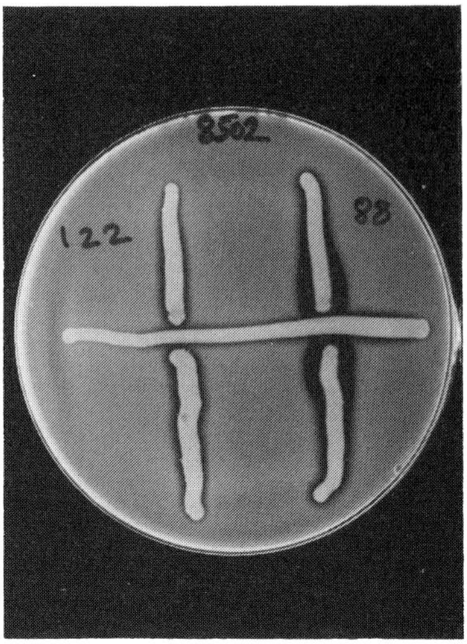

Fig. 1. Evidence for a diffusible regulatory substance produced by *X. campestris* pv. *campestris*. A strain of *X. campestris* pv. *campestris*, 8502, which is *rpfF* $^+$ was streaked horizontally across the middle of a 9 cm Petri dish containing skimmed-milk agar to detect protease production. Other strains, 88 and 122, were streaked at right angles to 8502, taking care that the streaks did not touch. Strain 88 (*rpfF*::Tn*5-lac*) produced much more protease activity, shown by the clear zone around the bacterial growth, adjacent to 8502. In contrast, strain 122, which is a *rpfC*::Tn*5-lac* mutant, is not 'cross-fed' by 8502. Note that, for clarity, 8502 carries a mutation in another gene which reduces protease production; this does not affect the cross-feeding phenomenon.

probable recognition site for a response regulator of the two-component class. Thus it is possible that the synthesis of the regulator RpfN is itself subject to regulation by another gene, as yet unknown. It is not yet known whether the positive (*rpfA–G*) and negative (*rpfN*) regulatory systems interact. For example, negative regulation could be accomplished by blocking positive regulation. However, preliminary experiments with double mutants suggest that the systems are independent.

The regulation of enzyme and xanthan synthesis is further complicated because a gene called *clp* has been shown to influence their production [38]. This gene was cloned by virtue of its ability to partially complement an *E. coli crp* mutant (defective in the catabolite activator protein) by restoring ability to grow on certain sugars. A *clp* mutant of *X. campestris* pv. *campestris* was constructed. Synthesis of xanthan, endoglucanase and pectinase was reduced, and the residual xanthan had altered properties compared with the polymer produced by the wild-type bacteria. Protease production was increased in the *clp* mutant, and amylase production was unchanged. Virulence was reduced. In *E. coli*, the *crp* gene is believed to regulate expression of many hundreds of genes. It is not known whether the *X. campestris* pv. *campestris clp* gene is similarly catholic in its action. The mutation does not affect the ability of *X. campestris* pv. *campestris* to utilize a wide range of carbon sources tested, but transcription of a reporter gene from the *E. coli lac* promoter is reduced in a *clp* mutant background, in comparison with wild-type *X. campestris* pv. *campestris* (S. Soby and M.J. Daniels, unpublished work). The Clp protein is very similar to Crp [38,39] and probably binds to DNA. If it is a catholic regulator in *X. campestris* pv. *campestris*, it may play an important role in 'tuning' virulence functions to match the physiological state of the bacteria.

It was noted above that RpfC and RpfG are members of the two-component family of regulators. Bacteria are believed to contain many regulatory systems of this class, perhaps as many as 50 [9]. Osbourn *et al.* [3] used oligonucleotides which would code for conserved peptide domains as hybridization probes to isolate further two-component regulatory genes from *X. campestris* pv. *campestris* When mutated, many of these genes gave no obvious phenotypic change. However, one caused a reduction of 70% in xanthan production. This reduction was, however, not sufficient to have a noticeable effect on virulence.

hrp genes

hrp genes are so called because mutagenesis causes a loss of ability to incite both a hypersensitive response and pathogenicity. That is to say, both compatible and incompatible bacteria–plant interactions are affected. Mutants of this class have been found in *E. amylovora*, *P. solanacearum* and *P. syringae* pathovars, and in *X. campestris* pathovars. In all cases, the *hrp* genes are organized in clusters of 20 kb or more, probably with some additional unlinked loci. Early studies on *hrp* genes were reviewed by Willis *et al.* [40]. Recent work has revealed striking similarities between the *hrp* genes of several plant pathogens and certain pathogenicity genes of animal pathogens. For this reason, *hrp* genes are discussed together, rather than under the headings of individual bacterial genera. The relatedness was first suspected when it was found that *P. solanacearum hrp* genes hybridized to sequences in the genome of a range of *X. campestris* pathovars [41]. The homologous DNA was cloned from *X. campestris* pv. *campestris* and shown to contain *hrp* genes in this organism, with a similar organization to the *P. solanacearum* cluster [4]. Sequencing has subsequently confirmed the relationships between *hrp* genes of different bacteria [42–45]. The organization of *hrp* genes and features of their expression was determined by genetic methods. In all cases the *hrp* gene clusters consist of at least six transcriptional units. There is general agreement that expression of *hrp* genes is repressed in rich media and induced during bacterial growth *in planta* [46]. The nature of the controlling factors may, however, vary. In *P. syringae* pv. *phaseolicola*, *hrpA, B, C* and *D* are regulated by osmolarity of the medium, pH and type of carbon source; *hrp* E is affected by pH and carbon source, but relatively little by osmolarity, and *hrpF* is unaffected by any of these factors. *hrpL* and *S* are scarcely expressed *in vitro*, but expressed at high levels *in planta*. *hrpS* and *hrpR* encode transcriptional regulators of the other *hrp* loci and are, in turn, influenced by the unlinked genes *hrpQ* and *hrpT*. *hrpQ* and *T* mutants are also auxotrophic for methionine and tryptophan, respectively [44]. In *X. campestris* pv. *vesicatoria* [47,48] a low-molecular-mass plant factor has been implicated in regulation of *hrp* expression, and levels of phosphate, sucrose, methionine and salt are also important. In *E. amylovora* [49], ammonium, nicotinic acid, temperature, pH and carbon sources were implicated as controlling factors. In contrast, the major factor in stimulating expression of *X. campestris* pv. *campestris hrp* genes appears to be starvation (S.A. Liddle and M.J. Daniels, unpublished work).

Some hints about the functions of *hrp* gene products are beginning to emerge. Genetic evidence indicates that *hrpR, S, Q* and *T* products are transcriptional regulators of other *hrp* genes in *P. syringae* pv. *phaseolicola* [44], and in *P. solanacearum hrpB* encodes a positive regulator of the XylS/AraC class [50]. The products of the *hrpN* gene of *E. amylovora* [51] and of an *hrpB*-dependent gene of *P. solanacearum* [50] are proteinaceous elicitors of plant cell necrosis. The necrosis is observed with both host and non-host plants [51], and has many of the characteristics of cell collapse and necrosis observed in a hypersensitive response caused by intact bacteria. Incompatible interactions at the level of pathogen race and plant cultivar require the presence of avirulence genes in the pathogen and matching resistance genes in the host. The transcription of some *P. syringae* avirulence genes is *hrp*-dependent [52,53], whereas in *X. campestris* pv. *vesicatoria* and *X. campestris* pv. *campestris* this is not the case [54,55]. Even in these cases where the avirulence gene products are presumably synthesized, the characteristic incompatible interaction phenotype is dependent on *hrp* gene function.

Recent results from sequencing of *hrp* gene clusters of *P. solanacearum* and *X. campestris* pv. *vesicatoria* have indicated striking relatedness of some deduced *hrp* products with pathogenicity gene products of animal pathogens (*Yersinia enterocolitica*, *Y. pestis*, *E. coli*, *Shigella flexneri*, *Salmonella typhimurium*) [42,43]. Although the functions of the animal pathogenicity gene products are not fully understood, it is believed that they are required for secretion of other factors. The similarities between the proteins of the animal and plant pathogens promises to open up new areas of research in pathogenesis. Animal and plant systems each have advantages and disadvantages for experimental analysis of microbial pathogenicity. The related pathogenicity gene products suggest that common mechanisms may be operating and will encourage investigators to search for further similarities.

Conclusions

In this paper we have reviewed only a fraction of the literature on bacterial pathogenicity to plants. In particular we have not considered important work on host specificity determinants, and on mechanisms by which plants resist infections. We have tried to demonstrate the similarities in the pathogenicity mechanisms of a range of bacteria, which incite quite different diseases. Production of pathogenicity factors is regulated in a complex manner, in response to changes in both the external environment of the bacteria (including plant tissues), and the internal physiological and metabolic state of the cells. This indicates that many biochemical pathways need to be considered in relation to host–pathogen interaction. In the early days of genetic analysis of pathogenicity it was expedient to regard pathogens as 'consensus bacteria', with some additional pathogenicity genes added. We have now moved beyond this stage; the challenge for the coming years is to integrate specific studies of pathogenicity with general bacterial and plant cell biology.

The Sainsbury Laboratory is supported by the Gatsby Charitable Foundation. Research in the authors' laboratory has also been supported by the Agricultural and Food Research Council, the Commission of the European Communities and the Royal Society.

References

1. Bradbury, J.F. (1986) Guide to Plant Pathogenic Bacteria, CAB International, Farnham Royal
2. Osbourn, A.E., Barber, C.E. and Daniels, M.J. (1987) EMBO J. **6**, 23–28
3. Osbourn, A.E., Clarke, B.R., Stevens, B.J.H. and Daniels, M.J. (1990) Mol. Gen. Genet. **222**, 145–151
4. Arlat, M., Gough, C.L., Barber, C.E., Boucher, C. and Daniels, M.J. (1991) Mol. Plant–Microbe Interact. **4**, 593–601
5. Ruvkun, G.B. and Ausubel, F.M. (1981) Nature (London) **289**, 85–88
6. Osbourn, A.E., Clarke, B.R. and Daniels, M.J. (1990) Mol. Plant–Microbe Interact. **3**, 280–285
7. Beijersbergen, A. and Hooykaas, P.J.J. (1993) Adv. Mol. Genet. Plant–Microbe Interact. **2**, 37–49
8. Winans, S.C. (1992) Microbiol. Rev. **56**, 12–31
9. Parkinson, J.S. and Kofoid, E.C. (1992) Annu. Rev. Genet. **26**, 71–112
10. Jin, S., Roitsch, T., Ankenbauer, R.G., Gordon, M.P. and Nester, E.W. (1990) J. Bacteriol. **172**, 525–530
11. Jin, S., Prusti, R.K., Roitsch, T., Ankenbauer, R.G. and Nester, E.W. (1990) J. Bacteriol. **172**, 4945–4950
12. Jin, S., Roitsch, T., Christie, P.J. and Nester, E.W. (1990) J. Bacteriol. **172**, 531–537
13. Pazour, G.J. and Das, A. (1990) J. Bacteriol. **172**, 1241–1249
14. Binns, A.N., Joerger, R.D., Banta, L.M., Lee, K. and Lynn, D.G. (1993) Adv. Mol. Genet. Plant–Microbe Interact. **2**, 51–61
15. Nasser, W., Reverchon, S. and Robert-Baudouy, J. (1992) Mol. Microbiol. **6**, 257–265
16. Lojkowska. E., Dorel, C., Reignault, P., Hugovieux-Cotte-Pattat, N. and Robert-Baudouy, J. (1993) in Mechanisms of Plant Defense Responses (Fritig, B. and Legrand, M., eds.), pp. 72–75, Kluwer, Dordrecht
17. Chatterjee, A., Liu, L., Murata, H., Souissi, T. and Chatterjee, A.K. (1993) Adv. Mol. Genet. Plant–Microbe Interact. **2**, 241–251
18. Pirhonen, M., Saarilahti, H., Karlsson, M.B. and Palva, E.T. (1991) Mol. Plant–Microbe Interact. **4**, 276–283
19. Flego, D., Pirhonen, M. and Palva, E.T. (1992) Abstr. 6th Int. Symp. Mol. Plant–Microbe Interact. 170
20. Chatterjee, A., McEvoy, J.L., Chambost, J.P., Blasco, F. and Chatterjee, A.K. (1991) J. Bacteriol. **173**, 1765–1769
21. Williams, P., Bainton, N.J., Swift, S., Chhabra, S.R., Winson, M.K., Stewart, G.S.A.B., Salmond, G.P.C. and Bycroft, B.W. (1992) FEMS Microbiol. Lett. **100**, 161–168
22. Zhang, L., Murphy, P.J., Kerr, A. and Tate, M.E. (1993) Nature (London) **362**, 446–448
23. Piper, K.R., von Bodman, S.B. and Farrand, S.K. (1993) Nature (London) **362**, 448–450
24. Sauvage, C., Mahé, B., Laulhre, J.P., Masclaux, C. and Expert, D. (1992) Abstr. 6th Int. Symp. Mol. Plant–Microbe Interact. 167
25. Coleman, M., Pearce, R., Hitchin, E., Busfield, F., Mansfield, J.W. and Roberts, I.S. (1990) J. Gen. Microbiol. **136**, 1799–1806
26. Poetter, K. and Coplin, D.L. (1991) Mol. Gen. Genet. **229**, 155–160

27. Roberts, I.S., Taylor, N., Rigg, G. and Eastgate, J. (1992) Abstr. 6th Int. Symp. Mol. Plant–Microbe Interact., 203
28. Hrabak, E.M. and Willis, D.K. (1992) J. Bacteriol. **174**, 3011–3020
29. Tang, J.L., Liu, Y.N., Barber, C.E., Dow, J.M., Wootton, J.C. and Daniels, M.J. (1991) Mol. Gen. Genet. **226**, 409–419
30. Barta, T.M., Kinscherf, T.G. and Willis, D.K. (1992) J. Bacteriol. **174**, 3021–3029
31. Rich, J.J., Hirano, S.S. and Willis, D.K. (1992) Appl. Environ. Microbiol. **58**, 1440–1446
32. Denny, T.P., Carney, B.F. and Schell, M.A. (1992) Mol. Plant–Microbe Interact. **3**, 293–300
33. Allen, C., Simon, L. and Sequeira, L. (1992) Abstr. 6th Int. Symp. Mol. Plant–Microbe Interact. 209
34. Schell, M.A., Denny, T.P., Clough, S.J. and Huang, J. (1993) Adv. Mol. Genet. Plant–Microbe Interact. **2**, 231–239
35. Daniels, M.J., Barber, C.E., Turner, P.C., Cleary, W.G. and Sawczyc, M.K. (1984) J. Gen. Microbiol. **130**, 2447–2455
36. Daniels, M.J., Barber, C.E., Turner, P.C., Sawczyc, M.K., Byrde, R.J.W. and Fielding, A.H. (1984) EMBO J. **3**, 3323–3328
37. Tang, J.L., Gough, C.L. and Daniels, M.J. (1990) Mol. Gen. Genet. **222**, 157–160
38. de Crécy-Lagard, V., Glaser, P., Lejeune, P., Sismeiro, O.,Barber, C.E., Daniels, M.J. and Danchin, A. (1990) J. Bacteriol. **172**, 5877–5883
39. Dong, Q. and Ebright, R.H. (1992) J. Bacteriol. **174**, 5457–5461
40. Willis, D.K., Rich, J.J. and Hrabak, E.M. (1991) Mol. Plant–Microbe Interact. **4**, 132–138
41. Boucher, C.A., van Gijsegem, F., Barberis, P.A., Arlat, M. and Zischek, C. (1987) J. Bacteriol. **169**, 5626–5632
42. Gough, C.L., Genin, S., Zischek, C. and Boucher, C.A. (1992) Mol. Plant–Microbe Interact. **5**, 384–389
43. Fenselau, S., Balbo, I. and Bonas, U. (1992) Mol. Plant–Microbe Interact. **5**, 390–396
44. Miller, W., Mindrinos, M.M., Rahme, L.G., Frederick, R.D., Grimm, C., Gressman, R., Kyriakides, X., Kokkinidis, M. and Panopoulos, N.J. (1993) Adv. Mol. Genet. Plant–Microbe Interact. **2**, 267–274
45. Beer, S.V., Wei, Z., Laby, R.J., He, S.Y., Bauer, D.W., Collmer, A. and Zumoff, C. (1993) Adv. Mol. Genet. Plant–Microbe Interact. **2**, 281–286
46. Xiao, Y., Lu, Y., Heu, S. and Hutcheson, S.W. (1992) J. Bacteriol. **174**, 1734–1741
47. Schulte, R. and Bonas, U. (1992) J. Bacteriol. **174**, 815–823
48. Schulte, R. and Bonas, U. (1992) Plant Cell **4**, 79–86
49. Wei, Z., Sneath, B.J. and Beer, S.V. (1992) J. Bacteriol. **174**, 1875–1882
50. Genin, S., Gough, C.L., Arlat, M., Zischek, C., van Gijsegem, F., Barberis, P. and Boucher, C.A. (1993) Adv. Mol. Genet. Plant–Microbe Interact. **2**, 259–266
51. Wei, Z., Laby, R.J., Zumoff, C.H., Bauer. D.W., He, S.Y., Collmer, A. and Beer, S.V. (1992) Science **257**, 85–88
52. Huynh, T.V., Dahlbeck, D. and Staskawicz, B.J. (1989) Science **245**, 1374–1377
53. Keen, N.T., Sims, J.J., Midland, S., Yoder, M., Jurnak, F., Shen, H., Boyd, C., Yucel, I., Lorang, J. and Murillo, J. (1993) Adv. Mol. Genet. Plant–Microbe Interact. **2**, 211–220
54. Knoop, V., Staskawicz, B. and Bonas, U. (1991) J. Bacteriol. **173**, 7142–7150
55. Parker, J.E., Barber, C.E., Fan, M.J. and Daniels, M.J. (1993) Mol. Plant–Microbe Interact. **6**, 216–224

Biochem. Soc. Symp. **60**, 241–248
Printed in Great Britain

Molecular mechanisms underlying induction of plant defence gene transcription

Christopher J. Lamb* and Richard A. Dixon†

*Plant Biology Laboratory, Salk Institute for Biological Studies, 10010 North Torrey Pines Road, La Jolla, CA 92037, U.S.A. and
†Plant Biology Division, Samuel Roberts Noble Foundation, P.O. Box 2180, Ardmore, OK 73402, U.S.A.

Introduction

Our overall objective is to dissect the mechanisms involved in the perception of, and response to, microbial attack leading to activation of inducible defences. Our strategy is to trace a causal sequence of events by working back from the final induced defence responses toward initial perception. Plants respond to fungal elicitor or microbial attack by transcribing defence genes encoding products involved in the elaboration of inducible protective responses, such as phytoalexin synthesis and production of lytic enzymes. Thus defence gene promoters are targets of stress signal pathways, and as a model we are characterizing *trans*-factors that mediate elicitor induction of genes encoding the key phytoalexin enzyme chalcone synthase. Our studies are revealing how these *trans*-factors function as terminal components of the signal pathway, and provide the basis for stepwise delineation of earlier components at the molecular level.

Plants exhibit inducible defence responses

Plants elaborate inducible defences following microbial attack including: synthesis of phytoalexins, deposition of lignin, hydroxyproline-rich glycoproteins and callose, and deployment of lytic enzymes [1]. Defence induction associated with expression of localized hypersensitive resistance is observed in the early stages of attempted infection by a non-pathogen or by an avirulent pathogen in an incompatible interaction [1]. In some cases, these responses are observed in distant tissue, associated with the expression of acquired resistance [2]. These defences can also be induced by

*To whom correspondence should be addressed.

microbial glycan and glycoprotein elicitors [3–5], or by metabolites such as arachidonate and glutathione [6,7], as well as chemically diverse abiotic elicitors that appear to act by non-specific traumatization [5,8].

The spatial and temporal correlation between induction of defences and expression of resistance in many systems argues for a functional relationship [1,9–11]. Direct evidence comes from: (i) the relation of phytoalexin detoxification in *Nectria hematococca* to pathogenicity on pea [12]; (ii) the susceptibility to avirulent races of *Phytophthora megasperma* f.sp. *glycinea* of soybean seedlings treated with an inhibitor of phenylalanine ammonia-lyase (PAL), the first enzyme in lignin and phytoalexin synthesis [13]; and (iii) the protective effect of constitutive expression of chitinase [14,15].

Defence gene transcription

With the major exceptions of callose production [16] and oxidative cross-linking of cell wall structural proteins [17], defence induction generally involves transcription of the corresponding genes as part of a massive switch in gene expression [1,18–20]. For example, in suspension-cultured bean cells, genes encoding chitinase and enzymes of phenylpropanoid biosynthesis involved in phytoalexin and lignin production are activated within 5 min of elicitor treatment [20], and in elicited parsley cells there is rapid induction of certain pathogenesis-related (PR) protein genes [21]. These kinetics suggest that there are few steps between elicitor binding to a receptor and transcriptional activation. In contrast, elicitor induction of hydroxyproline-rich glycoprotein (HRGP) transcription in bean cells exhibits a 1 h lag [20], implying the existence of more than one stimulus or a single stimulus leading to either sequential effects or divergent pathways. Moreover, different elicitor fractions from the same fungus can induce different qualitative patterns of phytoalexin accumulation and translatable mRNA populations in equivalent cells [22], further suggesting the operation of more than one intracellular signal pathway.

Defence gene induction is also observed in plant tissues in response to attempted infection [1,9,10,18,20]. A number of race-cultivar specific interactions, e.g. between bean and the fungus *Colletotrichum lindemuthianum* [9,10], demonstrate clear differences in the temporal and/or spatial patterns of defence gene activation between incompatible and compatible interactions, correlated with early expression of hypersensitive resistance and attempted lesion limitation, respectively [1,8,18]. In some cases, e.g. induction of bean HRGP genes (REFS), transcript accumulation is observed not only in directly infected tissue but also in adjacent uninfected tissue, implying the operation of endogenous intercellular signals [23]. Members of the bean PAL and HRGP gene families exhibit markedly different patterns of expression, indicating that different signal pathways operate in the early stages of the incompatible interaction, the later stages of the compatible interaction and in uninfected wounded tissue [23,24]. Establishment of acquired resistance is associated with the systemic expression of a subset of defence genes, notably various classes of PR proteins, including extracellular glucanases and chitinases [25,26].

Recent data have shown that induction of PAL transcripts and those of a novel defence gene designated ELI3 in *Arabidopsis* by *Pseudomonas syringae* pv. *maculicolu* carrying the *avrRpm1* avirulence gene is strictly dependent on the presence of dominant alleles at the *RPM1* locus [27]. This demonstrates a direct relationship between molecular recognition of an avirulent pathogen mediated by a disease resistance gene and defence gene activation, as inferred from previous studies [1]. However, plants have a battery of inducible defence mechanisms and, while these may have synergistic effects, specific mechanisms may have complementary functions and the protective contribution of individual mechanisms will probably be different in different disease and physiological states [1,15]. This flexible functional deployment is important in assessing apparent exceptions to the general concept, which should also be evaluated in relation to the complex temporal and spatial patterns of activation of defence gene families associated with other forms of disease resistance, in addition to the development of a visible, macroscopic, hypersensitive response to an incompatible pathogen applied at high inoculation densities [23,24].

Chalcone synthase induction as an experimental system

The above observations focus attention on the events underlying defence gene activation as early steps in response to perception of biological stress. We have studied genes encoding chalcone synthase (CHS), which catalyses the first step in phenylpropanoid metabolism specific for isoflavonoid phytoalexin synthesis [28]. Fungal elicitor stimulates CHS transcription in bean cells within 5 min [20]. CHS is encoded by a family of six to eight genes in bean, and for analysis of gene activation mechanisms CHS15 and CHS8 were studied, since the corresponding transcripts are the most abundant CHS transcripts in elicited cells [29]. CHS15–chloramphenicol acetyltransferase (CAT) gene fusions are induced by fungal elicitor and glutathione in electroporated soybean protoplasts [30], and there is a rapid induction of glucuronidase (GUS) activity in cell suspensions derived from callus transformed with CHS15–GUS gene fusions [31]. CHS15–GUS is induced in transgenic tobacco plants by wounding and $HgCl_2$, and induction is associated with a hypersensitive response to tobacco mosaic virus (TMV) [32] (C.J. Lamb and R.A. Dixon, unpublished work). CHS8–GUS is likewise inducible by fungal elicitor, TMV and an incompatible strain of *P. syringae* [32–34]. Stress induction in many cell types adjacent to an applied stress stimulus is superimposed on a tissue- and cell-type-specific developmental programme of expression in the corolla, inner epidermis and root tips [33].

Functional analysis of the CHS15 promoter

A 5′ deletion of CHS15, from −130 to −72, abolishes elicitor inducibility in soybean protoplasts [30]. This region contains two CCTACC sequences (designated H-boxes), a third version of which is located between −61 and −56 [30,35]. The 3′ H-box is the site of an elicitor-inducible DNase I hypersensitive site in chromatin from suspension cultured bean cells, and an elicitor inducible footprint maps to this motif in the parsley PAL-1 promoter *in vivo* [36,37]. H-box function was confirmed by specific mutations in these *cis*-elements. For example, mutation of H-box III (TATA-proximal version) markedly inhibited CHS15–GUS induction by TMV in transgenic tobacco plants (C.J. Lamb and R.A. Dixon, unpublished work).

A G-box (CACGTG) is located at −74 to −69 in CHS15. The G-box has been implicated in the activation of a wide range of inducible promoters, including those regulated by abscisic acid (e.g. Em), light (e.g. rbcS-3A), and u.v. (e.g. parsley CHS) [38–42]. The CHS15 G-box is disrupted by 5′ deletion to −72, which destroys induction of CHS15–CAT in soybean protoplasts [30]. Developmental expression and stress induction of CHS15 in tobacco is abolished by 5′ deletion from −140 to −72, and specific mutation of the G-box markedly reduces induction by TMV (C.J. Lamb and R.A. Dixon, unpublished work). These observations implicate the G-box in combination with the H-box in CHS15 expression, and this was confirmed by analysis of constructs containing three or four tandem repeats of the −80 to −40 sequence in CHS15 (identical sequence in CHS8) upstream of CaMV 35S and CHS15 (−33) minimal promoters. Mutation of either the set of G-boxes or the set of H-boxes abolished developmental and stress-induced expression. Interestingly, the G-box *cis*-element is also involved in wound induction of a proteinase inhibitor gene [43].

Purification and characterization of H-box binding factors

An H-box consensus sequence in a 28-mer oligonucleotide was used in gel shift assays of binding activity to monitor H-box factor purification. Two protein factors, KAP-1 and KAP-2, were purified to homogeneity from bean cells by ion-exchange and DNA-affinity chromotography with yields of 1.5–3% [35]. KAP-1 and KAP-2 are polypeptides of 97 kDa and 56 kDa, respectively. KAP-2 has been resolved from a 76 kDa polypeptide that co-purified in earlier experiments. Assuming a binding stoichiometry of 1 mol of activity to 1 mol of DNA, 10–20 pmol of KAP-1 and KAP-2 were obtained from one 1 kg of cells. These purified factors showed the same binding specificity, with respect to natural H-box sequences and mutant sequences, as that initially observed in crude extracts.

Properties of KAP-1 and KAP-2 as elicitor signal pathway components

Dephosphorylation of purified KAP-1 or KAP-2 does not inhibit H-box binding, but there is a marked increase in the mobilities of the respective binding complexes in gel shift assays, suggesting that the functional properties of these *trans*-factors may be modulated by reversible phosphorylation [35]. Elicitation caused a striking change in the cellular distribution of H-box factor binding activities. Thus, while no change in total binding activity of KAP-1 or KAP-2 in whole-cell extracts was observed following exposure of cells to the elicitor glutathione for 30 min, elicitation caused a 7.5-fold increase in the specific activity of KAP-1 plus KAP-2 in nuclear extracts [35].

These results provide further evidence for the function of these H-box binding factors in the elicitor signal pathway, and there may be a separate pool of KAP-1 and KAP-2 in the nucleus that responds to the elicitation signal. However, binding activity recovered in the nuclear fraction isolated from elicited cells accounted for almost 10%

of that present in crude whole-cell extracts. During isolation of nuclei in cell fractionation procedures, there are rapid, substantial losses of nuclear-localized proteins [44], and recovery of proteins localized exclusively in the nucleus is typically of the order of that observed for KAP-1 and KAP-2. Hence, our data indicate that a substantial fraction of the total cellular KAP-1 and KAP-2 binding activity is localized in the nucleus of elicited cells, and this — together with the observation of no change in total cellular binding activity — suggests that elicitation may cause migration of these *trans*-factors from the cytosol into the nucleus [35].

Analysis of CHS15 transcription *in vitro*

Soybean whole-cell and nuclear extracts were obtained that efficiently transcribed the CHS15 promoter [44]. However, these standard transcription reactions also generated non-template-dependent radiolabelled products which hindered quantification of CHS15 template-specific transcripts. This problem was circumvented by assembly of transcription complexes on a CHS15 promoter template coupled to agarose beads. Washing such immobilized transcription complexes resulted in the almost quantitative recovery of CHS15 template-dependent transcriptional activity in the absence of background products. Transcription from the CHS15 promoter template was dependent on RNA polymerase II, and was initiated from the *in vivo* transcription start site as judged by the size of the product, primer extension mapping and generation of correspondingly smaller transcripts from 3′ truncated templates.

To delineate the functional organization of the CHS15 promoter we examined whether extracts pre-incubated with DNA fragments of the CHS15 promoter were then competent to transcribe the immobilized CHS15 promoter template in a subsequent transcription assay *in vitro* [45]. Thus the function of specific *cis*-elements in the CHS15 promoter could be determined by assaying the effects of depletion of the cognate *trans*-factors from the cell extract. *Trans*-competition with the −80 to −40 region containing mutations in either the G-box or H-box, and *trans*-competition with the individual G-box or H-box sequences, showed that *trans*-factors that bind to these *cis*-elements make major contributions to CHS15 transcription *in vitro* and that both *cis*-element/*trans*-factor interactions in combination are required for maximal activity.

Elicitation mechanisms were investigated by analysis of matched extracts from control cells and cells induced by treatment with a defined heptaglucoside fungal elicitor from *P. megasperma* f.sp. *glycinea* for 1 h [46,47]. Whole-cell extracts from elicited cells supported 2–3-fold higher rates of transcription from the CHS15 promoter than equivalent extracts from control cells, whereas no difference was observed with control promoter templates, e.g. the minimal CHS15 (−33) promoter. Likewise, 3–4-fold higher rates of CHS15 transcription were observed with nuclear extracts from elicited cells compared with control cells. Thus, the transcription system retains key physiological attributes *in vitro*. Extract depletion by *trans*-competition with the G-box almost completely inhibited CHS15 transcription in whole-cell extracts from elicited cells, confirming a direct function for this *cis*-element in elicitor activation. Depletion of H-box factors from these whole-cell extracts caused only a partial inhibition of transcription, and actually increased the ratio of CHS15 transcription in depleted whole-cell extracts from elicited cells compared with uninduced cells from 1.8 (no depletion) to >5 (H-box factor depletion). These results suggest the presence of saturating levels of H-box factors in whole-cell extracts, and are consistent with the presence of a pool of these factors in the cytosol, with elicitor-induced relocation to the nucleus from this cytosolic pool as part of the elicitor signal mechanism, as inferred from the parallel studies of *trans*-factor binding [35], rather than activation of the H-box *trans*-factors *per se*.

Molecular cloning of H-box binding factors From 6 kg of bean cells, 50 μg (80 pmol) of the 56 kDa KAP-2 polypeptide was purified. This quantity is sufficient for peptide microsequencing for the design of primers to isolate H-box binding factor cDNAs, and to confirm the identity of clones isolated by South-western screening of expression libraries. A clone has been isolated by screening a cell suspension culture cDNA library in λgt11 (8×10^5 plaques screened) with the H-box consensus 28-mer oligonucleotide. The plaque-purified clone (1.7 kb insert) does not bind mutant H-box sequences or other *cis*-elements (e.g. G-box), and nucleotide sequencing predicts a protein product with features suggesting possible mechanisms for factor function in relation to the concepts developed from the prior analysis of CHS15 promoter architecture.

Tracing an elicitor signal pathway

The elicitor signal pathway has also been characterized in terms of elicitor binding sites, rapidly induced physiological changes and effects of pharmacological reagents. High-affinity, plasma-membrane-binding sites for the heptaglucoside elicitor have been characterized [46–48] and a 76 kDa putative receptor protein purified by affinity chromatography [48]. Rapid changes in H^+ and Ca^{2+} influx, and K^+ efflux, have been demonstrated [49,50], and inhibitor studies implicate Ca^{2+} in defence gene activation [51]. Jasmonic acid, recently implicated in signalling wound induction of proteinase inhibitors [52,53], also induces CHS transcripts [54], suggesting that jasmonic acid may be a pathway component for internalization of fungal elicitor signals. Elicitor induces rapid changes in the phosphorylation of functionally uncharacterized proteins [55], and protein kinase inhibitors block defence activation in tomato [56], suggesting that protein phosphorylation is an early stage in the elicitor signal pathway. However, in soybean cells, such inhibitors apparently potentiate elicitor activation of the phytoalexin defence response [57].

The identification of G-box and H-box binding factors as terminal components of an elicitor signal pathway may advance the investigation of reversible protein phosphorylation in elicitor signalling — from observations of general physiological and pharmacological effects to the analysis of the regulation of defined pathway components and the development of specific functional assays for the corresponding protein kinases and phosphatases. The rapid increase in H-box factor binding activity in the nuclear fraction without a change in total cellular activity, together with modulation of the DNA-binding complex by changes in the status of factor phosphorylation, suggests a possible mechanism for transmission of the elicitor signal from the cytosol to the nucleus — with striking parallels to interferon signalling of the nucleus by phosphorylation-induced migration of the ISGF3 transcription factor complex [58]. A second striking parallel with animal systems pertains to the elicitor- or pathogen-induced oxidative burst [17,60,61], which closely resembles early events in macrophage activation and egg protection following fertilization [62,63]. Recent data indicate that the elicitor-induced oxidative burst may be involved not only in providing H_2O_2 at the surface for rapid wall toughening by cross-linking of cell wall structural proteins [17], but also in signalling the nucleus for defence gene activation [60,61] (M. Mehdy, personal communication). Interestingly, NF-*K*B activation appears to be mediated by various stimuli which act through the generation of activated oxygen species [64].

We thank the Samuel Roberts Noble Foundation for support and encouragement.

References

1. Lamb, C.J., Lawton, M.A., Dron, M. and Dixon, R.A. (1989) Cell **56**, 215–224
2. Sequeira, L. (1984) Trends Biotechnol. **2**, 25–29
3. Darvill, A. and Albersheim, P. (1986) Annu. Rev. Plant Physiol. **35**, 243–275
4. Dixon, R.A., Dey, P.M. and Lamb, C.J. (1983) Adv. Enzymol. **55**, 1–135
5. Ebel, J. (1986) Annu. Rev. Phytopathol. **24**, 235–264
6. Dean, R.A. and Kuc, J. (1985) Trends Biotechnol. **3**, 125–129
7. Wingate, V.P.M., Lawton, M.A. and Lamb, C.J. (1988) Plant Physiol. **87**, 206–210
8. Bailey, J.A. (1983) in The Dynamics of Active Defense (Bailey, J.A. and Deverall, B.J., eds.), pp 1–32, Academic Press, Sydney
9. Bell, J.N., Dixon, R.A., Bailey, J.A., Rowell, P.M. and Lamb, C.J. (1984) Proc. Natl. Acad. Sci. U.S.A. **81**, 3384–3388
10. Bell, J.N., Ryder, T.B., Wingate, V.P.M., Bailey, J.A. and Lamb, C.J. (1986) Mol. Cell. Biol. **6**, 1615–1623
11. Hahn, M.G., Bonhoff, A. and Grisebach, H. (1985) Plant Physiol. **77**, 591–601
12. Van Etten, H.D., Matthews, D.E. and Matthews, P.G. (1989) Annu. Rev. Phytopathol. **27**, 143–164
13. Moesta, P. and Grisebach, H. (1982) Physiol. Plant Pathol. **21**, 65–70
14. Broglie, K., Chet, I., Holliday, M., Cressman, R., Biddle, P., Knowlton, C., Mauvais, C.J. and Broglie, R. (1991) Science **254**, 1194–1197
15. Lamb, C.J., Ryals, J.A., Ward, E.R. and Dixon, R.A. (1992) Bio/technology **10**, 1436–1445
16. Koehle, H., Jeblick, W., Poten, F., Blaschek, W. and Kauss, H. (1985) Plant Physiol. **77**, 544–581
17. Bradley, D.J., Kjellbom, P. and Lamb, C.J. (1992) Cell **70**, 21–30
18. Dixon, R.A. and Harrison, M.J. (1990) Adv. Genet. **28**, 165–234
19. Cramer, C.L., Edwards, K., Dron, M., Liang, X., Dildine, S.L., Bolwell, G.P., Dixon, R.A., Lamb, C.J. and Schuch, W. (1989) Plant Mol. Biol. **12**, 367–383
20. Lawton, M.A. and Lamb, C.J. (1987) Mol. Cell. Biol. **7**, 335–341
21. Somssich, I.E., Schmelzer, E., Bollmann, J. and Hahlbrock, K. (1986) Proc. Natl. Acad. Sci. U.S.A. **83**, 2627–2630
22. Hamadan, M.A.M.S. and Dixon, R.A. (1987) Physiol. Mol. Plant Pathol. **31**, 105–121
23. Corbin, D.R., Sauer, N. and Lamb, C.J. (1987) Mol. Cell. Biol. **7**, 4337–4344
24. Liang, X., Dron, M., Cramer, C.L., Dixon, R.A. and Lamb, C.J. (1989) J. Biol. Chem. **264**, 14486–14492
25. Enyedi, A.J., Yalpani, N., Silverman, P. and Raskin, I. (1992) Cell **70**, 879–886
26. Ward, E.R., Uknes, S., Williams, S.C., Dincher, S.S., Wiederhold, D.L., Alexander, D., Métraux, J.-P. and Ryals, J.A. (1991) Plant Cell **3**, 1085–1094
27. Kiedrowski, S., Kawalleck, P., Hahlbrock, K., Samssich, I.E. and Dangl, J. (1992) EMBO J. **11**, 4677–4684
28. Hahlbrock, K. and Scheel, D. (1989) Annu. Rev. Plant Physiol. Plant Mol. Biol. **40**, 347–369

29. Ryder, T.B., Hedrick, S.A., Bell, J.N., Liang, X., Clouse, S.D. and Lamb, C.J. (1987) Mol. Gen. Genet. **210**, 219–233
30. Dron, M., Clouse, S.D., Lawton, M.A., Dixon, R.A. and Lamb, C.J. (1988) Proc. Natl. Acad. Sci. U.S.A. **85**, 6738–6742
31. Franklin, C.I., Trieu, T.N, Cassidy, B.G., Dixon, R.A. and Nelson, B.S. (1992) Plant Cell Rep. **12**, 74–79
32. Stermer, B.A., Schmid, J., Lamb, C.J. and Dixon, R.A. (1990) Mol. Plant-Microbe Interact. **3**, 381–388
33. Schmid, J., Doerner, P.W., Clouse, S.D., Dixon, R.A. and Lamb, C.J. (1990) Plant Cell **2**, 619–631
34. Doerner, P.W., Stermer, B., Schmid, J., Dixon, R.A. and Lamb, C.J. (1990) Bio/technology **8**, 845–848
35. Yu, L., Lamb, C.J. and Dixon, R.A. (1993) Plant J. **3**, 805–816
36. Lawton, M.A., Clouse, S.D. and Lamb, C.J. (1989) Plant Cell Rep. **8**, 504–507
37. Lois, R., Dietrich, A., Hahlbrock, K. and Schulz, W. (1989) EMBO J. **8**, 1641–1648
38. Marcotte, W.R., Jr., Russell, S.H. and Quatrano, R.A. (1989) Plant Cell **1**, 969–976
39. Schindler, U., Menkens, A.E., Beckmann, H., Ecker, J.E. and Cashmore, A.R. (1992) EMBO J. **11**, 1261–1273
40. Weisshaar, B., Armstrong, G.A., Block, A., da Costa e Silva, O. and Hahlbrock, K. (1991) EMBO J. **10**, 1777–1786
41. Oeda, K., Salinas, J. and Chua, N.-H. (1991) EMBO J. **10**, 1793–1802
42. Williams, M.F., Foster, R. and Chua, N.-H. (1992) Plant Cell **4**, 485–486
43. Kim, S.-R., Choi, J.-L., Costa, M.A. and An, G. (1992) Plant Physiol. **99**, 627–631
44. Paine, P.L., Austerberry, C.F., Desjarlais, L.J. and Horowitz, S.B. (1983) J. Cell Biol. **97**, 1240–1242
45. Arias, J.A., Dixon, R.A. and Lamb, C.J. (1992) Plant Cell **5**, 485–496
46. Cheong, J.-J., Birberg, W., Fügedi, R., Pilotti, A., Garegg, P.J., Hong, N., Ogawa, T. and Hahn, M.G. (1991) Plant Cell **3**, 127–136
47. Cheong, J.-J. and Hahn, M.G. (1991) Plant Cell **3**, 137–147
48. Cosio, E.G., Frey, T. and Ebel, J. (1992) Eur. J. Biochem. **204**, 1115–1123
49. Scheel, D., Colling, C., Hedrick, R., Kawallcek, P., Parker, J.E., Sacks, W.R., Somssich, I.E. and Hahlbrock, K. (1991) Adv. Mol. Genet. Plant-Microb. Interact. **1**, 373–380
50. Mathieu, Y., Kurkdjian, A., Xia, H., Gueru, J., Koller, A., Spiro, M.D., O'Neill, M., Albersheim, P. and Darvill, A. (1991) Plant J. **1**, 333–343
51. Stab, M.R. and Ebel, J. (1987) Arch. Biochem. Biophys. **257**, 416–423
52. Farmer, E.E. and Ryan, C.A. (1992) Trends Cell Biol. **2**, 236–241
53. Raskin, I. (1992) Plant Physiol. **99**, 799–803
54. Creelman, R.A., Tierney, M.L. and Mullet, J.E. (1992) Proc. Natl. Acad. Sci. U.S.A. **89**, 4938–4941
55. Dietrich, A., Mayer, J.E. and Hahlbrock, K. (1989) J. Biol. Chem. **265**, 6360–6368
56. Felix, G., Grosskopf, D.G., Regengass, M. and Boller, T. (1991) Proc. Natl. Acad. Sci. U.S.A. **88**, 8831–8834
57. Kauss, H., Theisinger-Hinkel, E., Minderman, R. and Conrath, U. (1992) Plant J. **2**, 655–660
58. Schindler, C., Shuai, K., Prezioso, V.R. and Darnell, J.E. (1992) Science **257**, 809–813
59. Reference deleted
60. Apostol, I., Heinstein, P.F. and Low, P.S. (1989) Plant Physiol. **90**, 109–116
61. Sutherland, M.W. (1991) Physiol. Mol. Plant Pathol. **39**, 79–94
62. Baggiolini, M. and Wyman, M.P. (1990) Trends Biochem. Sci. **15**, 69–75
63. Shaprio, B.M. (1991) Science, **252**, 533–536
64. Schreck, R., Rieber, P. and Banerle, P.A. (1991) EMBO J. **10**, 2247–2258

Biochem. Soc. Symp. **60**, 249–256
Printed in Great Britain

Photomorphogenic mutants of tomato

R.E. Kendrick*†§, J.L. Peters†, L.H.J. Kerckhoffs†, A. van Tuinen‡ and M. Koornneef‡

†Department of Plant Physiology, Wageningen Agricultural University, Arboretumlaan 4, NL-6703 BD Wageningen, The Netherlands; ‡Department of Genetics, Wageningen Agricultural University, Dreijenlaan 2, NL-6703 HA Wageningen, The Netherlands and §Laboratory for Photoperception and Signal Transduction, Frontier Research Program, Institute of Physical and Chemical Research (RIKEN), Hirowsawa 2-1, Wako City, Saitama 351-01, Japan

Introduction

Photomorphogenesis is the process by which light regulates aspects of plant growth and development [1]. Throughout the life cycle of a plant, from germination to flowering, light has been shown to play an important role. Some examples are: (i) the triggering of germination; (ii) the process of de-etiolation, which results in the transition from the strategy of dark growth (while below the ground, typified by poorly developed leaves and rapid elongation growth while living heterotrophically on the store food reserves within the seed), to the strategy of light growth, as a green, photosynthetically self-sufficient seedling; and (iii) shade avoidance (near-neighbour detection) owing to the perception of change in spectral quality of the light environment as a consequence of transmission and reflectance from other plants. To achieve these and other processes, plants utilize at least three classes of photoreceptor: phytochrome(s) which absorbs predominantly in the red (R) and far-red (FR) region of the spectrum, blue light (B)/u.v.-A photoreceptor(s) and u.v.-B photoreceptor(s). Phytochrome is the most extensively studied, existing in two interconvertible forms: one R-absorbing (Pr) and the other FR-absorbing (Pfr). Phytochrome responses can be further subdivided into response modes on the basis of the amount of light required: very low fluence responses (VLFRs), low fluence responses (LFRs) and high irradiance responses (HIRs). There is also evidence that the different classes of photoreceptors act in concert. It is our aim to unravel this complexity by studying photomorphogenic

*Address for correspondence: Department of Plant Physiology, Wageningen Agricultural University, Arboretumlaan 4, NL-603 BD Wageningen, The Netherlands.

mutants, in which elements of the system are modified, using tomato *(Lycopersicon esculentum* Mill.) as a model species.

Why tomato?

Tomato offers an alternative to *Arabidopsis* as a model plant, and has one notable advantage in that its seeds and seedlings are relatively large and amenable to both biochemical and biophysical studies (Table 1). Tomato is an economically important species and, therefore, many groups are engaged in developing molecular biological techniques for ultimate crop improvement (e.g. disease resistance). In addition, there is a large collection of mutants available, among which photomorphogenic mutants already exist, although they were not initially recognized as such. In a broader context, while *Arabidopsis* has become the universal model plant, it should be remembered that a crucifer with its overwintering rosette is just one example of the many different growth strategies exhibited by plants.

Phytochrome genes

The first phytochrome gene to be cloned was from the monocotyledon oat *(Avena sativa)*, in which multiple copies of the gene designated *phyA* exist [2]. A small family of phytochrome genes, *phyA, phyB, phyC, phyD* and *phyE* [2,3] has been identified in *Arabidopsis.* These genes encode the apophytochromes A, B, C, D and E which, after insertion of the linear tetrapyrrole chromophore, result in the holophytochromes A, B, C, D and E, respectively. Genes corresponding to *phyA* and *phyB* have been identified in all species of higher plant, so far studied, both monocotyledons and dicotyledons [2]. There are multiple phytochrome genes in tomato but, to date, none has been completely sequenced [2,4,5]; however, one *phy* gene resides on chromosome 10 [6]. The cloning of phytochrome genes in tomato is currently in progress in collaboration with L.H. Pratt and M.-M. Cordonnier-Pratt. Only when this has been done, and specific DNA probes and antibodies for each member of the gene family are available, will it be possible to fully characterize the molecular nature of the mutants discussed below.

Table 1. Comparison of tomato and *Arabidopsis* as model plants.

	Arabidopsis	Tomato
Haploid chromosome number	5	12
Haploid genome (kb)	7×10^4	7.1×10^5
Gene maps available	+	+
Generation time (months)	2	6
Transformation possible	+	+
Tagging possible	+	+
Seeds and seedlings	Small	Large

Types of photomorphogenic mutant

Mutants deficient in phytochrome action

During selection of gibberellic acid (GA)-responsive mutants (dwarfs), a mutant was isolated that required GA for germination but, in contrast to GA-deficient mutants, had a long hypocotyl and a marked reduction in chlorophyll content when grown in white light [7]. A genetic analysis revealed that this recessive mutant was allelic with a previously described *aurea* (*au*) mutant [8] located on chromosome 1 [9]. Another mutant at the *au* locus has been isolated in the progeny of tomato plants derived from tissue culture by Lipucci di Paola *et al.* [10]. A similar mutant to *au*, although less extreme is called yellow-green-2 (*yg-2*) (allelic with *auroid*) and is located on chromosome 12 [11–14].

One aspect of the *au* phenotype is its reduced germination in darkness compared with wild type. [8]. The freshly harvested seeds, which are dormant, can be induced to germinate after treatment with a combination of chilling and nitrate [15]. Moreover, exposure to continuous R, an effect which could be replaced by R pulses, led to an increase in germination of *au*-mutant seed batches. Interestingly, functional phytochrome must be present, since the effect of R pulses was reversible by FR pulses. However, no inhibitory effect of continuous FR was observed in older seed batches, with appreciable dark germination, in contrast to wild-type seeds which exhibit a strong FR-irradiance-dependent inhibition of germination [8]. Lipucci di Paola *et al.* [10] have found a promotion of seed germination by FR for *au* mutants and suggested that this is the consequence of the absence of an inhibitory FR–HIR.

At the etiolated seedling stage, compared with wild-type, the *au* mutant is characterized by a reduction in: (i) hypocotyl growth inhibition in white light [8], R, B and u.v.-A [16]; (ii) chlorophyll and chloroplast development [8,17], appearing to lack the VLFR component in the fluence-response curve for greening; (iii) anthocyanin content [18]; and (iv) the photoregulation of the transcript levels of chlorophyll *a*/*b*-binding proteins of photosystems I and II, plastocyanin and subunit II of photosystem I [6,19,20]. This pleiotropic phenotype, coupled with lack of phytochrome in etiolated *au*-mutant tissues, is precisely that predicted for a phytochrome-deficient mutant.

Adult, light-grown plants (both wild-type and *au* mutant) exhibit a quantitatively similar elongation growth response to end-of-day FR treatment [16,21,22] and changes in the R:FR photon ratio during the daily photoperiod [23,24], indicating the presence of functional phytochrome in light-grown *au*-mutant plants. Casal [25] has presented evidence which suggests that the *au* mutant is less sensitive to detection of small changes in the R:FR photon ratio, indicating that de-etiolated plants have a partially aberrant shade detection mechanism. The yellow colour of the leaves indicates a retained problem with respect to chlorophyll accumulation in de-etiolated plants. In the *au*-mutant, nitrate reductase and nitrite reductase levels are not enhanced by R in etiolated seedlings (in contrast to wild type). This resulted in 70–80% lower levels in the *au*-mutant after 3 days continuous R [26]. However, in darkness on nitrate, levels are similar in the *au*-mutant and wild-type plants. In addition, the mRNA transcript levels of chlorophyll *a*/*b*-binding protein and ribulose-bisphosphate carboxylase (small subunit) are similar to wild type in older plants. Surprisingly, despite the difference in chlorophyll content, the *au*-mutant leaves show net photosynthesis rates comparable to wild type [21,26].

Whereas etiolated seedlings of the *au* mutant contain less than 5% (detection limit) of the spectrophotometrically detectable phytochrome found in wild-type seedlings, light-grown tissues of this mutant contain about 60–70% of the phytochrome present in the wild type [16,21]. Using monocotyledon antibodies, Parks *et al.* [27] and Oelmüller *et al.* [19] found no immunologically detectable phytochrome in etiolated *au*-mutant tissue. However, preliminary experiments using dicotyledon antibodies raised against *phyA* and *phyB* gene products show that the phytochrome A apoprotein in etiolated *au*-mutant seedlings is about 20% of that in the wild type [28,29], whereas a comparable level of phytochrome B apoprotein is detected in etiolated wild-type and *au*-mutant seedlings. In contrast to wild type, *in vitro* proteolysis after R and FR irradiation fails to show any conformational change in the phytochrome A apoprotein of the *au* mutant. This suggests that etiolated *au*-mutant seedlings accumulate low levels of phytochrome A apoprotein, which is largely photochemically inactive, and normal levels of phytochrome B apoprotein. In light-grown, wild-type and *au*-mutant seedlings, comparable levels of phytochrome B are found. These results indicate that the *au* mutant is deficient in phytochrome A at the seedling stage, but do not give a complete explanation of the nature of the *au* mutation. However, it is unclear whether phytochromes other than phytochrome A and phytochrome B are deficient in etiolated seedlings.

The *au* mutant produces wild-type levels of phytochrome mRNA which is fully translated *in vitro* to yield a polypeptide product of the same abundance, size and immunochemical properties as the wild-type translation product [6]. Therefore, the deficiency of spectrophotometrically active phytochrome in the *au* mutant appears to result not from a lack of phytochrome gene expression, but from instability of the protein. Explanations for the phytochrome polypeptide instability *in vivo* include (i) a phytochrome structural gene (*phyA*) mutation, (ii) an aberrant proteolytic degradation and (iii) a defect in chromophore biosynthesis or attachment to the protein [6]. If this latter possibility is correct then tomato is different to *Arabidopsis*, where the chromophore mutants *hy1* and *hy2* both accumulate wild-type levels of phytochrome A [30].

Since one phytochrome coding sequence identified by Sharrock *et al.* [6] is located on chromosome 10 and the *au* locus resides on chromosome 1, they concluded that the mutation is not a mutation of that particular structural gene [6]. However, this phytochrome coding sequence, in principle, may encode a phytochrome type different from that absent in the *au* mutant. Transformation of the *au* mutant with *phyA* cDNA has so far not been reported, but experiments are underway to combine, by crossings, a tomato that overexpresses the oat *phyA3* gene [31] with the *au* and *yg-2* mutants (A. van Tuinen, unpublished work). Results suggest a lack of rescue of the au phenotype at the seedling stage.

Attempts to rescue the *au* mutant by feeding with the chromophore precursor biliverdin (P.H. Quail, personal communication; R.P. Sharma, personal communication) have been unsuccessful. This result must be considered inconclusive, since a test to restore photochemical activity to phytochrome by feeding biliverdin to wild-type seedlings grown in the presence of gabaculine, an inhibitor of tetrapyrrole biosynthesis, was also unsuccessful (P.H. Quail, personal communication; R.P. Sharma, personal communication). Therefore, the possibility that biliverdin is not taken up by the plant cannot be excluded. Only further studies will resolve this problem. What is without doubt is that etiolated seedlings and seeds exhibit R/FR reversible responses and have photochemically active holophytochrome. This must mean that *au*-mutant plants, at least after de-etiolation, can produce sufficient chromophore to fulfil some phytochrome synthesis.

A recessive mutant selected for its slightly longer hypocotyl than wild type under white light, has recently been partially characterized physiologically (L.H.J. Kerckhoffs and A. van Tuinen, unpublished work). While continuous low fluence rate irradiation with R is relatively ineffective in hypocotyl inhibition, both B and FR are just as effective as in the wild type. Short pulses of R, or R followed immediately by FR, given every 4 h demonstrate that phytochrome response via its low fluence mode on hypocotyl growth is not detectable in the mutant during the first 2 days of treatment. We have called this mutant temporarily red light insensitive (*tri*). This mutant accumulates some anthocyanin during the period of no inhibition of hypocotyl elongation, but the level is reduced compared with the wild type. In experiments where the R/FR-reversible anthocyanin synthesis during a 24 h dark period was studied after a 12 h pretreatment with R or B of different fluence rates, the response was dramatically less in the case of R pretreatments in the *tri* mutant. Fluence rate-response curves for anthocyanin accumulation in R indicate that the LFR component is present, but the R–HIR at higher fluence rates is absent. Adult, light-grown, *tri*-mutant plants respond to reduction in the R:FR photon ratio during the daily photoperiod and end-of-day FR treatments by stimulation of elongation growth. In contrast to the R-insensitive, phytochrome-deficient, tomato *au* mutant, white-light-grown *tri*-mutant plants are not yellow and have a similar chlorophyll level to wild-type plants. Preliminary measurements of phytochrome in etiolated seedlings indicate a normal light-labile phytochrome pool. A detailed study of all phytochrome pools in this mutant is eagerly awaited.

Mutants exhibiting exaggerated phytochrome responses

A spontaneous mutant at a high pigment (*hp*) locus was found as early as 1917 [32]. The monogenic recessive *hp-1* mutants are characterized by features such as dark-green foliage and immature fruit colour due to high chlorophyll levels [33], higher lycopene and carotene content resulting in deep-red fruits [34] and high levels of anthocyanin [35,36]. Mochizuki and Kamimura [37] observed that *hp-1*-mutant hypocotyls had more anthocyanin than wild type when grown in yellow light, and they used this as a selection criterion. Using continuous R to select *hp* mutants, several new *hp* mutants have been selected. Plant height is somewhat reduced in *hp-1* mutants: (i) hypocotyl growth is more inhibited than in wild-type when the seedlings are grown in R or yellow light [37]; (ii) hypocotyl dry mass is lower than in wild type when the seedlings are grown in white light [36]. Thompson *et al.* [34] reported that the seed germination of *hp-1* mutants was lower than wild type and that the stems of *hp-1* mutant plants were more brittle — resulting in a higher mortality than wild type. The pleiotropic nature of the *hp-1* mutant suggests that it has a modification of a basic process affecting plant morphogenesis rather than being a specific response mutant affecting pigment synthesis only.

There are also mutants which are similar in some aspects to the mutant phenotype but are non-allelic with *hp-1*, such as *hp-2* [38], *atroviolatia* (*atv*) [14] and intensive pigment (*Ip*) [39] mutants. Furthermore, plants with *hp-1*-like characteristics at their seedling stage were obtained when high levels of the oat *phyA3* gene were expressed in tomato [31]. These mutants all had short hypocotyls and more anthocyanin than wild type, although only some retained a strong dwarf phenotype as adult plants. Combining this information with the finding that the phytochrome regulating the anthocyanin synthesis must be relatively stable, the capacity for anthocyanin synthesis could be established by the bulk phytochrome A pool, while the anthocyanin synthesis is actually photoregulated via a stable phytochrome pool (phytochrome B?). Detailed fluence rate-response curves for anthocyanin biosynthesis in *hp-1* suggest that both the LFR and HIR components in R are amplified [40].

The *hp-1* mutant of tomato exhibits exaggerated phytochrome responses, whereas the phytochrome content of etiolated seedlings and the characteristics of the phytochrome system are similar to those in wild type [18,41]. Therefore, there is no evidence so far to suggest that the *hp-1* mutant is a photoreceptor mutant. In contrast to wild type, the *hp-1* mutant does not require co-action of the B photoreceptor and phytochrome for normal development and exhibits maximum anthocyanin synthesis and hypocotyl growth inhibition in R alone, i.e. it mimics the action of B. On the basis of its recessive (loss-of-function) nature, it is proposed that the phytochrome action in etiolated seedlings is under the constraint of the *hp-1* gene product (HP-1) [40]. Both exposure to B and the *hp-1* mutation appear to result in reduction of the level of HP-1 or its effectiveness. The exaggerated response of the *hp-1* mutant compared with wild type fits the definition of 'responsiveness amplification' proposed by Mohr [42] to describe the amplification of a phytochrome response as a result of pre-irradiation, which excites either the B photoreceptor or phytochrome. We propose that the *hp-1* mutation is associated with this amplification step in the phytochrome transduction chain. It is proposed that the amplification affects phytochrome-A action, since (i) the phytochrome A-lacking *au,hp-1* double mutant shows severely reduced or no anthocyanin accumulation, and (ii) the overexpression of *phyA* exhibits similar exaggerated phytochrome responses [31]. A study of the photoregulation of phenylalanine ammonia-lyase, a key enzyme in flavonoid biosynthesis, showed a higher level in the *hp-1* mutant when compared with the *au* mutant, *au,hp-1* double mutant and wild-type level [43]. Interestingly, an R/FR reversible effect on phenylalanine ammonia-lyase activity was shown in all these genotypes, indicating that etiolated seedlings of the *au* mutant do contain some functional phytochrome. Adult plants of the *hp-1* and the *au,hp-1* double mutant show a quantitatively similar elongation response to reduction in the FR:FR photo ratio during the daily photoperiod and end-of-day FR treatments [22,24].

While it is premature to make firm conclusions, it has become apparent that the process of de-etiolation is complex. Perhaps the reason for this is that the selection pressure for this critical process in the life of a plant is so strong that several different photoreceptors and modes of action function together to achieve this goal. In the laboratory we have the opportunity to reveal apparent response characteristics which in nature never appear independently. For example, the inhibition of elongation growth of a hypocotyl can be achieved by different wavelengths of light in a number of ways. The application of FR could function via the FR–HIR mode of phytochrome A, whereas R might function via an R–LFR and an R–HIR in which both phytochrome A and phytochrome B may play a role. Perhaps in nature none of these processes is saturated by the low light levels below the soil surface, but collectively they enable the selective advantage of perception of the light environment (soil surface) to be anticipated. An additional complication is that the process of de-etiolation is not the same in all seedlings. If we take anthocyanin biosynthesis as an example, in some seedlings there is a very strong FR–HIR, but in tomato this is not the case; yet in the same hypocotyl there is a strong FR–HIR for inhibition of elongation growth.

Supported by the Foundation for Fundamental Biological Research (BION) which is subsidized by the Netherlands Organization for the Advancement of Research (NWO). The authors are most grateful to Dr. E. López-Juez for his comments on the manuscript.

References

1. Kendrick, R.E. and Kronenberg, G.H.M. (1986) Photomorphogenesis in Plants, Martinus Nijhoff, Dordrecht
2. Quail, P.H. (1991) Annu. Rev. Genet. **25**, 389–409
3. Sharrock, R.A. and Quail, P.H. (1989) Genes Dev. **3**, 1745–1757
4. Hauser, B. and Pratt, L. (1990) Plant Physiol. Suppl. **93**, 137
5. Hauser, B. and Pratt, L. (1991) Beltsville Symp. Agric. Res. **16**, 70
6. Sharrock, R.A., Parks, B.M., Koornneef, M. and Quail, P.H. (1988) Mol. Gen. Genet. **213**, 9–14
7. Koornneef, M., van der Veen, J.H., Spruit, C.J.P. and Karssen, C.M. (1981) in Induced Mutations: A Tool in Plant Breeding, pp. 227–232, International Atomic Energy Agency, Vienna
8. Koornneef, M., Cone, J.W., Dekens, R.G., O'Herne-Robers, E.G., Spruit, C.J.P. and Kendrick, R.E. (1985) J. Plant Physiol. **120**, 153–165
9. Khush, G.S. and Rick, C.M. (1968) Chromosoma **23**, 452–484
10. Lipucci di Paola, M., Collina Grenci, F., Caltavuturo, L., Tognoni, F. and Lercari, B. (1988) Adv. Hortic. Sci. **2**, 30–32
11. Burdick, A.B. (1958) Tomato Genet. Coop. Rep. **8**, 9–11
12. Kerr, E.A. (1979) Tomato Genet. Coop. Rep. **29**, 27–28
13. Kerr, E.A. (1981) Tomato Genet. Coop. Rep. **31**, 8
14. Rick, C.M., Reeves, A.F. and Zobel, R.W. (1968) Tomato Genet. Coop. Rep. **18**, 34–35
15. Georghiou, K. and Kendrick, R.E. (1991) Physiol. Plant. **82**, 127–133
16. Adamse, P., Jaspers, P.A.P.M., Bakker, J.A., Wesselius, J.C., Heeringa, G.H., Kendrick, R.E. and Koornneef, M. (1988) J. Plant Physiol. **133**, 436–440
17. Ken-Dror, S. and Horwitz, B.A. (1990) Plant Physiol. **92**, 1004–1008
18. Adamse, P., Peters, J.L., Jaspers, P.A.P.M., van Tuinen, A., Koornneef, M. and Kendrick, R.E. (1989) Photochem. Photobiol. **50**, 107–111
19. Oelmüller, R., Kendrick, R.E. and Briggs, W.R. (1989) Plant Mol. Biol. **13**, 223–232
20. Oelmüller, R. and Kendrick, R.E. (1991) Plant Mol. Biol. **16**, 293–299
21. López-Juez, E., Nagatani, A., Buurmeijer, W.F., Peters, J.L., Kendrick, R.E. and Wesselius, J.C. (1990) J. Photochem. Photobiol. B **4**, 391–405
22. Peters, J.L., Schreuder, M.E.L., Heeringa, G.H., Wesselius, J.C., Kendrick, R.E. and Koornneef, M. (1992) Acta Hortic. **305**, 67–77
23. Whitelam, G.C. and Smith, H. (1991) J. Plant Physiol. **139**, 119–125
24. Kerckhoffs, L.H.J., Kendrick, R.E., Whitelam, G.C. and Smith, H. (1992) Photochem. Photobiol. **56**, 611–616
25. Casal, J.J. (1991) Beltsville Symp. Agric. Res. **16**, S1
26. Becker, T.W., Foyer, C. and Caboche, M. (1992) Planta **188**, 39–47

27. Parks, B.M., Jones, A.M., Adamse, P., Koornneef, M., Kendrick, R.E. and Quail, P.H. (1987) Plant Mol. Biol. **9**, 97–107
28. López, E., Sharma, R., Nagatani, A., Kendrick, R.E. and Furuya, M. (1991) Beltsville Symp. Agric. Res. **16**, 79
29. Sharma, R., López, E., Nagatani, A., Kendrick, R.E. and Furuya, M. (1992) Proc. Int. Congr. Photobiol. **11**, 347
30. Parks, B.M., Shanklin, J., Koornneef, M., Kendrick, R.E. and Quail, P.H. (1989) Plant Mol. Biol. **12**, 425–437
31. Boylan, M.T. and Quail, P.H. (1989) Plant Cell **1**, 765–773
32. Raynard, G.B. (1956) Tomato Genet. Coop. Rep. **6**, 22
33. Sanders, D.C., Pharr, D.M. and Konsler, T.R. (1975) Hortic. Sci. **10**, 262–664
34. Thompson, A.E., Hepler, R.W. and Kerr, E.A. (1962) Proc. Am. Soc. Hortic. Sci. **81**, 434–442
35. Kerr, E.A. (1965) Can. J. Plant Sci. **45**, 104–105
36. von Wettenstein Knowles, P. (1968) Hereditas **61**, 255–275
37. Mochizuki, T. and Kamimura, S. (1985) Tomato Genet. Coop. Rep. **35**, 12–13
38. Soressi, G.P. and Salamini, F. (1975) Tomato Genet. Coop. Rep. **25**, 21–22
39. Rick, C.M. (1974) Hilgardia **42**, 493–510
40. Peters, J.L., Schreuder, M.E.L., Verduin, S.J.W. and Kendrick, R.E. (1992) Photochem. Photobiol. **56**, 75–82
41. Peters, J.L., van Tuinen, A., Adamse, P., Kendrick, R.E. and Koornneef, M. (1989) J. Plant Physiol. **134**, 661–666
42. Mohr, H. (1986) in Photomorphogenesis in Plants (Kendrick, R.E. and Kronenberg, G.H.M., eds.), pp. 547–563, Martinus Nijhoff, Dordrecht
43. Goud, K.V., Sharma, R., Kendrick, R.E. and Furuya, M. (1991) Plant Cell Physiol. **32**, 1251–1258

Biochem. Soc. Symp. **60**, 257–263
Printed in Great Britain

Genes controlling *Arabidopsis* photomorphogenesis

Joanne Chory*, Tedd Elich, Hsou-min Li, Alan Pepper, Daniel Poole, Jason Reed, Ronald Susek, Veronique Vitart, Tracy Washburn, Masaki Furuya† and Akira Nagatani‡

Plant Biology Laboratory, The Salk Institute for Biological Studies, San Diego, CA 92186-5800, U.S.A.; †Advanced Research Laboratory, Hitachi Ltd, Hatoyama, Saitama 350-03, Japan and ‡Laboratory of Plant Biological Regulation, Frontier Research Program, RIKEN Institute, Wako-shi, Saitama 351-01, Japan

Introduction

Light plays a critical role in the development of seedlings of dicotyledonous plants, such as *Arabidopsis thaliana.* Upon seed germination, seedlings that fail to perceive light undergo a specific mode of development, called etiolated growth, that emphasizes hypocotyl or epicotyl elongation at the expense of leaf development [1]. Upon perceiving light, a developmental transition called 'de-etiolation' occurs, and leaf development is initiated while the rate of hypocotyl elongation slows. Underlying the development of leaves are changes in patterns of gene expression and the differentiation of etioplasts into functional chloroplasts. Both red and blue light have a stimulatory role. Red light perception is mediated by a red/far-red reversible photoreceptor phytochrome [2,3]. Phytochrome is composed of a linear, tetrapyrrole chromophore and an apoprotein that is encoded (in *Arabidopsis*) by one of five genes, designated *PHYA*, *PHYB*, *PHYC*, *PHYD* and *PHYE* [4]. Light signals are integrated with intrinsic developmental programming that defines the temporal and spatial specificity of gene expression and cell differentiation. Aside from what is known about phytochrome, little is known about the signal transduction and morphogenetic pathways which mediate light-regulated development. In an effort to study these pathways, our laboratory has utilized a combined molecular and genetic approach to identify and study the molecules that mediate the photomorphogenetic transition.

*To whom correspondence should be addressed.

Photomorphogenetic mutants

Mutants that are defective in photomorphogenesis are valuable tools in the study of seedling development for several reasons. First, mutants help to functionally define the number and types (positively acting/negatively acting) of regulatory molecules in a morphogenetic pathway. Analysis of the phenotypes of double mutants (epistasis) can define the order of gene action within a pathway, and helps to elucidate the organization and complexity of branched and/or parallel pathways. Moreover, mutants are valuable biological reagents for biochemical and physiological experiments, since *a priori* they perturb one component of the system while holding all others static. Morphological mutations that are shown to occur within a gene that has been cloned (e.g. by biochemical approaches or by sequence similarity) can be used to positively assign a function to the cloned gene. Multiple alleles at this locus can be studied phenotypically and their corresponding molecular lesions can be determined by DNA sequencing. This combined approach provides a powerful method to address questions of structure and function. Conversely, genes which have been defined mutationally can be cloned by complementation of the mutant phenotype with wild-type sequences. In higher plants, because of larger genome sizes and the lack of a high efficiency transformation method, this approach requires that sequences likely to contain the gene of interest must first be delimited to a small defined region. Typically, the method of choice for such an experiment is chromosome walking.

Photomorphogenetic mutants of *Arabidopsis* fall into two broad classes: those with reduced responses to light, and those that have a constitutive response in the absence of light. The latter class of mutant displays many of the characteristics of light-grown plants when grown in the dark. These mutants, designated *det* (for de-etiolated) fall into four complementation groups: *det1*, *det2*, *det3* and *det4* [5,6] (H. Cabrera and J. Chory, unpublished work). When grown in the dark, the mutant plants have short hypocotyls, expanded cotyledons and, with the exception of *det3*, the expression of photo-inducible genes is greatly elevated. For *det1*, true leaf development, as well as the partial differentiation of chloroplasts, is initiated in the dark. All alleles at the *det1*, *det2*, *det3* and *det4* loci are recessive, suggesting that they may play a role as negative regulators of de-etiolation. Epistasis analysis suggests that the *DET1* and *DET2* genes act via independent pathways and downstream from phytochrome, and are, therefore, likely to be signal transduction elements [7]. Two additional *Arabidopsis* loci, *cop1* and *cop9* (constitutively photomorphogenic), have a similar phenotype to the *det* loci [8,9].

Mutants that are deficient in the photomorphogenetic response display a long hypocotyl when grown in continuous white, blue or far-red light. These are designated *hy* (long hypocotyl), *fre* (far-red elongated) or *blu* (blue light unresponsive). These mutants fall into more than 11 complementation groups, designated *hy1–8*, *blu1*, *blu2* and *blu3*, and *fre1* [10–15]. The *hy1*, *hy2* and *hy6* mutants of *Arabidopsis* make phytochrome which is lacking in spectrophotometric activity, although the apoprotein is still present. Based on feeding experiments with chromophore intermediates, the mutations in these lines appear to affect biosynthesis of the tetrapyrrole chromophore of phytochrome, suggesting that these mutations probably affect all types of phytochrome [16] (J. Reed and J. Chory, unpublished work).

The *hy4*, *blu1*, *blu2* and *blu3* mutants are deficient in blue light, but not red light, inhibition of hypocotyl elongation [11,13]. Given the number of genes that affect either red light (*HY1*, *HY2*, *HY3* and *HY6*) or blue light (*HY4*, *BLU1*, *BLU2* and *BLU3*) inhibition of hypocotyl elongation, it is likely that the blue and red light signal transduction pathways are at least partially independent of one another. Supporting this idea, the phenotypes of plants homozygous for both *blu1* and *hy6* are much more severe than either mutant alone [13]. However, epistasis studies suggest that red and blue light signals converge on pathways defined by *det1* and *det2* [7]. The *hy5* mutant is deficient in red light inhibition and has a lesser, yet reproducible, deficiency in blue light inhibition of hypocotyl growth [11]. The *hy5* phenotype cannot be suppressed by chromophore intermediates, has normal levels of PHYA, PHYB and PHYC mRNAs and proteins, and has wild-type levels of phytochrome spectral activity. Based on these observations and epistasis studies [7,11], it appears that the *HY5* gene product is a downstream element in a light signal transduction pathway that leads to inhibition of hypocotyl growth [7].

Mutations in the PHYB gene: molecular analysis of the *HY3* locus

In addition to playing a role in stimulating de-etiolation, the *HY3* gene product also functions as a regulatory photoreceptor in the light-grown plant. Preliminary experiments utilizing monoclonal antibodies to phytochrome B suggested that the *hy3* defect specifically affects the accumulation of the PHYB apoprotein [17,18]. Recent experiments in our laboratory have shown that a *hy3* mutation is tightly linked (within 0.2 cM) to a restriction fragment length polymorphism (RFLP) detected by a cloned *PHYB* gene probe [19]. In Northern and Western blotting experiments, severe *hy3* alleles displayed a dramatic decrease in PHYB message and protein accumulation. A less severe allele, *hy3-4-117*, displayed normal levels of mRNA and protein. Since most of the mutant alleles of *hy3* were isolated in the Landsberg ecotypic background, we cloned and sequenced the wild-type Landsberg genomic *PHYB* gene. To rapidly sequence the *PHYB* gene from *hy3* mutant alleles, we used an array of *PHYB* specific oligonucleotides to amplify segments of the *PHYB* gene. The amplified products were then subjected to asymmetric PCR sequencing [20]. Severe alleles of *hy3* were found to induce stop codons in the *PHYB* coding sequence, thus establishing that the *HY3* locus encodes the PHYB apoprotein. Allele *hy3-4-117* contained a His to Tyr mis-sense mutation at codon 283. Interestingly, His-283 is absolutely conserved among both monocotyledons and dicotyledons. Thus, both evolutionary conservation and mutational analysis suggest a critical role for this residue in phytochrome function. Additional *hy3* alleles, which have their basis in point mutations in the *PHYB* gene, will help to elucidate structure and function relationships within the PHYB molecule. The phenotypes of the *hy3* (*phyB*) null mutants include early flowering, reduced specific chlorophyll content and a general defect in the inhibition of cell elongation in several tissues of seedling and adult plants [19]. PHYB thus controls *Arabidopsis* development at numerous stages and in multiple tissues.

Phytochrome-A-deficient mutants

Other *Arabidopsis* mutations appear to specifically disrupt phytochrome A accumulation, although it remains to be shown whether these mutations reside in the

PHYA gene itself [14,15]. These mutants, called *hy8* [14] or *fre1* [15] are characterized by insensitivity to inhibition of hypocotyl elongation by continuous far-red light, a response that had previously been attributed to the light-labile phytochrome, PHYA [21]. Severe alleles of *hy8* and *fre1* also do not show phytochrome spectrophotometric activity in etiolated seedlings, nor do they accumulate phytochrome A. It has not been determined whether *hy8* and *fre1* mutations define the same gene, or whether that gene encodes PHYA. However, our recent genetic data show tight linkage of *fre*1 to the cloned *PHYA* gene. Therefore, the *FRE1* gene is an excellent candidate for encoding the PHYA apoprotein. *hy8* and *fre1* mutants appear completely normal under continuous white or red light, suggesting that PHYA is less important than PHYB under normal growth conditions. PHYA may play a highly specialized role in allowing seedlings to germinate in extreme shaded environments highly enriched in far-red light (J. Reed and J. Chory, unpublished work).

Positional cloning of the photomorphogenetic loci *DET1* and *DET2*

To study the mechanism of action of the *DET1* and *DET2* gene products, we are in the process of cloning the *DET1* and *DET2* genes, by chromosome walking followed by complementation of the mutant phenotype with cloned wild-type DNA sequences. The *det1* mutation was mapped to a position near 26 cM on chromosome 4, and is flanked by the RFLP markers cos2616 and λAt518. These two RFLP probes were then used to screen a yeast artificial chromosome (YAC) library with *A. thaliana* genomic DNA inserts (average insert size 150 kb). Probes containing *Arabidopsis* DNA from the ends of the YAC have been obtained either by inverse PCR, utilizing vector-specific primers, or by 'plasmid rescue' using *Escherichia-coli*-derived plasmid sequences contained within one of the vector ends. These YAC end-probes often constitute novel RFLP markers and can also be used to re-screen the YAC library to isolate overlapping YAC clones. Two sets of overlapping YACs, surrounding the cos2616 and λAt518 markers, were obtained. RFLP analysis has shown that the *DET1* gene is within the λAt518-derived YAC contig. A 160–240 kb interval containing the *DET1* gene has been defined as a region that is flanked by RFLP probes for which no recombination has been detected between the probe and the *det1* mutation. A YAC which spans this interval was purified by contour-clamped homogenous electric field (CHEF)-gel electrophoresis, labelled by random hexamer priming and used to screen an *Arabidopsis* genomic library constructed in an *Agrobacterium*-transformable cosmid vector. These cosmids are being introduced into the *det1* mutant to test for complementation. To identify the transcribed *DET1* sequences, complementing cosmids that restore the wild-type *DET1* phenotype will be used to screen a cDNA library from etiolated seedlings and to probe Northern blots containing RNA samples from wild-type *DET1* plants and from several *det1* mutant alleles.

Utilizing identical methods we have mapped the location of *det2* to near 75 cM on chromosome 2. The *det2* gene is flanked by RFLP markers λAt429 and cos4514 and has been localized to a 50–150 kb interval on the λAt429-derived YAC set. Transformable cosmid clones containing *Arabidopsis* genomic inserts from this interval are currently being introduced into the *det2* mutant to test for functional complementation.

Suppressors of *det1*

In an effort to identify additional genes that act in the photomorphogenetic pathway, including those which interact directly with the *DET1* gene product, we have isolated second-site genetic suppressors of the *det1* mutation. A homozygous *det1-1* strain was mutagenized and screened in the dark for seedlings with some or all of the characteristics of the etiolated wild-type plant. A total of 120 000 M_2 seedlings from 10 000 M_1 plants were screened. Of the plants that appeared etiolated in these screens, a subset de-etiolated normally in response to light and were characterized further. These recessive and dominant mutations were designated *ted* or *TED*, respectively, since they reverse the *det1* seedling phenotype. Of the four recessive mutants, two are allelic to *hy5*. Based on preliminary data, four of the dominant mutations map to a position near 5 cM on chromosome 4, distinct from the *DET1* locus. An additional dominant allele shows no linkage to the *DET1* locus. These mutations are likely to be extragenic suppressors of the *det1* mutation, and may identify novel regulatory elements in the photomorphogenetic pathway.

Downstream events in light signal transduction

Though the photomorphogenetic mutations have been extremely useful in the study of light-regulated development, they are pleiotropic, each affecting a large number of downstream light-regulated processes. We have utilized a second genetic approach that focuses on one particular downstream light-regulated response, the transcription of a chlorophyll *a*/*b*-binding protein promoter, *cab3*. This molecular genetic approach allows us to identify mutants by aberrant gene expression patterns, rather than by predicted phenotype, and should be applicable to the study of other signal transduction pathways.

Numerous studies have shown that *cab* gene transcription is regulated by light, intrinsic developmental signals, and is also sensitive to signals originating from the chloroplast itself (reviewed in [22]). The *cab* promoter–marker gene chimera (pOCA107) needed for these studies was introduced into plants at a single site in a transgenic line called pOCA107-2. The pOCA107 construction contains a fully regulated *cab3* promoter sequence fused to the *hph* (hygromycin phosphotransferase) gene, which confers hygromycin resistance allowing for positive selection strategies [23]. The construct also contains a second *cab3* promoter fused to a screenable marker *uidA* gene (β-glucuronidase, GUS) [24]. We mutagenized the transgenic pOCA107-2 line and selected for plants that aberrantly express the marker transgenes under a variety of conditions. After selection on hygromycin, we also screen for GUS activity, which is under the control of the second *cab3* promoter. This step is important so that true signal transduction mutants can be distinguished from *cis*-acting promoter mutations.

Using this line, we have identified a number of mutant lines in which the *cab3* promoters are expressed at aberrant high levels in the dark, though the seedlings remain etiolated in appearance. To date, the mutations define at least five complementation groups that we have designated *doc* (for dark overexpression of cab). All *doc* mutations examined are recessive, with the exception of one mutant which is dominant. Some of the mutations also affect the levels of *rbc*S (ribulose-bisphosphate carboxylase, small subunit) mRNAs in dark-grown mutant seedlings. Others affect the accumulation of *cab* mRNA, but not *rbc*S mRNA. Thus, the *doc* mutations define loci that act late in the light signal transduction pathways. Moreover, the mutations identify a branch point in which the control of *cab* gene expression can be genetically separated from *rbc*S expression.

Conclusion

We have demonstrated the utility of morphological and molecularly marked *Arabidopsis* mutants to study seedling development in response to light. Mutations at the *fre1* and *hy3* loci define the overall functions of PHYA and PHYB in *Arabidopsis* development. *PHYB* sequences from a number of *hy3* alleles have begun to address structure and function relationships within the PHYB molecule. Mutations have also identified a novel class of signal transduction molecules that may act to negatively regulate photomorphogenesis. Positional cloning, utilizing the *det1* and *det2* mutations, is being used to identify these genes. Finally, analysis of mutations that affect the expression of the *cab3* promoter *in trans* will define downstream branches of the light signal transduction pathways.

Our work on photomorphogenetic mutants of *Arabidopsis* is supported by grants from the National Science Foundation, the Department of Energy, the USDA, the International Human Frontier Science Program Organization and the Samuel Roberts Noble Foundation.

References

1. Cosgrove, D.J. (1986) in Photomorphogenesis in Plants (Kendrick R.E. and Kronenberg, G.H.M., eds.), pp. 341–366, Martinus Nijhoff, Dordrecht
2. Quail, P.H. (1991) Annu. Rev. Genet. **25**, 389–409
3. Furuya, M. (1989) Adv. Biophys. **25**, 133–167
4. Sharrock, R.A. and Quail, P.H. (1989) Genes Dev. **3**, 1745–1757
5. Chory, J., Peto, C., Feinbaum, R., Pratt, L. and Ausubel, F. (1989) Cell **58**, 991–999
6. Chory, J., Nagpal, P. and Peto, C.A. (1991) Plant Cell **3**, 445–459
7. Chory, J. (1992) Development **15**, 337–355
8. Deng, X.W., Caspar, T. and Quail, P.H. (1991) Genes Dev. **5**, 1172–1182
9. Wei, N. and Deng, X.W. (1992) Plant Cell **4**, 1507–1518
10. Redei, G.P. and Hirono, Y. (1964) *Arabidopsis* Information Service **1**, 9–10
11. Koornneef, M., Rolff, E. and Spruitt, C.J.P. (1980) Z. Planzenphysiol. **100**, 147–160
12. Chory, J., Peto, C.A., Ashbaugh, M., Saganich, R., Pratt, L. and Ausubel, F.M. (1989) Plant Cell **1**, 867–880
13. Liscum, E. and Hangarter, R.P. (1991) Plant Cell **3**, 685–694
14. Parks, B.M. and Quail, P.H. (1993) Plant Cell **5**, 39–48
15. Nagatani, A., Reed, J.W. and Chory, J. (1993) Plant Physiol. **102**, 269–277
16. Parks, B. M. and Quail, P.H. (1991) Plant Cell **3**, 1177–1186
17. Somers, D.E., Sharrock, R.A., Tepperman, J.M. and Quail P.H. (1991) Plant Cell **3**, 1263–1274
18. Nagatani, A., Chory, J. and Furuya, M. (1991) Plant Cell Physiol. **32**, 1119–1122
19. Reed, J.W., Nagpal, P., Poole, D.S., Furuya, M. and Chory, J. (1993) Plant Cell **5**, 147–157

20. Beitel, G.J., Clark, S.G. and Horvitz, H.R. (1990) Nature (London) **348**, 503–509
21. Smith, H. and Whitelam, G.C. (1990) Plant Cell Environ. **13**, 695–707
22. Gilmartin, P.M., Sarokin, L., Memelink, J. and Chua, N.H. (1990) Plant Cell **2**, 369–378
23. Gritz, L. and Davies, J. (1983) Gene **25**, 179–188
24. Jefferson, R.A., Burgess, S.M. and Hirsh, D. (1986) Proc. Natl. Acad. Sci. U.S.A. **83**, 8447–8451

Biochem. Soc. Symp. **60**, 265–275
Printed in Great Britain

Cloning and characterization of cDNAs encoding oat PF1: a protein that binds to the PE1 region in the oat phytochrome A3 gene promoter

Jorge Nieto-Sotelo† and Peter H. Quail*

University of California at Berkeley/USDA, Plant Gene Expression Center, 800 Buchanan Street, Albany, CA 94710, U.S.A.

Synopsis

In monocotyledons, the expression of the oat phytochrome A gene (*PHYA*) is down-regulated by phytochrome itself. This autoregulatory repression is the most rapid light-induced effect on gene expression reported in plants to date. A functional analysis of the oat *PHYA3* gene minimal promoter in a rice transient expression assay has identified two promoter elements, PE1 and PE3, that interact synergistically in positive regulation. We have isolated an oat cDNA clone (pO2) that encodes a DNA-binding protein that binds to the PE1 region of the oat *PHYA3* gene promoter. The *in vitro* binding properties of the pO2-encoded protein, towards DNA probes containing either the PE1 sequence or linker-substitution mutations in PE1, correlate with the activity of these DNA elements in the rice transient expression assay. These mutations are known to abolish expression of a reporter gene *in vivo*. Binding of these linker-substitution mutants to the pO2-encoded protein *in vitro* was lower by one to two orders of magnitude than the binding of the native PE1 region. We suggest, therefore, that the pO2 clone may encode the putative nuclear factor, oat PF1, that is involved in positive regulation of *PHYA3* by binding to PE1 *in vivo*. pO2 encodes a 170-amino-acid-long protein that contains three repeats of the 'AT-hook' DNA-binding motif found in high mobility group I-Y (HMGI-Y) proteins. Oat PF1 is highly similar to rice PF1 and to the protein encoded by soybean cDNA SB16. They all have a strong similarity in their *N*-terminus to the pea H1 histone, and the presence of several AT-hook DNA-binding motifs in their *C*-terminal halves.

*To whom correspondence should be addressed.
†Present address: Instituto de Biotecnología, Universidad Nacional Autónoma de México, apdo. postal 510-3, Cuernavaca, Mor. 62271, México.

Introduction

In plants several photoreceptors, such as the blue-light receptor, the u.v. receptor and phytochrome, control growth and development. Of these receptors, phytochrome is by far the best characterized. Phytochrome is a polypeptide with a chromophore covalently linked to it. In *Arabidopsis thaliana*, five different genes encoding phytochrome, designated *PHYA*, *PHYB*, *PHYC*, *PHYD* and *PHYE* have been found so far [11a]. In rice, three genes encoding phytochromes A, B and C have been found [2,3] (K. Dehesh and P. Quail, unpublished work) and preliminary evidence suggests the existence of two more genes in the rice genome, encoding phytochromes D and E (K. Dehesh and P. Quail, unpublished work). In addition, sequences with similarity to members of the *Arabidopsis* and rice phytochrome subfamilies have been detected in a very diverse group of plants and algae [4]. This photoreceptor has the capacity to exist in two photo-interconvertible forms: Pr, which absorbs maximally in the red region (666 nm), and Pfr, which absorbs maximally in the far-red region (730 nm) of the spectrum. Once a photon has been absorbed by Pr, Pr is converted into Pfr and vice versa. The key regulatory function of phytochrome as a molecular switch is based upon this photoconvertibility. Phytochrome is synthesized *de novo* as Pr, the inactive form, which accumulates in etiolated tissues. The photoconversion to the active Pfr form induces substantial changes in the patterns of gene expression [5,6]. This induction triggers distinctive patterns of development in some plants, such as flowering, seed germination, and stem elongation. Many of these events can be prevented if Pfr is immediately converted to the Pr form [7].

The rapid transcriptional repression by phytochrome of its own *PHYA* gene has been used as a model system to understand how phytochrome controls gene transcription. This response is the fastest light-mediated transcriptional event reported in plants [8]. Functional analysis of the *PHYA* promoter has been made by means of a transient expression assay in rice seedlings using oat *PHYA3* promoter/chloramphenicol acetyltransferase (CAT) fusion constructs [9]. This approach identified three positive elements, which were designated PE1 (−367 bp to −346 bp), PE2 (−635 bp to −489 bp) and PE3 (−111 bp to −81 bp), that support high-level expression in low-Pfr cells, and one negative element, designated RE1, that represses transcription in high-Pfr cells [10,11]. The PE1 and PE2 elements are functionally redundant and the synergistic interaction of either one of these elements with PE3 is required for maximum expression.

Earlier studies of protein–DNA interactions involved in the transcriptional regulation of the *PHYA* gene resulted in the cloning and characterization of a rice cDNA clone encoding the GT-2 protein which is implicated in the positive regulation of the rice *PHYA* gene [11–13]. Here, we present the isolation and initial characterization of the DNA-binding activity of oat cDNAs encoding proteins that bind to the PE1 region in the oat *PHYA3* gene promoter.

Methods

Plant material and growth

Soaked oat (*Avena sativa* L. cv. *Gary*) seeds were grown on moist termiculite at 26°C in the dark. The harvesting of tissue took place under green safe light.

RNA isolation

Total RNA was isolated from 6-day-old, dark-grown seedlings. Only the top 2–3 cm of shoot tissue was harvested using scissors, immediately weighed, frozen under liquid nitrogen, and kept at −80°C until RNA isolation. Total RNA was isolated by the method of Rochester *et al.* [14]. Poly(A^+) RNA was selected by oligo (dT) cellulose chromatography as described [15].

Construction and screening of cDNA library

A cDNA library was constructed using a commercially available cDNA synthesis kit (Promega), *Eco*RI adaptors (Promega) and λgt11 arms (Stratagene). Starting with 4 μg of poly(A^+) RNA, 2.6 μg of cDNA were obtained. From 200 ng of cDNA ligated to λgt11 arms, 13.3×10^6 recombinant phage were obtained.

The method of Vinson *et al.* [16] was used for screening the cDNA library with a double-stranded O-PE1 probe that was prepared by annealing two complementary synthetic oligonucleotides 5′-GGCTGGAAATAGCAAATGTTAAAAATAAA-3′ and 5′-AGCCTTTATTTTTAACATTTGCTATTTCC-3′ that had been gel purified. After annealing, oligonucleotides were kinased, ligated and radiolabelled, with ^{32}P and a nick translation kit (Amersham) using standard techniques, to a specific activity of at least 1×10^8 c.p.m./μg of DNA [17]. A total of 750 000 recombinant phage were screened and five positive clones were obtained after three rounds of screening. Characterization of the DNA binding of protein made by recombinant phage was determined in a filter-binding assay by comparing the binding of O-PE1 probe with mutant probes that were radiolabelled at equal specific activity. Mutant probes are synthetic, double-stranded oligonucleotides O-614 (5′-GGCTGCTCGAGTCAAATGTTAAAAA-TAAA-3′ and 5′-AGCCTTTATTTTTAACATTTGACTCGAGC-3′), O-615 (5′-GGCTGGAAATAGGCTCGAGTAAAAATAAA-3′ and 5′-AGCCTTTATTTTTA-CTCGAGCCTATTTCC-3′), and O-616 (5′-GGCTGGAAATAGCAAATGTGCT-CGAGTAA-3′ and 5′-AGCCTTACTCGAGCACATTTGCTATTTCC-3′) prepared similarly to O-PE1. DNA bound to filters containing blotted phage protein was estimated by liquid scintillation counting and by autoradiography. cDNA inserts were subcloned as *Eco*RI fragments in two orientations: in the *Eco*RI site of $pBSSK^+$ vector (Stratagene) for sequence analysis, and in the pPO-9 vector for overexpression of the encoded protein in *E. coli* (for more details see section on Expression of recombinant PF1 protein in bacteria).

RNA blot analysis

Analysis was performed as described previously [18], using 1.25 μg of poly(A^+) RNA per lane.

Expression of recombinant PF1 protein in bacteria and protein extract preparation

The cDNA insert from recombinant λgt11 bacteriophage pO2 was subcloned as an *Eco*RI fragment into plasmid vector pPO-9 [18a]which is a modified version of the pET-3a vector originally described by Rosenberg *et al.* [19]. Vector pPO-9 keeps the *Eco*RI cDNA insert in the same frame as it was in the original λgt11 phage and, like its parent plasmid pET-3a, expression of the cDNA is under the control of a T7 promoter. Sense and antisense orientation versions of pO2 cDNA insert subcloned in pPO-9 were transformed into *E. coli* BL21(DE3) cells. Transformed bacteria were induced with

1 mM IPTG at an A_{600} between 0.7 and 0.9 for 3 h, harvested, and resuspended in 1X binding buffer [20 mM Hepes (pH 7.9), 3 mM $MgCl_2$, 40 mM KCl, 1 mM DTT, 20% glycerol, 1 μg/ml leupeptin, 1 μg/ml antipain]. Protein extracts were obtained by freezing and thawing three times between liquid N_2 and room temperature followed by sonication on ice. The suspension was centrifuged at 80 000 r.p.m. for 20 min at 4°C (Beckman TL 100 ultracentrifuge) and the supernatant fraction was saved and stored in small aliquots at −80°C. Protein concentration was measured by the Bradford assay [20].

DNA-binding assays *in vitro*

South-western blot assays Protein extracts were resolved in 11% polyacrylamide/SDS gels and transferred to nitrocellulose. Blots were denatured/renatured with Guanidine–HCl, hybridized to nick-translated, O-PE1, concatamerized, double-stranded oligonucleotides for 3–4 h at 4°C, washed and autoradiographed as described [16].

Mobility shift analysis Protein extracts were incubated in 1X binding buffer, 1–4 μg of salmon sperm DNA and an end-labelled probe at room temperature for 15 min. Samples were analysed by electrophoresis in 6% polyacrylamide gels in 1X Tris-Borate-EDTA (TBE) buffer. End-labelled probes were a 300 bp fragment of the oat *PHYA3* gene promoter (−416 to −116 relative to transcription start site) or linker-substitution mutants of this same 300 bp promoter fragment. The 300 bp wild-type probe corresponds to the *Bam*HI–*Sal*I fragment of clone 449 [10], whereas the linker-substitution mutants correspond to the *Bam*HI–*Sal*I 300 bp fragments of clones 614, 615 and 616 [11].

Results

Isolation of oat cDNA clones encoding proteins that bind to an AT-rich DNA element

We prepared a λgt11 cDNA expression library from poly(A^+) RNA that was purified from etiolated oat seedlings. This library was then screened with double-stranded synthetic oligonucleotides containing the sequence of the PE1 region of the oat *PHYA3* gene promoter (O-PE1). Starting with a total of 750 000 recombinant phages, we isolated five independent clones expressing proteins that bound to the ^{32}P-labelled O-PE1 probe after three rounds of screening. The five positive clones were designated pO1, pO2, pO3, pO4 and pO9. We carried out a filter DNA-binding assay to confirm the specificity of the positive clones towards the O-PE1 probe. We compared the binding of O-PE1 with the binding of synthetic oligonucleotides containing linker-substitution mutations in PE1, named O-614, O-615 and O-616, whose sequences are shown in Fig. 1*c*. We hypothesized that the binding to DNA of factors involved in transactivation of the oat *PHYA3* gene through the PE1 region in oat seedlings should be affected by these mutations, since these linker-substitution mutations caused the loss of expression of a reporter gene in a rice seedling transient expression assay [11].

In Fig. 1 we can see the binding of O-614, O-615 and O-616 to the recombinant protein expressed by pO2. This binding was 18%, 23% and 1%, relative to the level of

(a)

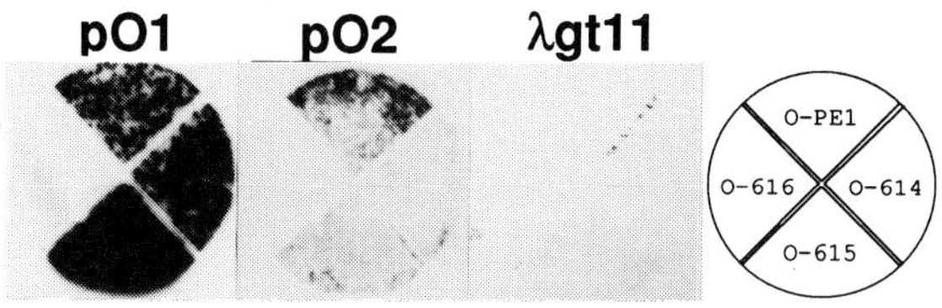

(b)

Oligonucleotide probe	% Binding pO1	pO2	pO3	pO4	pO9
O-PE1	100	100	100	100	100
O-614	187	18	12	11	12
O-615	517	23	13	13	6
O-616	0.6	1.1	0.5	1.4	0.8

(c) **Sequence of oligonucleotide probes**

5' GGC TGG AAA TAG CAA ATG TTA AAA ATA AA 3' O-PE1

5' GGC TGc tcg agt CAA ATG TTA AAA ATA AA 3' O-614

5' GGC TGG AAA TAG g c t c g a gTA AAA ATA AA 3' O-615

5' GGC TGG AAA TAG CAA ATG Tgc tcg Agt AA 3' O-616

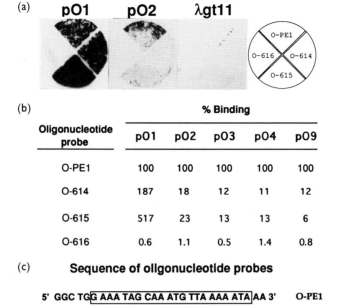

Fig. 1. A comparative filter binding assay of recombinant proteins encoded by λgt11-pO1, -pO2, -pO3, -pO4, -pO9 and a non-recombinant λgt11 phage to synthetic oligonucleotides. (a) Autoradiograms of the filter-binding assay made to test the binding of proteins expressed by λgt11-pO1, λgt11-pO2 and a λgt11 non-recombinant phage to synthetic oligonucleotides whose sequences are indicated in (c). λgt11-recombinant or a λgt11 non-recombinant phage that were grown to confluence on *E. coli* Y1090 cells were plaque-lifted on nitrocellulose filters. Each filter was cut into quarters and further processed by the filter-binding assay, as described in the Methods section. The diagram shows the oligonucleotides used to probe each quarter filter. (b) Quantification of the DNA-binding activity for O-PE1 and mutant oligonucleotides of PE1 of recombinant proteins expressed by λgt11-pO1, -pO2, -pO3, -pO4 and -pO9. Activity is expressed as a percentage of the amount of radiolabelled oligonucleotide bound to each filter divided by the amount of radioactivity bound to filter probed with O-PE1. (c) Oligonucleotide probe O-PE1 comprises the sequence of PE1, a *PHYA3* gene promoter region previously identified in transient expression assays to be required for positive expression of a reporter gene in rice seedlings [11]. Oligonucleotide probes O-614, O-615 and O-616 contain the sequences of linker-substitution mutants in the PE1 region that disrupt positive expression of a reporter gene *in vivo* [11]. Boxes outline the PE1 region in O-PE1, and the nucleotide substitutions in O-614, O-615 and O-616.

binding of O-PE1, respectively (Fig. 1*b*). We obtained comparable results for binding to recombinant proteins encoded by clones pO2, pO3 and pO4. In contrast, binding of O-614, O-615 and O-616 to recombinant protein expressed by pO1 was 187%, 517% and 0.6%, relative to the level of binding to O-PE1, respectively. These results indicate that pO2, pO3, pO4 and pO9 clones encode DNA-binding proteins with the properties expected for PF1, the nuclear factor acting *in trans* with the PE1 region, while pO1 does not meet these requirements. Further studies showed that the pO4 cDNA insert is identical to pO2 but smaller in size. We characterized the pO2 clone further.

Characterization of the pO2 cDNA clone encoding oat PF1

The insert of the pO2 cDNA clone is approximately 1.1 kb in length. It was subcloned into the expression vector pPO-9 and total extracts were made from isopropyl β-D-thiogalactoside (IPTG)-induced or non-induced *Escherichia coli* cells transformed with the recombinant plasmid in both orientations. We determined the size of the recombinant protein encoded by pO2 by South-western analysis. The apparent molecular mass of the recombinant protein expressed in the sense orientation that bound O-PE1 was 24 kDa on SDS/PAGE (Fig. 2).

The expression of pO2 was studied by Northern blot analysis. A single 1.1 kb transcript was obtained with poly(A^+) RNA isolated from oat seedlings using the pO2 cDNA insert as probe. This finding indicates that the cDNA insert of the recombinant clone is near to full length (data not shown).

Characterization of oat recombinant PF1-binding activity

The binding of oat recombinant PF1 with a 300 bp oat *PHYA3* gene promoter fragment containing the PE1 region [P-PE1 (positions −416 to −116)] was compared

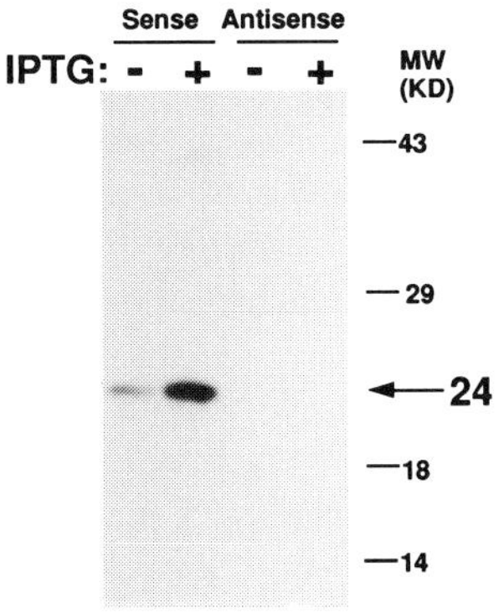

Fig. 2. South-western blot analysis of recombinant PF1 encoded by pO2 cDNA. Total protein extracts were prepared from *E. coli* BL21 (DE3) cells transformed with vector pPO-9 containing the pO2 cDNA insert in sense or antisense orientation. Samples were separated by SDS/PAGE and electroblotted to nitrocellulose. The blot was denatured/renatured and incubated with radiolabelled O-PE1 concatamerized probe, as described in the Methods section.

with the binding of oat recombinant PF1 with 300 bp fragments of the same region containing linker-substitution mutations within PE1 by gel mobility shift analysis. These promoter fragments contain the same mutations in PE-1 as depicted in Fig. 1*c*. In Fig. 3 we can see that complex formation between oat recombinant PF1 and P-PE1 needs lower protein concentration than complex formation with any of the promoter mutations. There is no complex formation at all with P-614 (Fig. 3, lanes 5 and 6). However, P-615 and P-616 complex with oat recombinant PF1 although to a lesser extent than the recombinant PF1 complexes with P-PE1. These results indicate that, in the context of the 300 bp promoter region, oat recombinant PF1 requires an intact PE1 region for efficient binding. Nevertheless, it also suggests that some binding may occur outside the PE1 region, since mutations P-615 and P-616 are still capable of complex formation with recombinant PF1.

Nucleotide sequence analysis of PF1

We sequenced the entire pO2 cDNA insert on both strands by the dideoxy-nucleotide chain-termination method. Analysis of the sequence indicates that the cDNA insert has a length of 970 bp, and that the largest open reading frame found was in frame with the λgt11 β-galactosidase gene. Assuming that the initiation codon is at position 83, this open reading frame spans 510-nucleotides, followed by a stop codon and 365 nucleotides of 3 non-coding region and a poly(A^+) tail that is 10 nucleotides long [21].

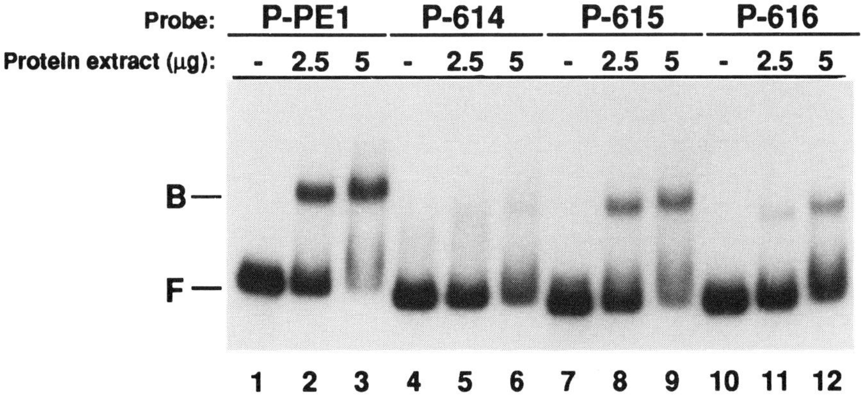

Fig. 3. Binding assay with recombinant PF1 and an oat *PHYA3* gene promoter fragment containing the PE1 region or promoter fragments containing linker substitutions within the PE1 region. A mobility shift analysis was performed with end-labelled 300 bp probes containing the oat *PHYA3* gene region between −416 bp and −116 bp (P-PE1) or 300 bp probes of the same promoter region containing linker substitutions within PE1 (P-614, P-615 and P-616) plus or minus soluble protein extracts from bacteria overexpressing recombinant oat PF1 protein. Linker-substitution mutants of PE1 are identical to the ones shown in Figure 1*c*. Abbreviations used: F, free probe; B, protein bound to DNA complex.

The predicted amino acid sequence contains 170 amino acids with a calculated molecular mass of 18 kDa. This value is discrepant from that of 24 kDa, which is the apparent molecular mass determined by SDS/PAGE for oat PF1 recombinant protein expressed in *E. coli.* The cause of this discrepancy is not clear, but it may be due in part to the unusual composition of the protein. The polypeptide is 13% alanine and 19% proline, with a calculated pI of 10.8 due to the abundance of lysine and arginine residues. Dot plot analysis of the protein sequence compared with itself revealed the presence of three nonapeptide repeats at the *C*-terminal half of the protein. The consensus sequence for all three internal repeats is K/GRG/PRGRPP/AK.

Comparison of oat PF1 with rice PF1 and with the database

A rice cDNA clone encoding PF1 that shows DNA-binding properties similar to those of oat PF1, was independently isolated using the same library screening method used to obtain the oat PF1 clone [22]. A comparison of the structure of oat PF1 with its homologous protein, rice PF1, indicated a 79% similarity and 67% identity in the primary sequences and showed a high degree of conservation in the secondary structure of the two proteins. The similarity within all the nonapeptide repeats is 100% [21]. When the entire oat PF1 sequence was used for comparison with the GenEMBL database, a 50% similarity (29% identity) to histone H1 from pea was found [23]. However, none of the similarities is within any of the three internal repeats of oat PF1; instead they are spread along the entire length of the protein, with a maximum percentage of invariable amino acids in the *N*-terminal end. When the sequence of the two identical nonapeptide repeats 1 and 3 of oat PF1 (amino acid residues 87–95 and 153–161) was compared with the database, we found a 100% match to the sequence of two asparagine-rich antigens of unknown function from *Plasmodium falciparium* [Schreiber *et al.* (1989), unpublished work, in GenEMBL accession numbers X17485 and X17488]. Eight out of nine amino acids of the nonapeptide were identical to a sequence within the protein of *Saccharomyces cerevisiae* encoded by the SNF2 gene that shows functional interdependence to SNF5 and SNF6 proteins in transcriptional activation [24], to a developmentally regulated mRNA of *Dictyostelium discoideum* [25], and to the mammalian HMGI and HMGY proteins [26,27]. However, this region shows a 100% match in a recent search in the database to the soybean HMGY-a and HMGY-b sequences [28]. This nonapeptide repeat, which is present in all HMGI and HMGY proteins, constitutes their DNA-binding domain. Since this region binds to AT-rich DNA, it has been named the AT hook [29]. A comparison of the amino acid sequences of oat and rice PF1 with the human and soybean HMGI and HMGY proteins revealed the following: (i) oat PF1 is more similar to rice PF1 (79% similarity, 67% identity); (ii) the soybean HMGY proteins show higher similarity to oat PF1 (65% similarity, 51% identity) than their mammalian counterparts (41% similarity; 22% identity). The soybean HMGYs and oat and rice PF1s have additional *N*- and *C*-terminal sequences absent in the mammalian proteins. The alignment of all sequences showed three small gaps of 10, 9 and 14 amino acids in length in the oat PF1 and soybean HMGYs relative to the rice PF1 sequence (amino acids 6–15, 109–117, and 166–179 of the rice PF1 sequence, respectively). The second gap spans a short hydrophobic region characteristically present in mammalian HMGI proteins and absent in HMGY. The absence of this hydrophobic region in oat PF1, as in soybean HMGYs, suggests that oat PF1 is an HMGY-like protein.

Discussion

We have isolated five different cDNA clones encoding DNA-binding proteins by screening an oat expression library with synthetic oligonucleotides containing the sequence of the oat *PHYA3* gene PE1 region. In this work, we have described the characterization of clone pO2 which encodes a protein referred to as oat PF1. The size of the cDNA in pO2 is similar to the size of its mRNA, indicating that the insert is close to full length.

The protein encoded by the pO2 clone meets the criteria expected for the nuclear factor that interacts with PE1, based on its DNA-binding properties. Two kinds of experiment indicated the aforementioned properties. The first involved a filter-binding assay. When we compared the binding to recombinant oat PF1 of synthetic oligonucleotides containing the sequence of PE1 (O-PE1) with mutations of PE1 that were known to disrupt expression of a reporter gene in a rice transient expression assay (O-614, O-615 and O-616) [11], the binding of O-PE1 was one or two orders of magnitude higher than the binding of the mutant DNA probes. In a gel mobility assay we compared the binding to recombinant PF1 of a 300 bp oat *PHYA3* gene fragment containing PE1 (P-PE1) with the binding of mutants of the same promoter fragment containing linker substitutions in the PE1 region (P-614, P-615 and P-616). The results indicated that, although recombinant PF1 displays a higher affinity towards P-PE1, there is still some binding at high concentrations of protein extract to P-615 and P-616, implying that additional PF1-binding sites may be found outside PE1. Direct evidence that additional sites for PF1 do indeed exist in the P-PE1 promoter fragment comes from experiments using recombinant rice PF1 in footprinting experiments [22]. It was observed that recombinant rice PF1 protected PE1 and two other AT-rich regions upstream of PE1 at positions -372 to -385 and -387 to -399 of the oat *PHYA3* gene. A similar pattern of protection to the one found with recombinant rice PF1 was observed when oat nuclear protein extracts were used in footprinting experiments [11]. The same two AT-rich regions upstream of PE1 were protected in addition to PE1 and other elements within the promoter. The experiments performed with the rice PF1 suggest that at least one nuclear factor interacts with PE1 and two other AT-rich regions upstream of PE1. We are not certain about the functionality of these two AT-rich regions upstream of PE1 in the intact plant in terms of *PHYA3* gene expression. However, under the conditions of the rice transient expression assay, they did not seem to be functional in the presence of PE1 [10,11]. It is possible that they are functionally redundant to PE1.

A comparison of the predicted amino acid sequence of oat PF1 encoded by the pO2 clone and the sequence of rice PF1 showed that they have a high degree of similarity. Rice PF1 is encoded by pR4, a rice PE1-binding protein, independently isolated by screening a rice expression library with O-PE1 oligonucleotides [22]. Both sequences contain several repeats of the so-called AT-hook DNA-binding motif along the protein (three repeats in oat PF1 and four in rice PF1). This DNA-binding motif is repeated three times in mammalian HMGY and HMGI proteins [26,27] and four times in the protein encoded by the soybean SB16 cDNA [28] that also binds AT-rich DNA. To the same extent as the soybean SB16 cDNA, oat and rice PF1s are similar to pea H1 histone at their *N*-terminal 50 amino acids. Oat PF1 lacks a stretch of 11 amino acids that is present in mammalian HMGI but absent in HMGY proteins. Although the function of this domain has not yet been established, its possible involvement in the formation of homo- and/or heterodimers, owing to its hydrophobic character, has been proposed [26]. The synthesis of HMGY and HMGI proteins arises from the alternative splicing of the encoding mRNAs in the open reading frame, which results in the deletion of the 11 amino acid region in HMGY. This implies that these two proteins are the products of the same gene [26,27]. It is not known whether this happens in the case of oat and rice PF1s or the soybean SB16 gene products. Finally, we can only speculate on the significance of the similarity of rice PF1, oat PF1 and soybean SB16 proteins to pea H1 histone. It is possible that the presence of a histone H1 domain may allow these proteins to interact with other proteins and/or DNA in chromatin as H1 histone does.

We thank Dr. Katayoon Dehesh for her advice on protocols for screening expression libraries with oligonucleotides. Also, we are thankful to Dr. Paul Oeller for providing us with vector pPO-9. Finally, we would like to express our appreciation to the members of our laboratory for their helpful criticism while this work was in progress. This work was supported by National Science Foundation grant number MCB-9220161 and USDA/ARS CRIS number 5335-21000-006-00D.

References

1. Sharrock, R.A. and Quail, P.H. (1989) Genes Dev. **3**, 1745–1757
1a. Clack, T., Mathews, S. and Sharrock, R.A. (1994) Plant Mol. Biol., in the press
2. Dehesh, K., Tepperman, J., Christensen, A.H. and Quail, P.H. (1991) Mol. Gen. Genet. **225**, 305–313
3. Kay, S.A., Keith, B., Shinozaki, K., Chye, M.-L. and Chua, N.-H. (1989) Plant Cell **1**, 351–360
4. Quail, P.H. (1991) Annu. Rev. Genet. **25**, 389–409
5. Gilmartin, P.M., Sarokin, L., Memelink, J. and Chua, N.-H. (1990) Plant Cell **2**, 369–378
6. Thompson, W.F. and White, M.J. (1991) Annu. Rev. Plant Physiol. Plant Mol. Biol. **42**, 423–466
7. Kendrick, R.E. and Kronenberg, G.H.M. (1986) Photomorphogenesis in Plants, Martinus Nijhoff, Dordrecht
8. Lissemore, J.L. and Quail, P.H. (1988) Mol. Cell. Biol. **8**, 4840–4850
9. Bruce, W.B., Christensen, A.H., Klein, T., Fromm, M. and Quail, P.H. (1989) Proc. Natl. Acad. Sci. U.S.A. **86**, 9692–9696
10. Bruce, W.B., and Quail, P.H. (1990) Plant Cell **2**, 1081–1089
11. Bruce, W.B., Deng, X.-W. and Quail, P.H. (1991) EMBO J. **10**, 3015–3024
12. Dehesh, K., Bruce, W.B. and Quail, P.H. (1990) Science **250**, 1397–1399
13. Dehesh, K., Hung, H., Tepperman, J.M. and Quail, P.H. (1992) EMBO J. **11**, 4131–4144
14. Rochester, D.E., Winter, J.A. and Shah, D.M. (1986) EMBO J. **5**, 451–458
15. Theologis, A., Huynh, T.V. and Davies, R.W. (1985) J. Mol. Biol. **183**, 53–68
16. Vinson, C.R., LaMarco, K.L., Johnson, P.F., Landschulz, W.H. and McKnight, S.L. (1988) Genes Dev. **2**, 801–806
17. Sambrook, J., Frisch, E.F. and Maniatis, T. (1989) Molecular Cloning: A Laboratory Manual, 2nd edn, Cold Spring Harbor Laboratory Press, Cold Spring Harbor
18. Colbert, J.T., Hershey, H.P. and Quail, P.H. (1985) Plant Mol. Biol. **5**, 91–102
18a. Rottman, W.H., Peter, G.F., Oeller, P.W., Keller, J.A., Shen, N.F., Nagy, B.P., Taylor, L.P., Campbell, A.D. and Theologis, A. (1991) J. Mol. Biol. **222**, 937–961

19. Rosenberg, A.H., Lade, B.N., Chui, D.-S., Lin, S.-W., Dunn, J.J. and Studier, F.W. (1987) Gene **56**, 125–135
20. Bradford, M. (1976) Anal. Biochem. **72**, 248–254
21. Nieto-Sotelo, J., Ichida, A. and Quail, P.H. (1994) Nucleic Acids Res. **22**, 1115–1116
22. Nieto-Sotelo, J. Ichida, A. and Quail, P.H. (1994) Plant Cell **6**, 287–301
23. Gantt, J.S. and Key, J.L. (1987) Eur. J. Biochem. **166**, 119–125
24. Laurent, B.C., Treitel, M.A. and Carlson, M. (1991) Proc. Natl. Acad. Sci. U.S.A. **88**, 2687–2691
25. Shaw, D.R., Richter, H., Giorda, R., Ohmachi, T. and Ennis, H.L. (1989) Mol. Gen. Genet. **218**, 453–459
26. Ecker, R. and Birnstiel, M.L. (1989) Nucleic Acids Res. **17**, 5947–5959
27. Johnson, K.R., Lehn, D.A. and Reeves, R. (1989) Mol. Cell. Biol. **9**, 2114–2123
28. Laux, T., Seurinck. J. and Goldberg, R.B. (1991) Nucleic Acids Res. **19**, 4768
29. Reeves, R. and Nissen, M.S. (1990) J. Biol. Chem. **265**, 8573–8582

Biochem. Soc. Symp. **60**, 277–284
Printed in Great Britain

Elucidation of phytochrome signal-transduction mechanisms

Gunther Neuhaus*, Chris Bowler† and Nam-Hai Chua

Laboratory of Plant Molecular Biology, The Rockefeller University, 1230 York Avenue, New York, NY10021-6399, U.S.A. and *Institut für Pflanzenwissenschaften, ETH-Zentrum, Universitätstrasse 2, CH-8092 Zürich, Switzerland

Introduction

Plants must be able to detect changes in ambient light to optimize photosynthetic reactions and to regulate their growth and development. Light is perceived through three distinct light receptors that respond to particular wavelengths and intensities of light: phytochrome, which absorbs red and far-red light; cryptochrome, which absorbs blue and u.v.-A light; and the u.v.-B photoreceptor [1,2]. Of these, the best characterized biochemically and physiologically is phytochrome. Phytochrome consists of a polypeptide and a covalently bound chromophore, and light perception is mediated by the photo-isomerization of the chromophore. Etiolated and green tissue contain different pools of phytochrome, respectively referred to as type I and type II. Most plants contain small multigene families which encode the proteins that make up these two pools; *PHYA* encodes the type I phytochrome (also known as PHYA), while other genes (e.g. *PHYB*) encode type II phytochrome [2]. Two spectrophotometrically different, but photo-interconvertible, forms of these phytochromes exist in the plant: the red-light-absorbing Pr, and the far-red-absorbing Pfr. Pfr is generally thought to be the active form.

There has recently been much progress in the elucidation of *cis*- and *trans*-acting factors that mediate phytochrome-regulated gene expression [3]. In contrast, there has been little real progress in our understanding of the molecular transduction events that couple photoperception by phytochrome in the cytosol to changes in gene expression in the nucleus. In the current work, our aim was to develop a new approach for elucidating phytochrome signal-transduction intermediates. Specifically, we wanted to address the following questions. (i) What are the components of the phytochrome phototransduction machinery, and how are they ordered? (ii) Because phytochrome regulates such a large number of molecular targets, are the transduction chains separate or branched,

†To whom correspondence should be addressed.

and are any of the intermediates shared? (iii) Is phytochrome signalling cell autonomous or is there cell-to-cell signalling?

To address these questions, we developed single-cell assays, employing microinjection to introduce large or small molecules into cells in an attempt to identify components that could modulate reactions normally carried out by phytochrome. As recipient material, we chose to use hypocotyl cells of a PHYA-deficient mutant of tomato known as *aurea* [4,5]. We used this mutant because, while wild-type plant cells develop chloroplasts and anthocyanins in response to light, *aurea* hypocotyl cells do not.

In this report, we show that microinjection of an exogenous PHYA can restore chloroplast development and anthocyanin biosynthesis within an injected cell. Furthermore, we show that these responses require the participation of one or more heterotrimeric G-proteins, together with calcium and calmodulin acting further downstream. Although G-protein activation can mediate a full cellular response, equivalent to that initiated by PHYA, the efficacy of calcium and calmodulin is more restricted: there is no anthocyanin biosynthesis and chloroplast development is incomplete, owing to the absence of two of the five major photosynthetic complexes: photosystem I (PSI) and cytochrome $b_6 f$.

Results

Identification of phytochrome signalling intermediates

When dark-grown *aurea* seedlings are placed in the light for 48 h, there is neither significant development of chloroplasts nor anthocyanin pigment biosynthesis in hypocotyl cells, in striking contrast to wild-type cells treated in the same way. Since *aurea* lacks PHYA, we reasoned that it should be possible to biochemically rescue wild-type traits by microinjecting PHYA protein into *aurea* cells. Indeed, when oat PHYA was injected into such cells, we observed within 48 h the development of anthocyanin pigments within the vacuole and the appearance of mature chloroplasts (Table 1). Interestingly, the injection of PHYA into a cell resulted only in the expression of PHYA-regulated processes in that cell and not in neighbouring cells (data not shown). This demonstrates that phytochrome signalling is cell autonomous.

To use this system to biochemically dissect the signal-transduction pathways downstream of phytochrome, we used oat PHYA to initiate the molecular responses and attempted to identify pharmacological reagents that could antagonize these processes. A very common mechanism of signal transduction in animal cells involves the activation of a heterotrimeric G-protein. We speculated that an initial event in phytochrome signalling may be the activation of such a component, and this was tested by co-injecting inhibitors of G-proteins (either GDP-β-S or pertussis toxin) with PHYA. This resulted in the loss of all detectable phytochrome responses, ie. no anthocyanin or chloroplast development (Table 1). This result suggested that phytochrome signal transduction requires the participation of a G-protein very early in the signalling process, because its inhibition resulted in the loss of all phytochrome-mediated events observed in a single cell.

The importance of calcium as a second messenger for phytochrome was assessed by first analysing the effect of PHYA injection in cells that were subsequently bathed in

Table 1. Summary of microinjection experiments in *aurea* cells. All injections and subsequent incubations (for 48 h) were done in white light. All concentrations are expressed as final estimated intracellular concentrations. Concentrations are in number of molecules, unless indicated otherwise. G-protein inhibitors were GDP-β-S ($>1\ \mu M$) and pertussis toxin (>5000 molecules); Ca^{2+} inhibitors were nifedipine ($125\ \mu M$) and verapamil ($130\ \mu M$); calmodulin inhibitors were trifluoroperazine ($200\ \mu M$) and W7 ($210\ \mu M$). When a reaction was observed inside an injected cell, it was always the same, e.g. all positive cells injected with $>10\,000$ molecules PHYA in the light showed anthocyanin and chloroplasts, and never just one; all positive cells injected with >5000 molecules Ca^{2+}-activated calmodulin showed only chloroplasts and never anthocyanin. Abbreviations used: A, anthocyanin; BSA, bovine serum albumin; C, chloroplasts; CaM, calmodulin; CTX, cholera toxin.

Injected material	Conditions	Concentration	Number of injections	Number of activations of anthocyanine and chloroplasts	Efficiency (%)
PHYA		$>10\,000$	628	48 (A and C)	7.6
Injection buffer			1210	0	—
BSA		10 000	709	0	—
PHYA	+ G-protein inhibitors	$>10\,000$	891	0	—
PHYA	+ Ca^{2+} or CaM inhibitors	$>10\,000$	823	37 (A)	4.5
GTP-γ-S		30–100 μM	463	32 (A and C)	6.9
GTP-γ-S	+ Ca^{2+} or CaM inhibitors	30–100 μM	225	18 (A)	8.0
CTX	+ GTP-γ-S	>1000	280	15 (A and C)	5.3
Ca^{2+}		0.5–5 μM	415	32 (C)	7.8
Ca^{2+}	+ CaM inhibitors	0.5–5 μM	784	0	—
CaM	Ca^{2+} activated	>5000	620	37 (C)	5.9
CaM	Not activated	>5000	420	0	—

nifedipine, which blocks L-type calcium channels in animal cells. Interestingly, nifedipine was found to block chloroplast development but had no effect on anthocyanin biosynthesis, indicating specificity (Table 1). The same results were obtained with verapamil (another L-type calcium-channel blocker, with an unrelated chemical structure) and also with the calmodulin antagonists trifluoroperazine and W7 (which are also structurally distinct) (Table 1). These results suggested that calcium, probably acting through calmodulin, is essential for regulating events during chloroplast development but not anthocyanin biosynthesis. Hence, it appears that phytochrome signalling may involve at least two pathways, one of which requires calcium whereas the other does not. Since GDP-β-S and pertussis toxin are able to block all PHYA-regulated events, heterotrimeric G-protein activation probably precedes the branch point identified by its calcium requirements.

Although this antagonist data provided some hints as to the nature of the signal-transduction chain, it was necessary to perform positive 'gain-of-function' experiments to show that injection of proposed transduction intermediates could stimulate a response in the absence of the photoreceptor. This is especially important for plant cells because, although it is assumed that the effects of these inhibitory drugs parallel those in animal cells, their action has not been well characterized in plants.

If a heterotrimeric G-protein was involved early in the pathway we would predict that injection of G-protein activators, such as GTP-γ-S and cholera toxin, would potentiate expression of all the markers for phytochrome signalling. Indeed, this was found to be the case (Table 1). In addition, we found that cholera toxin and GTP-γ-S effects on chloroplast development were antagonized by nifedipine and trifluoroperazine, while anthocyanin production was unaffected by these treatments (Table 1). This again suggests that the calcium-defined branch point is downstream of the putative G-protein.

A suggestion following these antagonist results (using nifedipine, verapamil, W7 and trifluoroperazine) was that calcium and activated calmodulin might be active participants in a transduction pathway that leads from PHYA to chloroplast development, but play no signalling role in regulating PHYA-mediated anthocyanin biosynthesis. As such, injection of calcium or calcium-activated calmodulin ought to stimulate processes required for chloroplast development but not those required for anthocyanin biosynthesis. The results in Table 1 clearly show this to be true.

The observed responses of cells injected with calcium could be antagonized in the presence of trifluoroperazine or W7 (Table 1), but not by nifedipine or verapamil, suggesting that calcium probably acts by activating calmodulin. Indeed, we found that cells injected with calcium and incubated in trifluoroperazine or W7 could be rescued (i.e. to show chloroplast development) by removing the antagonist and injecting a second time with activated calmodulin (data not shown), albeit at low efficiency. The low efficiency is due to the low survival rate of cells that have been injected twice.

As shown previously, PHYA or GTP-γ-S injection could induce only anthocyanin biosynthesis in cells bathed in nifedipine, verapamil, trifluoroperazine or W7 (Table 1). If the antagonists were removed and the cells injected a second time with either calcium (in the case of nifedipine- or verapamil-treated seedlings) or activated calmodulin (in the case of trifluoroperazine- or W7-treated seedlings), chloroplast development could be restored, again at low efficiency (data not shown). This suggests that in the normal physiological response, PHYA, acting via a heterotrimeric G-protein, can somehow stimulate entry of calcium into the plant cells, which then activates calmodulin. Presumably, calmodulin then mediates other signal-transduction events, possibly involving phosphorylation.

Efficacy of calcium and calmodulin for stimulating chloroplast development

In the experiments described above, we used chloroplast development as a gross phenotype of the end products of phytochrome signalling. In experiments where calcium or activated calmodulin were injected into cells, however, the chloroplasts appeared to be slightly malformed (data not shown), and this prompted us to examine them in more detail. To do this we used immunofluorescence to study the appearance of individual chloroplast proteins. The proteins examined represented all five of the major photosynthetic complexes: RUBISCO, photosystem II (PSII), PSI, cytochrome b_6f and ATP synthase. Table 2 shows that all 15 proteins examined were synthesized in *aurea* cells injected either with PHYA or GTP-γ-S, although not all of them appear after calcium or calmodulin injection — we were only able to clearly detect RbcS (ribulose-bisphosphate carboxylase, small subunit) from RUBISCO; LHCII (light-harvesting complex II); OEE1 (oxygen-evolving enzyme I); D1; D2 (from PSII); and the α- and γ-subunits from ATP synthase. While most of the proteins examined are nuclear-encoded, the D1, D2 and ATP synthase α-subunit are chloroplast encoded. Hence, in addition to their role in controlling nuclear gene expression, calcium and calmodulin are also able to modulate the expression of chloroplast genes.

Interestingly, none of the proteins from PSI or the cytochrome b_6f complex was present in calcium- or calmodulin-injected cells (Table 2). Taken collectively, these results demonstrate that other signals, in addition to calcium and calmodulin, are required to initiate the formation of a fully mature chloroplast.

Discussion

In this work we have been able to identify several components that are likely to be necessary for the transduction and amplification of the signal initially generated by phytochrome following photoperception. An early event in the signal-transduction pathway is likely to be the activation of a heterotrimeric G-protein, identified on the basis of antagonist data from GDP-β-S and pertussis toxin, and agonist data from GTP-γ-S and cholera toxin. G-protein activation is able to mediate fully the appearance of all three of our cellular markers for PHYA effects, perhaps suggesting that this step is very close to the initial event of phytochrome activation.

Downstream of G-protein activation there appear to be two different pathways. One of these requires the participation of cytosolic calcium and activated calmodulin and results in (partial) chloroplast development, while the pathway leading to anthocyanin biosynthesis does not require calcium or calmodulin. In the normal cellular response, we might predict that calcium enters the cytoplasm via voltage-dependent L-type calcium channels (sensitive to both dihydropyridines, such as nifedipine, and phenylalkylamines, such as verapamil [6]) and the ensuing increase in cytosolic calcium (presumably equivalent to that measured by Shacklock *et al.* [7] in response to red light) acts through calmodulin activation. We know that calcium-activated calmodulin is essential for calcium signalling because the amount of calcium used to activate the

Table 2. Immunofluorescence analysis of polypeptides present in plasmids induced by microinjection of PHYA, GTP-γ-S, calcium and activated calmodulin. Abbreviations used: N, nuclear; C, chloroplastic; cyt., cytochrome. + denotes a strong positive reaction; +/− denotes a clearly positive reaction; −/+ denotes weak immunoreactivity; − denotes no observable staining. Antibody staining was examined in 10 independent cells injected with each compound and the staining pattern for each was reproducible in all cases.

Antibody	Chloroplast complex	Genome	Reaction in cells injected with PHYA	Reaction in cells injected with GTP-γ-S	Reaction in cells injected with Ca^{2+} or CaM
RbcS	RUBISCO	N	+	+	+
LHCII	PSII	N	+	+	+
OEE1	PSII	N	+	+	+
D1	PSII	C	+	+	+/−
D2	PSII	C	+	+	+/−
Ferredoxin	PSI	N	+	+	−
Plastocyanin	PSI	N	+	+	−
PsaD	PSI	N	+	+	−
PsaF	PSI	N	+	+	−
LHCI	PSI	N	+	+	−/+
Rieske Fe–S	Cyt b_6f	N	+	+	−
Cyt. b_6	Cyt b_6f	C	+	+	−
Cyt. f	Cyt b_6f	C	+	+	−
γ-subunit	ATP synthase	N	+	+	+/−
α-subunit	ATP synthase	C	+	+	+/−

calmodulin was not sufficient, by itself, to stimulate any cellular responses. Furthermore, calmodulin not activated with calcium was ineffective (Table 1). By analogy with signal-transduction mechanisms in animal cells, calmodulin is perhaps most likely to modulate subsequent events through phosphorylation reactions, either by activating a phosphatase, such as calcineurin, or by activating a kinase, such as Ca^{2+}/calmodulin-dependent protein kinase II, or both. There is currently no convincing evidence for the presence of a Ca^{2+}/calmodulin-regulated phosphatase, such as calcineurin, in plants; the only Ca^{2+}/calmodulin-activated kinases so far identified in plants contain a calcium-binding domain, similar to calmodulin, fused to the kinase domain [8]. Such a kinase can, therefore only be activated by calcium and not calmodulin. Hence, future research must be directed towards the identification of the cellular target(s) of the injected calmodulin.

The chloroplasts induced by calcium or calmodulin injection were not as well developed as those that were formed by PHYA or GTP-γ-S stimulation. The reason for this was deduced from immunofluorescence analysis of 15 different chloroplast proteins, members of the five photosynthetic complexes: RUBISCO, PSII, PSI, cytochrome b_6f, and ATP synthase. While GTP-γ-S could fully substitute for PHYA to trigger the synthesis of all these proteins, calcium and activated calmodulin could only promote the synthesis of RbcS (from RUBISCO), LHCII, OEE1, D1, D2 (from PSII) and the ATP synthase α- and γ-subunits. Other proteins examined, from the PSI and cytochrome b_6f complexes, were undetectable in calcium- or activated-calmodulin-injected cells. These results suggest that the calcium-dependent pathway is only able to direct the synthesis of the RUBISCO, PSII and ATP synthase complexes, and that the whole PSI and cytochrome b_6f complexes are absent in calcium- or activated-calmodulin-injected cells.

These results also reveal that the calcium-dependent pathway is able to stimulate the synthesis of chloroplast-encoded proteins (D1, D2 and the γ-subunit of ATP synthase), in addition to nuclear-encoded proteins. The fact that the synthesis of chloroplast-encoded proteins is largely dependent on post-transcriptional and translational regulatory mechanisms, whereas known nuclear genes for photosynthetic proteins appear to be regulated primarily at the level of transcription [9–12], demonstrates the astonishing effectiveness of calcium in being able to modulate cellular processes in plants at several fundamental levels. Equally impressive is the effectiveness of PHYA and GTP-γ-S in mediating chloroplast development. The biosynthesis of the photosynthetic apparatus alone requires the participation of several hundred genes, and is dependent upon the concerted action of the nuclear and chloroplast genomes. Functional complexes also require the binding of numerous cofactors, such as chlorophyll, carotenoids and plastoquinones. A long-standing question has been how light can generate signals that can initiate and control all these processes. Through the studies reported here we have shown that light acting through PHYA activation can be sufficient to mediate these events. Furthermore, since GTP-γ-S appears to be equally effective in stimulating this myriad of cellular responses, it would appear that PHYA mediates these processes in just one way — by first activating one (or more) heterotrimeric G-proteins.

Taken collectively, these data begin to show the complexity of the phototransduction events mediated by phytochrome, and reveal that single signal intermediates are not always sufficient to promote a full cellular response. Notwithstanding, we believe that the continued use of this system will yield further information about phytochrome signal-transduction pathways. This will be all the more valuable when mutants with defective phytochrome signalling can be characterized at a similar level.

We thank Professor Ingo Potrykus (ETH Zentrum, Zurich, Switzerland), in whose laboratory the microinjection experiments were performed. We are indebted to Alessandro Galli for excellent technical assistance with the injection experiments. We thank Maarten Koornneef for his invaluable gift of wild-type and *aurea* seeds, and the following persons for their kind gifts of polyclonal antisera: Dr S.P. Mayfield (*Chlamydomonas* OEE1, D1 and D2); Dr R. Malkin (spinach PsaF, PsaD and LHCI); Dr G. Hauska (spinach Rieske Fe–S, Cytochrome *f*, and Cytochrome b_6). G.N. was partially supported by the Huber Kudlich Stiftung (Switzerland). C.B. was supported by a NATO Postdoctoral Fellowship from SERC, U.K., and subsequently by a Fellowship from the Norman and Rosita Winston Foundation. This work was supported by NIH grant number 44640 to N.-H.C.

References

1. Thompson, W.F. and White, M.J. (1991) Annu. Rev. Plant Physiol. Plant Mol. Biol. **42**, 423–466
2. Quail, P.H. (1991) Annu. Rev. Genet. **25**, 389–409
3. Gilmartin, P.M., Sarokin, L., Memelink, J. and Chua, N.-H. (1990) Plant Cell **2**, 369–378
4. Koornneef, M., Van Der Veen, J.H., Spruit, C.J.P. and Karssen, C.M. (1981) in Induced Mutations: a Tool for Crop Plant Improvement, pp. 227–232, International Atomic Energy Agency, Vienna
5. Parks, B.M., Jones, A.M., Adamse, P., Koornneef, M., Kendrick, R.E. and Quail, P.H. (1987) Plant Mol. Biol. **9**, 97–107
6. Schroeder, J. I. and Thuleau, P. (1991) Plant Cell **3**, 555–559
7. Shacklock, P.S., Read, N.D. and Trewavas, A.J. (1992) Nature (London) **358**, 753–755
8. Roberts, D.M. and Harmon, A.C. (1992) Annu. Rev. Plant Physiol. Plant Mol. Biol. **43**, 375–414
9. Mullet, J.E. (1988) Annu. Rev. Plant Physiol. Plant Mol. Biol. **39**, 475–502
10. Rochaix, J.-D. and Erickson, J. (1988) Trends Biochem. Sci. **13**, 56–59
11. Gruissem, W. (1989) Cell **56**, 161–170
12. Rochaix, J.-D. (1992) Annu. Rev. Cell Biol. **8**, 1–28

Subject Index